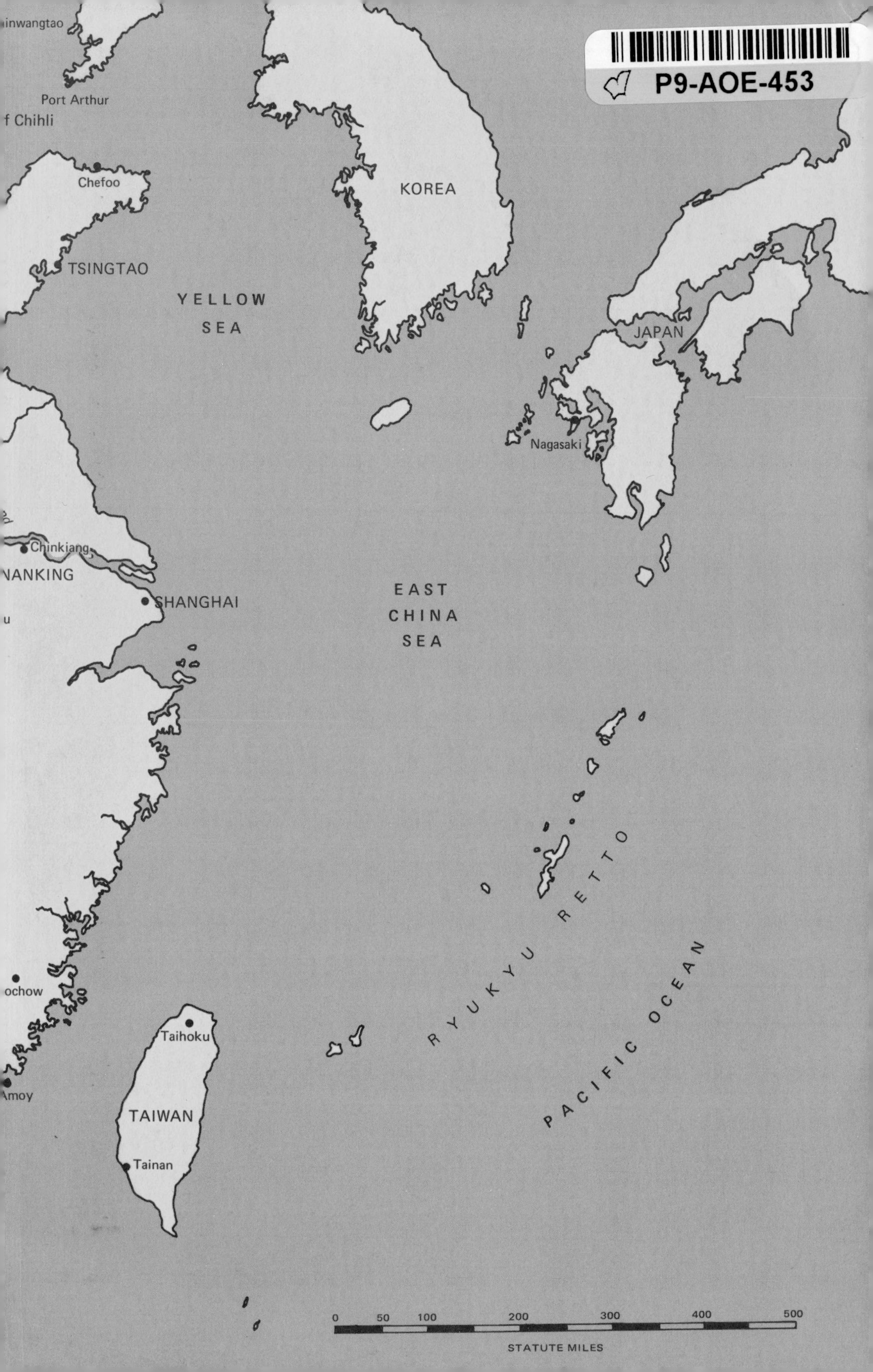
inwangtao
Port Arthur
f Chihli
Chefoo
TSINGTAO
KOREA
YELLOW
SEA
JAPAN
Nagasaki
Chinkiang
NANKING
SHANGHAI
EAST
CHINA
SEA
RYUKYU RETTO
PACIFIC OCEAN
ochow
Taihoku
Amoy
TAIWAN
Tainan
0 50 100 200 300 400 500
STATUTE MILES

Yangtze Patrol

The U.S. Navy in China

by Kemp Tolley
Rear Admiral, U.S. Navy (Retired)

NAVAL INSTITUTE PRESS
Annapolis, Maryland

Library of Congress Catalog Card Number: 73–146534
ISBN: 0–87021–797–6

Printed in the United States of America

To a very special dragon

Foreword

China, the oldest nation on the globe, a society that has endured for over two millennia, has been chronicled, analyzed, and dissected in a thousand volumes in a thousand ways.

Nevertheless, the China story has never been told completely, and accounts of the Chinese character and Chinese behavior earn and enjoy readership in whatever form they appear.

The practical reason is that the wisdom of ancient China— 規矩 *kuei-chü,* literally "old custom"—has never changed, and the ways Chinese people think and act are fascinating to the western world.

The Yangtze River has, for two thousand years, been simultaneously the spine and the central nervous system of the Chinese society. An understanding of how this 3,500-mile artery fits into the events of even two hundred brief years is a picture of China in microcosm, and a precious lesson in Chinese history. The Yangtze scene, as Admiral Tolley says, is timeless—the same today as it was in the time of Christ.

It is not odd that this great river should have served as the main stage for modern western adventure in China. The story of gunboat diplomacy on the rolling Yangtze is a portrait of the growth of colonialism in its simplest and most understandable form, as well as the interaction, conflict, and cupidity of the Western powers as they participated in the death throes of a three-hundred-year Manchu dynasty and the emergence of the Republic of China.

But all of this is essentially historical dividend. The real—and wholly unique—character of this volume is in its painstaking portrayal of the life of unfettered and high-hearted westerners in the most improbable and exotic frontier this world has ever seen.

Entertaining anecdotal writings of American adventures in China from the Civil War to World War II abound, but nowhere has there been a definitive history of our impact on the opening of the Yangtze River as seen severally through the eyes of the captain of a wood-burning gunboat, an expatriate American, a merchant sailor, a sloe-eyed Russian night club hostess, the bamboo remittance man, or the officer of the deck of the *Panay* under Japanese air attack.

It is all here in this wholly unique volume. A trip up the Yangtze rapids, a battle with bandits at Kiukiang, a pirate attack off Wuhu, a night on the town at Joe Farren's or Shanghai's Cercle Sportif Français—all will bring the reader a sense of having been there.

As one who was blessed with exposure to the Chinese life for a few precious years, I sense the care with which the account is put together. It is history, but it is history with an incomparable flair.

And it is all presented with indefatigable attention to factual and technical detail on the one hand and exquisite attention to humor on the other. To old China Hands and neophytes alike, the volume must provide an unforgettable experience.

Victor H. Krulak
Lieutenant General
U.S. Marine Corps, Retired

Preface

The Yangtze Patrol was a unique offshoot of the old U.S. Asiatic Fleet, itself a last remnant of those American squadrons which a century ago roamed the world's oceans under sail and steam. During its active life on China's greatest waterway, the Patrol carried out the longest single uninterrupted military operation in U.S. history, enduring without significant break for only twelve years short of a century.

Little is generally known of the adventures of the Americans who sailed merchantmen and warships on the turbulent Yangtze. Even less, perhaps, is known of "YangPat's" colorful history and exotic background. The "gunboaters" and "Old River Rats" who made up the Yangtze Patrol enjoyed a life such as the U.S. Navy will never know again. Here, along with the history, are the memories of one such Old River Rat, and the many anecdotes related by others who once served under the command of "ComYangPat."

Kemp Tolley

Kemp Tolley
Rear Admiral, U.S. Navy, Retired

Monkton, Maryland

Acknowledgments

Three factors have combined to make possible this first comprehensive coverage of U.S. naval penetration of China's great Yangtze River: the outstanding cooperation and expertise of Mr. Harry Schwartz of the National Archives in locating material; the generous offers by hundreds of individuals of old documents, diaries, journals, letters, clippings, and photographs; my own good fortune in having been able to serve in five river gunboats for a total of more than four years.

Of the many contributors, I am particularly grateful to Vice Admiral T. G. W. Settle for his encouragement and loan of documents, to Rear Admiral Roger E. Nelson for his delightful tales of *Elcano,* to Chief Commissary Steward Frank A. Brandenstein for his amusing anecdotes and donation of very useful books, and to Rear Admiral Stanley Haight and Rear Admiral Scott Umsted for their lengthy contributions on the upper river. Captain J. W. Morcott supplied the story of the Baker kidnapping and a number of other choice items. Colonel Adolph B. Miller put me onto excellent material covering the 1911 Revolution. Mr. Henry Barton, formerly of Standard Oil, was most helpful in the coverage of Changsha. Commander Andrew McIntyre let me use his father's diary for the 1927 Times of Trouble, followed by the Nanking Incident, my story derived from the voluminous papers of Commander Roy C. Smith, Jr., lent by his son, Captain Roy C. Smith III. Captain R. A. Dawes, USNA class of 1904, and Commander J. M. Doyle, 1909, furnished stories of the old Navy. Rear Admiral B. O. Wells and Lieutenant Commander H. M. Kieffer sent stories of much interest. Captain Philip Osborn furnished information on the last days of Hankow, and from Captain David Nash's diary came the story of the end of the Patrol's last ships. There are many, many others who were kind and helpful and their identity will be apparent as their contributions appear in the text.

Several of my China tales and one of Admiral Settle's had previously appeared in the U.S. Naval Academy Alumni Association's magazine *Shipmate.* Editor Captain R. B. Derickson kindly allowed me to use adaptations of mine, and of Admiral Settle's with his approval.

Much of the research material, a number of photographs, and the ship characteristics were supplied by the Naval History Division, Navy Department. At that time, Naval History was under the directorship of Rear Admiral E. M. Eller, U.S. Navy (Retired), ably seconded by Assistant Director Rear Admiral F. Kent Loomis, U.S. Navy (Retired), and Dr. Dean C. Allard; they strongly encouraged and assisted me in my previous writing, out of which this book has evolved.

Some of the photographs were provided by Mr. Andrew C. Fleming, Rear Admiral C. E. Coffin, U.S. Navy (Retired), and Captain R. C. Sutliff, U.S. Navy (Retired). Lieutenant Commander C. C. Hiles, U.S. Navy (Retired) not only lent photographs but throughout has been a much appreciated source of information, encouragement, and advice.

Contents

In the midnight blackness of 29 November 941, the U.S. gunboats *Oahu* and *Luzon* dropped down the Whangpoo, under the sleeping guns of Japanese warships. By dawn they had cleared the Yangtze River. A few days later nearly a century of U.S. naval activity in China ended in a terse radio obituary: YANGPAT DISSOLVED.

YANGPAT—Yangtze Patrol—was a logical result of increasing American commerce with China after the opening of the clipper ship era. The Americans were johnny-come-latelies in the Orient—the Portuguese, Dutch, British, French, and Russians were there first—but they oon saw that the commercial integrity of the actories and godowns along the foreshore was greatly benefited by the presence of a gunboat or so in the river. So in 1854 the USS *Susquehanna* arrived on the Whangpoo and the history of the Yangtze Patrol began.

This history of the Patrol, by a naval officer who was one of the "River Rats" and who served in the river boats until the bitter end, is far more than an account of naval operations on the river. While detailing the routine of gunboat life—months of tedium broken by moments of unbelievable uproar and confusion —the author also covers a century of Chinese history, replete with warlords and mandarins, bandits and kidnappers, missionaries and mercenaries, and riot, rape and revolution.

Along with a lively presentation of the Chinese political situation over the past century, an account of the changing international pressures which eventually gave Japan control of the and, and descriptions of the bombing of the *Panay,* the seige of Shanghai, the battle at Wanhsien, and the Nanking Incident, this book offers as *cumshaw* a liberal serving of colorful anecdotes concerning the River Rats, China, and the Chinese—the antics of the "two-side walkee" gunboat, how to make opium, the going price for "pigs" and mothers-in-law, what happens when the rice birds finally turn against men, and the fact that there was once in China a cook who could do two fried eggs for breakfast, "one fried on one side, one fried on the other."

China and the Yangtze Patrol will never be like that again.

$10.00

Rapids—Upper Yangtze River

, Shanghai

inity

Rear Admiral Kemp Tolley is well-qualified to write a book on the Yangtze Patrol, having spent three tours of duty in the Far East during his naval career. He first served in the gunboat *Mindanao* in 1932 and again in 1937, then put in a two-year stint aboard the gunboat *Tutuila* in remote Chungking before returning to the States in 1940. His last China assignment was to the Yangtze gunboat *Wake* in August 1941; a week before the Japanese bombed Pearl Harbor he departed for Shanghai as navigator aboard the *Oahu*, for a typhoon-beset trip to Manila, harassed by Japanese warships en route.

His China duty was interspersed with duty aboard the battleship *Florida*, Russian language study with the Fourth Marines in Shanghai, and eight more months of Russian in Riga, Latvia, followed by a year at the Naval Academy as an instructor of French. After Pearl Harbor, before Corregidor fell, he managed to slip out of Manila Bay in the schooner *Lanikai*, to arrive in Australia three months and 4,000 miles later. Eventually, naval orders took him to Moscow, where he served as assistant naval attaché for two years. The last year of World War II was spent as navigator on the battleship *North Carolina* off the Japanese coast.

Admiral Tolley, now retired, lives in Monkton, Maryland, near where his forebears settled 250 years ago. He says he is cook, gardener, carpenter, and plumber. He is also a writer, and has contributed numerous articles to the United States Naval Institute *Proceedings* and to *Shipmate*, the Naval Academy alumni magazine. He is presently working on another book—the story of the schooner *Lanikai*.

tell it as it is,
lly it becomes harder
an from Mars.
—Alexander Solzhenitsyn

1784-1865

"Red Hairs" on the China Coast
The Yangtze Rediscovered
Birth of the Yangtze Patrol
The Taiping Rebellion

Long before Marco Polo traveled the caravan route across central Asia to thirteenth-century China, Phoenicians, Carthaginians, and Syrians had found their way to Canton by sea. More than two thousand years ago the princes and wealthy merchants of Tyre and Sidon in the eastern Mediterranean dressed in Chinese silks, while those northern barbarians, the Britons, still danced around their camp fires in animal skins and blue paint. Pliny, the Roman historian, described the brisk trade in Chinese brocade, which his countrymen unraveled and rewove into something more diaphanous and revealing. When Roman power declined, the process perhaps hurried by silken-clad debauchees, Persians and Turks took over the Far Eastern commerce. With Arab ascendancy in the fifth century, they in turn became the masters of the China sea route and introduced to Europe that magnificent combination of medicine, rejuvenator, digestive, and aphrodisiac: tea.

All these Middle Easterners sailed to south China, where by the eighth century a maritime inspectorate had been established, so great was the foreign mercantile intercourse. The peripatetic Arab chronicler and traveler, ibn-Batuta, described south China's city of Amoy as the greatest port in the world. However, the Arabs and their predecessors came, not as swaggering would-be conquerors, with superior weapons and rude manners, but as traders. They were so accepted and allowed in their thousands to mingle freely with the Chinese populace.

While these Mediterranean merchants took the sea route to Canton, other Europeans came by land to China's back door. When the first Papal embassy reached Peking via central Asia in the seventh century, it found that Nestorian Christians from Asia Minor and a colony of Jews had arrived a century earlier by the same route. Kublai Khan's empress became a Christian convert. The great Emperor K'ang-hsi (reigning from 1662 to 1722) was cured by "Jesuits' bark"—quinine. His wife, mother, son, and half the court were baptized. The Jesuits were highly regarded in science, art, architecture; and astronomy until bickering and intrigue among Jesuits, Dominicans, and Franciscans raised the question which has confounded Chinese ever since

Christians first reached those shores: why, in the earthly realm of the Kingdom of Peace, do its vicars fight so viciously among themselves?

Long before Britain's colonies in North America had won independence, the British had found their way to China and were pounding on Canton's gates.* It was only natural that in due course, adventuresome Yankee traders from the former colonies would be on British heels.

Accordingly, the ink was barely dry on the treaty which proclaimed the United States independent when the little 370-ton ship *Empress of China* reached Whampoa, port of Canton, on 28 August 1784. Although ships from the American colonies had preceded her, she was the first to visit China flying the "Flowery Flag," as the Chinese called the new Stars and Stripes. The unfamiliar flag and the language of the newcomers caused the *Empress* to be widely mistaken for an "English country ship," as vessels from India were then termed.

Major Samuel Shaw, supercargo of the *Empress* and a highly educated, widely traveled man, was well equipped to record the arrival of the Americans at Canton.[1] In their treatment of the Americans, the English, not an overly demonstrative breed at best, were studiously correct but more than usually reserved as a result of the recent Revolutionary War. The more volatile French, patrons and military allies, were as firm friends of the new republic's representatives as they were suspicious of their traditional enemies, the English. "Merde pour le roi d'Angleterre, qui nous a déclaré la guerre!" runs the refrain of a popular old French Navy drinking song.

At that time, there were no foreign warships in China. Actually, the dividing line between warship and merchantman in those days was so fuzzy that sometimes only close scrutiny would reveal which was which. On arrival at Whampoa, as close to Canton as *fanquei* (foreign devil) ships were allowed to go, the *Empress* fired a 13-gun salute to the foreign vessels in port, returned by them gun for gun. It suggests that the *Empress* would not have found herself entirely helpless if attacked by Algerine pirates off West Africa en route the Cape of Good Hope—by which she had come—or by Malay freebooters in the South China Sea.

Tea had been known to the Arabs for over a thousand years before the English came to Canton, but the English had only recently developed a passion for this magnificent herb which helped disperse the gloom and chill of their bleak climate and which in time would become almost a euphemism for "British."

The demand for tea in Britain soon had British East India Company ships loading tea at Canton in exchange for silver specie, lead, and woolen cloth; the company was obliged by their national charter to export the cloth for the encouragement of the woolen industry. There was a heavy British import tax on tea, so the French, Danes, and Swedes, with little to trade that the Chinese wanted, smuggled tea wholesale into England. Then they brought the silver out to China to buy more tea.

The canny Dutch, long firmly established in their vast colonies in the East Indies, produced spices, tin, and quinine which they transported to Canton and traded to the

* The British reached Canton in 1635; the French in 1698; the Danes in 1731; the Swedes in 1732; the Russians in 1753. All were well after the Portuguese, who first came in 1514 and established Macao in 1557.

Chinese for silver. This was then exchanged locally with the English for bills payable in London. The Dutch thus avoided the long, dangerous haul of their hard money home, while the British immediately put it back into local circulation for more tea. Whatever way one looked at it, the English craving for tea was draining that country of scarce silver.

The East India Company, whose charter gave it a monopoly on all British trade in India and China, was in effect a government, with its own army, navy, and diplomatic service, the latter directing the company's foreign policy with little or no interference from a distant king and parliament. Faced with increasing demands for tea at home and dwindling silver, the Company was forced to turn to something else to trade. What better than the cheap opium grown in great quantity under Company auspices in India? Originally insinuated into China in small quantities, opium addiction and demand grew rapidly, paced by official Chinese alarm.

Major Shaw could view the situation with smug equanimity. There was no such demand on the small store of American silver. Americans paid for tea with ginseng, which was available in her forests in fair abundance, and, because the Chinese believed it effective in the maintenance of virility, worth its weight in gold. "On the whole," summarized Shaw, "it must be a satisfactory consideration for every American, that his country can carry on its commerce with China under advantages, if not in many respects superior, yet in all cases equal, to those possessed by other people." All this because of ginseng, that forest plant with humanlike root which the superstitious claim shrieks in agony as it is yanked out of the ground.

The Americans found the Chinese then as now bewildering, frustrating, occasionally insupportably overbearing, and frequently devilishly devious. But they were at the same time well organized for the business at hand: to make a mutually profitable exchange of goods. A *co-hong* of ten or twelve leading Chinese merchants provided the sole contact for "barbarian" trade. The contemptuously termed *fanquei* were confined to five hundred yards of fenced-in waterfront, each group of nationals in its high-walled "factory," wholly insulated from the Chinese public and forbidden to set foot in the city. A *comprador* attended to revictualling of ships. Superimposed over all was the *hoppo*, chief officer of customs. Dependent for his job on some mandarin higher up the totem pole, the *hoppo* was not reluctant to accept bribes in turn from the ships in the form of goods at ridiculously low prices. "Sing-songs," pidgin vernacular for clockworks, were a favorite item.

Like many newcomers to China, Shaw was at first inclined to be restrained and temperate in his treatment of Chinese merchants, some of whom by their persistence and occasional duplicity could be trying on the temper. This was taken note of by a Chinese, who, as recorded by Shaw in his journal, was astonished at Shaw's forbearance:

"You are an Englishman?" said he.

"No."

"But you speak English word, and when you first come, I can no tell difference; but now I understand very well. When I speak Englishman his price, he say, 'So much.—take it, let alone.' I tell him, 'No, my friend, I give you so much.' He look at me, 'Go to Hell, you damned rascal; what! you come here—set a price my goods?' Truly, Massa Typan, I see very well you no hap Englishman . . . All men come first time China very good gentlemen, all same you. I think two three times more you come Canton, you make all same Englishman too."

In the literal sense, that earliest Chinese contact made an accurate prediction. When Major Shaw returned to Canton the following year as first American consul, his journal suggests his attitude toward the Chinese had taken a more pragmatic turn.

In a figurative sense, Shaw's Chinese friend was equally right. The early Americans had come in an open, friendly way to trade. The Chinese were probably not far off in generalizing them as "maritime, uncultured, and primitive."[2] Americans took their lumps with fair grace, but little by little, they learned. The inescapable facts became clearer as the decades slipped by. Dealing with the Chinese required a magnificent degree of finesse—a combination of force, restraint, and delicacy—all exercised with a growing understanding of the Chinese mentality, without which the barbarian *fanquei* stood at least to be humiliated, at worst to be wholly thwarted.

By the time the *Empress* reached Whampoa, the European *fanquei* were not only demanding that their opium be received openly in exchange for silks, porcelain, and tea, but also were brazenly suggesting that they be represented at the Dragon Court in Peking. Understandably, the mandarins were shocked at the impertinence of the northern "Red Hairs," whom they considered to be a far cry from the learned monks or the mannerly Middle Easterners of an earlier day. When British envoys attempted to force their way to Peking, they were treated with contempt and derision. En route, Lord Macartney's houseboat carried a banner whose Chinese characters, unreadable to the haughty Britisher, proclaimed to the mandarins that here came "barbarians bearing tribute."

For their part the British Red Hairs were deeply pained by such humiliation. Many of them were men whose families in India wielded the power of princes, but in China they were treated as brigands posing as traders. An ever deeper pain was felt in their pocketbooks, through Chinese intransigeance in refusing to open wider the portals at Canton. The limited possibility for variety and volume of trade in such a one-port operation inevitably caused the Red Hairs to look farther afield. It was natural that their search should lead to the Yangtze.

The Yellow Sea was well named. The first "barbarian" adventurer who sailed it knew long before he met the Yangtze that he had discovered a mighty river. A hundred miles offshore the water is stained by the heavy burden of silt the river carries from east-central China, from the vast interior province of Szechwan, and from its source in distant Tibet, 3,500 miles from the sea. Each year the river pours out enough sediment to cover a square mile ninety feet deep. Over the millenia, billions of tons of silt have created a vast alluvial plain extending the original shoreline a hundred miles out into the shallow Yellow Sea.

On such ancient mudflats in the Yangtze delta, several sizeable cities have risen, the most familiar of which is Shanghai (Above-the-Sea). Twenty-two centuries before the Christian era, the Confucian Book of Classics mentioned the countryside where Shanghai is situated, and all through Chinese history the port has been an important center for waterborne commerce between the north and south China coasts and from the vast interior via the Yangtze and the Grand Canal.

Known in the time of Christ as "Lau in the Kingdom of Wu," and later as Hwating-hai, "Seaport of Hwating," the name "Shanghai" was first recorded at about the time Eric the Red sailed to the New World.

As Shanghai is not the most beautiful city in the world, neither is the Yangtze the widest, deepest, nor longest river in the world, but it is assuredly unique in its capriciousness, variety, and above all, its importance to the vast country through which it flows. There is a seasonal variation in water level at Chungking, some 1,300 miles from the sea, of as much as a hundred feet; half that at Hankow, only 595 miles from the sea. What is dry, level farm land at one time of the year can become a vast lake, the channel marked by an occasional navigational pole or groups of village rooftops. The population will have withdrawn behind a second line of dikes miles "inland" or moved to mounds and low hills to wait out the flood.

In the great gorges above Ichang, a thousand miles inland, a spring freshet can raise the water level a roaring 50 feet over night, 250 feet above winter low. At such times no ship or junk dares chute down the foaming cataract and certainly none could sail up against its racing, swirling current.

For navigational purposes the river falls into three parts—the 600 miles to Hankow, the 370 miles from Hankow to Ichang, and the 430 miles thence through the gorges to Chungking, the nominal end of the foreign exploited navigational road. (The river is navigable to Suifu, for 1,700 total miles.)

When the melting snows of the central Asian mountains and the monsoon rains from the south pour down the Yangtze in spring and summer, the heaviest draft ocean-going ships can reach Hankow and ships drawing up to twelve feet can reach Ichang. But winter's low water brings a drastic change. The limiting depth to Hankow becomes ten feet and to Ichang a mere six, leaving little enough water under even the flat keels of river gunboats.

Having described the Yangtze River, a little-known fact must now be revealed: to a Chinese, there is no such thing as the Yangtze River. To most literate Chinese, what foreigners know as the Yangtze has two names: from the mouth to Suifu (Ipin), a distance of about 1,700 miles, it is the Ch'ang Kiang (Long River); from that point on it becomes the Kin'sha Kiang (Gold Sand River; or more poetically, River of Golden Sand).[3]

The Min, which joins the Kin'sha at Suifu, is a large, long, and very important stream. These two form a *new* river, the "Long River." This is no more strange than the fact that because of incomplete knowledge of early explorers, the main stream of the Mississippi above St. Louis is called the Missouri.

China is a land of great regional independence. Her people, mostly uneducated, developed distinct local habits, customs, and nomenclature. It is not surprising to find informal, unofficial local names, and herein lies the confusion concerning the Yangtze.

Chen-chiang, first treaty port to be opened on the Yangtze, was known to gunboaters as Chinkiang. The ancient name for Chinkiang was *Yang-tzŭ*, a very important place historically. Thus the name for the river in its vicinity was *Yang-tzŭ Kiang*, the river *of*, or *at*, Yang-tzŭ. This local usage did not extend much above Chinkiang.

Undoubtedly, foreigners exploring the coast heard people using *Yang-tzŭ Kiang* and so picked up the informal, local name which they then mistakenly applied to the whole river, in early American accounts sometimes spelled *Yang-tzŭ*.

Another widely accepted bit of "old China hand" mythology was that many foreigners with a smattering of Chinese translated "Yangtze Kiang" as "Ocean River." The *yang* of the city Yang-tzŭ, and hence the stretch of river bearing its name, means "to scatter, to spread, to winnow, to praise." Though having identical pronunciation, it has no relation to the *yang* which means "ocean."

For the benefit of those who may aspire to write Chinese poetry, the Long River is sometimes referred to in that medium, as well as in history and literature, simply as *Kiang*, "The River." Also, in Chungking it may be called *Ta Kiang*, or *Ta Ho*, "Great River," to differentiate it from the *Kialing*, or "Little River," which joins it there.

According to Mr. Service:

> *There may be a wry footnote here to a history of the era of "gunboat diplomacy." In their ignorance, the "foreign barbarians" misnamed one of the great rivers of the world, and thinking it was the native name, stuck with it. The Chinese themselves do not use the name* Yangtze. *Unless they have had a lot of western contact or education, they are confused when they hear the name because they never think of it that way. The Yangtze Patrol obviously should have been called something else—Ch'ang Kiang Patrol, or Long River Patrol.*[3]

It is surprising that the "barbarians" discovered Shanghai so late in history. The general lay of the Chinese land had been well known to Europeans since the thirteenth century days of the Polos. European Christians and one band of Jews had entered China at least six hundred years earlier and had either been assimilated or, in the case of the Jesuits at Peking, lost their "franchise."

The first recorded European visit to Shanghai was that of the Prussian medical missionary, Karl Gutzlaff, in August 1831. He found the country roundabout highly fertile and the port magnificently situated for trade.

The next year, continuing near-insufferable Chinese intransigeance and harassment at Canton induced the East India Company to follow up Gutzlaff's visit with a more pretentious expedition. Consequently, Mr. Hugh Hamilton Lindsay, one of those men-of-all-accomplishments—diplomat, merchant, accountant, and amateur soldier known in those days as "supercargo"—appeared at Woosung in the Company's ship, the *Lord Amherst*. He was accompanied by Gutzlaff, who had for the moment left off employment with one of the largest British opium importing firms.

By dint of determination and strong-arm methods, Lindsay literally forced his way up the Whangpoo River to the city of Shanghai and into an audience with the wholly reluctant Chinese authorities. The expedition came to nothing. Nor was anything gained by a second thrust in 1835 to open Shanghai to foreign trade. It was this continued combination of Chinese bellicosity and British desire to expand trade and force opium on China which led to the War of 1841, with the capture of Shanghai on 16 June 1842 by Sir William Parker. The British then captured Chinkiang, the Tartar citadel, by which time it was crystal clear to even the most unreconstructed Chinese that firecrackers, gong beating, spears, and generals far to the rear in yellow mandarin coats were not the answer to Western arms. The war was brought to a close by the conciliatory August 1842 Treaty of Nanking, ignominiously signed by the Chinese aboard the British flagship, HMS *Cornwallis*, lying under the city's walls. The treaty at last opened Shanghai to foreign trade, along with Ningpo, Foochow, Amoy, and Swatow, plus ceding Hongkong to Britain. With the exception of Ningpo, all these ports were to become well known to U.S. Asiatic Fleet sailors of a later day.

The Chinese commissioners who put their "chop" on the Treaty of Nanking were dressed in their plainest clothes, because, as they explained, "Imperial commissioners are supposed to proceed in haste about their business, and have no time to waste on their persons."[4] To the British, this must have been the supreme irony; for two hundred years, they and the other "barbarians" had suffered suffocating exasperation at the hands of Chinese officials, who had compounded endless procrastination with infuriating condescension and intransigeance, mitigated only by cupidity over what could be squeezed from the "barbarian" trade.

The fictional time machine in the story of that name by H. G. Wells required only a twist of a knob to take its passenger into the infinite future, or back into the very dawn of history. An illusion of that sort appears to anyone in a position to observe the junk traffic on the Yangtze, where the scene is timeless—the same now as in the time of Christ.

So there was little change between the time Sir William Parker left Shanghai in 1842 and the day when Thomas W. Blakiston arrived to explore and write about China in 1861.[5] The river was as Parker had seen it, but Blakiston would see far more; he would be the first European to ascend the great gorges of the Yangtze to its remote navigable end. The trek began with an assist from the Royal Navy; on 12 February, Vice Admiral James Hope assembled a squadron, some paddle wheelers, others screw driven, and all rigged for sail, at the mouth of the Whangpoo. The ships, *Centaur, Waterman, Attalante, Couper, Banterer, Havoc,* and *Bouncer,* filled the Chinese with awe as they prepared to get under way in billowing clouds of smoke and showers of sparks. The first appearance of steam on the Yangtze was more impressive in appearance than performance; the engines were only marginally reliable, fuel supplies were few and far between and not much could be carried on board, and the boilers had an unfortunate proclivity for blowing up, even at a modest pressure of ten pounds.

Aboard the *Attalante* with author-explorer Blakiston were three other Europeans (euphemism for "white man" throughout the Far East), one of whom was an American missionary fluent in Chinese, the Reverend Mr. Schereschewsky. He would be the first American ever to view some of the most magnificent scenery on this planet—the rapids and gorges of the upper Yangtze.

The Blakiston party was made up of those typical intrepid explorer-type Britishers who with characteristic *sangfroid* have for centuries braved heat, cold, and the terrors of far places, placidly sucking away on cold pipes, carefully noting down in great detail what they saw and thought, adroitly but humanely handling the shifty natives, and all the while never missing the afternoon cup of tea. Americans, in all seriousness, owe much to this dedicated breed. The traditional British whim to climb a mountain "because it is there" or to trek through the remote interior of China merely to see what it contained struck the Chinese as symptoms of an insanity they found wholly inscrutable and as such, open to deep suspicion. Consequently, the local mandarins could only conclude that some secret mission was involved and that these people were important government officials, worthy of protection lest the foreign lightning flash as it had in the past.

Besides their tea, the explorers took three Chinese and four Sikh soldier guards, several hundred "Mexican" dollars, and four hundred English pounds worth of *taels,* boat-shaped lumps of silver. They also carried a sextant and artificial horizon for checking their position in the uncharted interior, pocket compasses, barometers, telescopes and binoculars, and the indispensable sporting guns which added an occasional snipe or partridge to the menu.

As the British squadron moved upriver, with water at the midwinter low, they constantly went aground and spent much time and effort pulling each other free.

At Chinkiang, Blakiston was horrified by what he saw. The once prosperous city was a blackened ruin. A few bony, half-starved survivors scuttled feebly for cover as the foreign party approached. There were no barking dogs or staring children. The dogs had long since been eaten and the children were dead. This was the work of the Taipings, whose once supposedly enlightened "Christian" movement had been the white hope of the missionaries as the saviors of a decadent and heathen Imperial China.

Admiral Hope remained at Chinkiang only long enough to install a British consul in the newly opened port, if one could be so charitable as to thus designate this blackened monument to Chinese fratricidal strife. Then the squadron chugged on upriver, supposedly charted in 1858 by Lord Elgin, but so capriciously changing its channels that yesterday's deep water became today's sand spit.

Arriving at the important trading city of Hankow, six hundred miles inland and the head of deep water navigation, Admiral Hope was chagrined to the quick to discover that in the lexicon of a later century, "Kilroy was here." On 26 February, the steamer *Yang-tze,* belonging to Messrs. Dent and Company of Shanghai, had arrived at Nanking, *sailing under American colors,* en route upriver to establish agents at new treaty ports. The Stars and Stripes, under what auspices unexplained, thus flew over the first merchant vessel to arrive at Hankow. "And I believe," wrote Blakiston, "that Mr. Webb received from the Viceroy some of the favors which were intended for Admiral Hope." It was not unusual in those days for civilians to affect gold braid and brass buttons when it seemed appropriate to impress the opposition. And in view of his mission to push his boss's business out ahead for Dent and Company, supercargo Webb, in a semimilitary version of full feathers and war paint, could be expected to put up something less than a struggle to disabuse the Viceroy of his importance. Nor did the execrable interpreters of the time serve to enhance full understanding.

Admiral Hope's once imposing squadron, reduced to the flagship *Attalante* and the gunboat *Bouncer* by groundings and one casualty or another familiar to any modern River Rat, staggered as far upriver as Yochow, taking sights to establish their position whenever the murk lifted sufficiently to bare the sun. They had shown the white ensign along seven hundred or more miles of the river for the first time. And also for the first time they had brought about contact between British and Chinese merchants. All these things were major benefits from the allied military effort at Tientsin, in the hostilities of 1858–60.

Thus, although the British failed to be the first *fanquei* to see the Yangtze, by a matter of several hundred years, it was the British, on occasion with a minor leg-up from the French, who furnished the initiative and blood and treasure to open the Great River to the foreign commerce which many countries later were to share. Americans must equanimically and sportingly accept the quote of a British officer of the times that "it became ingloriously, yet very profitably, the role of the United States pacifically to follow England to China in the wake of war, and to profit greatly by the victories of British arms."[6]

At Yochow the *Attalante* turned back downstream, leaving the Blakiston party on their own for the trip inland. They were not the first Europeans up the river; Jesuits and members of other orders had preceded them by four centuries. But the padres had in Oriental fashion "bent with the wind," adopting Chinese padded gowns, pigtails, and "football" moustaches—eleven hairs to a side—and thus managed to survive religious pogroms and political storms. Blakiston flaunted his "foreignness" and outraged Chinese decency by intruding into their very heartland, dazzling mandarins and generals with his Royal Engineer gold-encrusted full regimentals, complete with sword and feathered hat.

Blakiston harbored few of those tender feelings toward the Chinese which would smooth the way for many of those who followed: missionaries, quickie sightseers, remittance men, retired River Rats, and the professional Sinophiles. While still aboard *Attalante*, some patently fake fortifications had inspired Blakiston to something more than the usual saturnine English understatement, in which his views toward John Chinaman were made eminently clear: "It was one of those shams so constantly seen in China," he wrote, "from the government down to religion, or from religion down to government, whichever way one puts it, since it is impossible to say which is the more corrupt. In every institution, in the daily affairs of life, in business, in common conversation, is there not a vein of deception running through the whole? . . . A Celestial is a liar and the Central Flowery Land a myth."

Most earlier European travelers in China, such as Marco Polo, took a more charitable view. After all, the great contemporary cities of Europe were as fetid, flyblown, and pungent as those of China. Concurrently, in Polo's time, European political and public morals could scarcely be described as model. But by 1861 Europe had advanced, while China had lain weltering for half a millenium in her Confucian and Taoist and Buddhist introspective do-nothingness.

Nevertheless, Blakiston was studiously fair in his judgment. He noted the wretched coolie's generally cheerful good humor even under the most galling circumstances. He sniffed with righteous British disgust at the prediliction for opium from generals to sampan men, forgetting perhaps that it was through the force of British arms that the stuff was there in the first place.

By and large, Blakiston's party "roughed it" with the relish that only a nineteenth century British empire builder could muster. The junks on which they journeyed for the next several months were equipped with the proverbial magic chef, an ample number of servants, four devoted Sepoy guards, and generous supplies of tea, tobacco, and brandy, making the whole cross easier to bear. Only the ale had gone down the river with the *Attalante*. But its memory lingered on; Blakiston had seen a Chinese model of a British warship which was in action against the Peiho forts in 1859. The stern of the model carried the best the modelmaker could provide in the way of an English name: BASS&COPALEALE, undoubtedly copied from an empty bottle tossed overboard.

After endless palavar, the Reverend Schereschewsky hired a hundred-foot middle river junk for the trip to Ichang. The party stowed its baggage in the damp bilges and shook down in their cabin space aft, where the scuffle of the helmsman's feet overhead and the picturesque profanity of the skipper directed at passing junks tended to mask the jumble of noises emanating from the open amidships section. There, the crew ate, slept, and powwowed on a seemingly nonstop basis.

Alternately poling and rowing, hoisting the patchwork sail when there was a fair wind, the crew worried the clumsy craft upstream through flat countryside, denuded of trees, where only an occasional pinnacle or mound relieved the monotony of the brown winter landscape. The British stuffed cotton wadding in cracks to keep the chilling wind out of the cabin, and from time to time went ashore to stretch their legs and shoot a bird or two to vary the menu. Often they would hike across a neck of land and gain ten miles on the junk, beating around the loop. In fact, in the unlikely event he could have become airborne, Blakiston would have discovered that Ichang was only 150 miles from Hankow as compared to 370 miles by river.

At Shasi, the centuries-old transshipment port, junks from Szechwan transferred their cargoes to middle river junks. Some of the latter took the canal to Tungting Lake, thence south via river and canal to Canton. Others used the intricate canal system which laced the area north of the Yangtze between Shasi and Hankow rather than risk the open river with its twin hazards—navigational and piratical.

The upper river junks carried much less sail than the middle river craft, but they also carried up to a hundred crewmen for hauling up the rapids. It would have been uneconomical to maintain such a crew for continuing a middle river passage and in a paternalistic land such as China then was, explosive to lay them off.

Blakiston found that Shasi had a sort of split personality in that, separated by some distance from the Chinese part of the city, there existed that anachronism remaining from palmier days of the Manchu conquerors—a complete Tartar city. These enclaves had been established at important points all over China as garrisons of Manchu bannermen to maintain the Chinese in subjection. As peace prevailed over the centuries, the bannermen had degenerated into symbolic caricatures of warriors, more ornamental than useful.

The Tartars were from a western viewpoint more enlightened than their Chinese "subjects." Blakiston wrote of his liaison man aboard:

Our own mandarin also brought out his wife and family to see us. They were Tartars, or rather went under that name, and following the Tartar fashions in dress, and in not cramping the feet of the women. The ladies of the party were really pretty —if any European may be allowed to call any females pretty out of Europe—and were tastefully dressed in loose jackets and fancy trowsers, the younger ones having bright colored flowers set in their elaborately dressed hair. A Chinese would never have thought of introducing the females of his establishment, particularly on an occasion like this, when they would be brought face to face with 'the devils of the Western ocean,' for it is even considered bad manners to refer to females in ordinary conversation.

With tantalizing brevity and no documentation whatever he added that ". . . we never recollect to have heard that Chinese ladies were particularly noted for chastity."

In contrast to the Tartar reaction was that of a Chinese official of Canton, who included in a memorial to the emperor his views on foreign customs of the period 1850–60:[7]

Whenever there are honored guests, they [the foreigners] are sure to present the women. For instance the American barbarian and the French barbarian brought barbarian women with them. When your slave went to the barbarian houses to discuss matters, these barbarian women would suddenly appear to pay their respects. Your slave was composed and respectful but uncomfortable, while they were greatly honored. This is actually the custom of the various western countries and cannot be determined by Chinese standards of propriety. If we condemn them hastily there will be no means of dispelling their stupid ignorance, and at the same time we would arouse their doubts. . . .

There were other great gulfs of misunderstanding between the Chinese and Europeans. Most Chinese officials of Blakiston's time were extremely suspicious of anyone who would voluntarily subject himself to the rigors of an exploring trip up the Yangtze River. Such an individual simply *must* have a sinister motive and actually be engaged on some important secret government mission. When Blakiston and his equally bereft (from a Chinese point of view) companions announced their intentions to continue through the interior to Tibet and come out at Rangoon, the Chinese could only conclude that they were all valuable specimens to be protected at all costs in spite of their apparent folly.

Consequently, at Ichang, only minor problems were involved in turning in the middle-river junk for an upper-river craft. The new one was much more heavily constructed and carried its built-in "tug" in the form of some hundred-odd trackers, ready to jump ashore when the wind failed or the current got too strong. But the overland trek to Tibet and Rangoon proved beyond arranging, even for the redoubtable Captain Blakiston, and the party eventually returned, all intact, back down the Yangtze to Shanghai.

Despite the adventurous air of his expedition, Blakiston had been preceded by American warships which, eight years earlier, had gone up the Yangtze as far as Wuhu. By the time the U.S. Navy got around to definitely establishing the Yangtze Patrol, the routine on the river was no longer new.

By 1853, the United States was approaching the peak of its mercantile power on the sea, and, with close to five million tons of shipping, almost exclusively in sail, was nudging Britain for first place. There were twenty fast U.S. tea clippers on the run to London, getting double the rate paid English ships and taking the tea trade almost out of English hands.

Gold had been discovered in California in 1848, setting in motion the coolie trade to the Western Hemisphere. A good portion of it was in American hands, under conditions sometimes not much better than in the African slave trade. Ships now sailed via Cape Horn and California instead of Good Hope. They carried emigrants and supplies to the West Coast, loaded furs and New England manufactures for China, then packed tea or coolies for the return trip to the Americas and the Caribbean.

It was apparent to many that the Yangtze was the logical entrepot to replace Canton. But in 1853 the Yangtze Valley was in turmoil as a result of the Taiping Rebellion. The opposing forces battled on the Yangtze and on the land around Shanghai. American warships in numbers greater than ever before visited the port, the jumping-off place for Perry's operations directed toward opening first the Ryukyus and then Japan.

Along with their growing share of Chinese trade, the Americans were displaying the cockiness that comes with success in ventures on the sea. A stirring example of this attitude was displayed by Commander Kelly of the USS *Plymouth* at Shanghai in March 1853. A pilot boat showing the American flag was fired on by the *Sir H. Compton*, a vessel of the Chinese Imperial fleet, then boarded, and her Chinese crew removed and lashed to the *Compton*'s mast by their queues. When this was reported to the captain of the *Plymouth*, he immediately sent a boarding party to the *Compton*. After a short discussion hampered by the usual Oriental inertia and dissemblance, the Americans released the prisoners and removed them.

The next morning Commander Kelly had this insult to the American flag laid before the commander of the Imperial fleet and demanded public atonement in the form of a 21-gun salute, and when it was not forthcoming in a reasonable time, he moved the USS *Plymouth* to a position where her guns pointedly emphasized his demand. Almost immediately the captain of the *Compton* came on board to state that he would render the salute the following noon. Commander Kelly, in the words of the narrator, ". . . thought he might as well stay where he was and see it done, as he was determined it should be, and on the next day, at noon, it was done. . . ."[8]

The doughty Commander Kelly got his dander up once again during "The Battle of Muddy Flats," when Chinese soldiery moved in on the Shanghai race course, having first inflicted indignities on several foreign ladies and gentlemen in the vicinity.

A small party of British Marines sent to handle the situation found the odds too great, and went to earth behind some of the numerous grave mounds, thus losing considerable "face."

Reinforcements from British and U.S. ships in port were hurried to the scene and relieved the beleaguered Marines, but felt the Chinese should be taught a proper lesson. Consequently, next day the American and British consuls, seconded by Commander Kelly and Captain O'Callaghan, Royal Navy (who must have been a peppery Irishman of the same vein as Kelly), conferred on joint educational measures.

The preliminary step was to warn the local Chinese general and their fleet commander that their troops had better make themselves scarce on the sacred precincts of the race course by 4:00 P.M. or face the destruction of their breastworks. The Chinese, not having deigned to reply or move by 3:30, "the foreigners proceeded to their work." Captain O'Callaghan, with about a hundred and fifty British sailors and marines and the Shanghai volunteers occupied the right flank. On the left, Commander Kelly had about sixty *Plymouth* sailors and marines, half as many American merchant seamen, two "private" field pieces worked by American citizens, and a 12-pounder howitzer. At 4:00 P.M., the supposed deadline, the howitzer started lobbing shells into the Chinese encampment. After ten or fifteen minutes of this, with no opposition, Kelly and his sailors charged the barricades. To their surprise and discomfiture, they were met by a spirited fire as they struggled to cross a twenty-foot-wide, seven-foot-deep creek. A quick flanking movement to the left by Kelly and to the right by O'Callaghan put the Chinese soldiery between two fires. "In about eight minutes," the account concluded, "the Chinese fled in great disorder, leaving behind them a number of wounded and dead."

In short order, the ponies were pounding around the track and the ladies and gentlemen taking the spring air in peace, as "after this the Chinese behaved themselves quietly."

At the end of March 1853, the 2,500-ton *Susquehanna* was chunking her way up the Whangpoo, the largest steamer so far to reach Shanghai. Aboard her was Colonel Humphrey Marshall, U.S. commissioner to China, who, like most diplomats in those parts (to the discomfiture of the CinC) preferred to do his traveling by warship.

Like Ameratsu, the Sun Goddess, first forebear of the Japanese emperors, one might say that it was from *Susquehanna* that the ships of the Yangtze Patrol took their blood lines. She would be the first American man-of-war to ascend the Yangtze itself.

Commodore Perry, commanding the station, was not a China nor a Yangtze enthusiast. On visiting the river in the flagship *Mississippi,* he said that until proper landmarks and beacons were established, it was not a fit place for a naval depot. His opinion was reinforced when *Susquehanna, Plymouth,* and *Supply* grounded on the way in. "The wealthy foreign merchants established at Shanghai, who are gathering a plentiful harvest from the increasing trade of the place, should contribute some of their thousands toward rendering the navigation less dangerous," he complained.

Perry obviously longed for the beauty of Shimoda: "Nothing can be less picturesque than the scenery . . . in the approach to Shanghai. . . . The poetical observer is sadly disappointed in a view which represents a dead level of landscape, without a mountain, a hill-side, or even a tree to relieve the monotony." The Yangtze, if possible, was worse: "The muddy waters of the Yang-tse-Kiang, looking more muddy still in the yellow light of a foggy atmosphere, and the dull constraint of a tedious anchorage, presented a sad prospect to the eye, and a wearisome sensation to the feelings, which made all anxious for departure."

Perry's dim view of China's scenery was perhaps influenced by an even dimmer view of Chinese character: "It is true honesty is only a conventional virtue with the Chinese, but it can be obtained for money, like anything else among that nation of shopkeepers; and if a Chinese laborer stipulates to be honest for a consideration, he may, in ordinary cases, be depended upon. . . . If, however, honesty had not been made expressly a part of the bargain, a Chinaman thinks he retains the right of lying, cheating, and thieving. . . ."

In April 1854, as *Susquehanna* lay at Shanghai, the surrounding country was laced with battling "rebels"—Taipings, Triads, and "Imperials"—Peking oriented. Spillovers into the international settlement could be expected. *Susquehanna*'s 27 April log read that, "In case two rockets are fired from either station ashore, it means want of assistance. During the day, the English flag on the cathedral means the same."

The log also reported an expenditure of 3 6/32 gallons of whiskey for grog, with 448 5/32 gallons still in the casks. There were fourteen sick, most of them with that constant scourge, dysentery, or allied intestinal complaints. It was too early in the year for "fever."

On 3 May, the famous Samqua came aboard, being honored coming and going with the traditional 3-gun Chinese salute. Actually, he was Wu Chien-chang. "Sam" was a foreign nickname easier to remember and the "qua" was a sort of honorific "mister." There were strong rumors that Samqua was a silent partner in the powerful American firm of Russell and Company, which owned some of the fastest opium clippers in the business. As acting *taotai* of Shanghai, "Mister Sam" was an exceedingly useful man to have on the American side.

If it had been left to Perry, one doubts that the Yangtze would have been "opened" at all, as far as American effort was concerned. But both Commissioner Marshall and his successor, McLane, had strong enthusiasm for it. Marshall had made the proposal to the Chinese in 1853. The following spring, McLane arrived at Shanghai and demanded of Governor General I-liang that he memorialize the throne to allow Americans to trade along the entire course of the Yangtze.

Then, on 23 May, without any reply, or any treaty provisions allowing it, McLane shoved off in *Susquehanna* and chugged out of bounds up the Yangtze. In company was a convoying tug, aptly named, to soothe Chinese sensibilities, *Confucius*. Not relying wholly on the pacifying qualities of a name, however, principally in the case of the Taipings, who had scrapped the Chinese gods, *Confucius* was armed with *Susquehanna*'s field piece, manned by a squad of U.S. bluejackets.

Although *Susquehanna* was going only so far as Wuhu, it is appropriate here to name the Yangtze river ports which would become well known to the gunboat patrols in later years.

Those cities specified as "open" ports under the various agreements—the Treaty of Tientsin and codicils which followed—were in their order upwards: Chinkiang (To Pacify the River), Nanking (Southern Capital), Wuhu, Kiukiang (Nine Rivers Meeting at One Place), Hankow (At the Mouth of the Han River), Shasi (The Market on the Sand), Changsha (Sand Spit), Ichang (Can Be Prosperous) and Chungking (Happy Again). Actually, Changsha is not on the Yangtze, but about eighty-five miles up a tributary, the Siang.

It was at these places, some of them enormous centers of population with large foreign enclaves, that one was likely to find a foreign gunboat at anchor, swinging with the current or snugged down at a pontoon breasted out from the slimy stonework buttressing the bund. In addition, there were others, not treaty ports, at which foreign flag vessels might call and pick up or discharge passengers and freight. It would be highly unlikely to have unearthed a foreigner in such places, with the possible exception of a stray missionary. Indeed, only one with a passionate love for the human race could have endured such spots, unless he first had taken the precaution of having been born and raised there as a Chinese.

The *Susquehanna* had an uneventful, if slow, trip until Chinkiang, where a shot was fired at her. There would be none of *that* sort of foolishness! The *Susquehanna*'s log noted that "at 12:20, Lieut. Duer was sent on shore with a written communication from Captain Buchanan to the Commander in Chief of the Rebel Forces at Chin Kiang Fu to demand an apology for the Gun fired at the ship."

With most un-Chinese lack of protocol, "Minister Yew," from the "Rebel Forces" was out before breakfast the next day with an apology. Apparently all was forgiven; at 10:45, "Minister Yew," Lieutenant Duer amiably accompanying, went ashore on "official business," which one suspects included a thimbleful or two of "samshu," a 90-proof first cousin to shellac which the Chinese consider a delectable wine.

By 1:30 P.M., Duer was back aboard, relations so firmly cemented that "the Rebel General in Command of the Forces on the north side of the River visited the ship," and "a party of officers visited the shore." A final entry in the log suggested that amity was complete: "Furled sails."

Available charts being extremely sketchy, a Mr. Eyers had been commissioned as pilot, but apparently his recommendations exceeded his qualifications; the log records that "when he was found to know nothing of the river, he was not permitted to give any direction." In more ways than one, *Susquehanna* was discovering the facts of riverine life.

The Duer brand of diplomacy was repeated at Nanking, where amidst caulking the foredeck, Captain Buchanan "allowed a number of Rebels to visit the ships."

By 31 May, a week out of Shanghai, doubts began to creep in. They had got under way at 6:42 A.M. and made good fifty miles on anchoring at 6:00 P.M. Since noon, they had been off the chart, beyond the British survey. The next day, once more under way dead slow, *Confucius* probing ahead, they reached Wuhu, where "the ship was crowded until sunset with natives."

Wuhu was for McLane and *Susquehanna* the end of the line. By 6 June, they were back in Shanghai, with thirty-eight coolies busy coaling ship, and the 24-pounder howitzer and nine sailors returned from *Confucius*.

So, although the Navy would not know what to call it for many years, the Yangtze Patrol was born in June 1854.

Her place secure in history, *Susquehanna* settled down to routine Shanghai ship-keeping. On 17 June, the USS *Plymouth* was towed downriver by *Confucius* toward the freedom of the open seas. The crew of the *Susquehanna* "manned the rigging and gave three cheers, the other vessels of war in harbor following our actions." On 20 June, they fired a 21-gun salute "in honor of anniversary of accession of Queen Victoria, English flag at mizzen and American at fore."

There was another sort of salute on 23 June, this time with solid shot, when "the Flagship of the Imperial Fleet stood upriver and opened her fire on the Rebel battery in passing, which was returned." But this didn't spoil the day's fun: "From 4 to 8 lowered all our boats for a race. The first cutter won."

Sunrise of the Glorious Fourth brought the American flag to each masthead, the topgallant yards were crossed, a 21-gun salute was fired, and the band played "Hail Columbia." And the whiskey cask was depleted by twice the usual four gallons.

On 24 October 1854, the British and American envoys arrived at Taku Bar (port of Tientsin) to demand revision of the original 1842 treaty. The ship bearing the French envoy had got lost en route. The principal Anglo-American demands were for a plenipotentiary to reside at Peking, opening of the Yangtze to trade, and the right to build homes, buy land, and erect factories and godowns *anywhere in China*.

McLane must have liked what he saw up the Yangtze, for which Old River Rats should hold him in reverent memory as the first substantial mentor of their cause. On 4 November, according to Swisher, he told the Chinese representatives that "after having paid duty on American goods at Shanghai, merchants of the United States should be allowed, either using their own ships or leasing Chinese ships, to move them into the Yangtze basin without limitation or hindrance." Then he added the clincher: "The United States will provide them the protection herself."[7] The Yangtze Patrol, so recently born, was now being justified.

On 12 November, scarcely a tea break later, the emperor rejected McLane's "demand" out of hand, confiding to the Imperial subordinates that "these barbarians think only of profit; in their rushing back and forth, their purpose does not go beyond matters of trade and customs." The emperor comforted himself with the thought that McLane would not hold out much longer at Tientsin, thinking dangerous thoughts and making obnoxious requests. "Barbarians dread the cold!" the Chinese told each other; in December Tientsin would start to freeze up.

From the beginning of seventeenth century foreign penetration of China, the Celestials were hopeful of playing the powers off against each other. In 1851, Shanghai *taotai* Lin-k'uei (who had momentarily elbowed out acting-*taotai* "Mister Sam"), memorialized the throne. According to Swisher, he explained that "as to the United States, they do no more than follow in England's wake and utilize her strength. As feelings are not really cordial between them, they suspect and dislike each other. The Americans are not in the least worthy of our concern."

The 1854 negotiations and hopes for opening the Yangtze came to naught. It was therefore with some relief to the "allies" that a squabble over Chinese seizure of a British-owned *lorcha*, the *Arrow*, near Hongkong in 1856, once more precipitated hostilities. Variously called the Second Opium War, or the Arrow War, it led to the Anglo-French capture of Canton in December 1857, and of the Taku forts, below Tientsin, in May 1858.

Contrary to Lin's view that the Americans were not in the least worthy of concern, they had in fact at that time reached the pinnacle of their influence in Asia. This was in part due to Anglo-French preoccupation with the Crimean War, but there were positive factors on the U.S. side: The United States merchant marine was still relatively huge, and its ships were the fastest on the seas. Perry had just opened Japan, where Consul Townsend Harris had negotiated a brilliant commercial treaty. The British and French had been led to relinquish their exclusive pretensions at Shanghai, where the Chinese Maritime Customs had been set up under foreign, albeit British, control. All these things were in accord with American policy.

Consequently, Ho Kuei-ch'ing, an emerging statesman of vision, began to realize that China must depend on her own strength rather than on futile attempts to play off foreigners one against the other. But this change in tactics in no way suggested an increasing endearment toward the *fanquei,* whose ". . . mentality really is inscrutable." His 5 April 1858 memorial to the throne, quoted by Swisher, revealed that sentiment, and added that "the nature of dogs and sheep is difficult to analyze." But the Chinese were grateful that the Americans had not joined in the hostilities.

The Chinese soon reneged on the treaty forced on them in 1858, so in 1859 the Anglo-French once more moved against the Taku forts. This time they were in for a surprise. Trying to force obstructions in the channel with eleven gunboats, they lost three of them. In addition, eighty-nine troops were killed and several hundred, including Rear Admiral Sir James Hope, the CinC, were wounded.

Although the Americans took a back seat, it was here that peppery Commodore Josiah Tattnall, commanding the East India Squadron, coined the famous phrase: "Blood is thicker than water!" Hearing that Hope had been wounded, Tattnall bent neutrality to the point of towing a number of British troop-laden barges into action. Then he went aboard the British flagship to see Hope. "Blood is thicker than water!" exclaimed Tattnall, in explanation, adding that he'd be damned if he'd stand by and see white men butchered before his eyes. (Tattnall was a southerner, and perhaps the matter of color bore some influence.) "No, Sir!" he said. "Old Tattnall isn't that kind, Sir. This is the cause of humanity!"[9]

In 1860, the Allies returned with 25,000 men. The "war" was soon over and the treaty put into effect. It was to be the end of any Chinese pretense of sovereignty for the next seventy years.

"Two powers [Great Britain and France] had China by the throat while the other two [the United States and Russia] stood by to egg them on, so that all could share the spoil," said a British officer in 1858, just after the treaty signing.[10]

This the Russians straightaway did, in the form of what is now part of the Maritime Provinces of eastern Siberia. As for the Americans, their actions were stated in the words of Flag Officer Stribling of the East India Squadron, in a 29 March 1861 letter to the Secretary of the Navy:

*The opening of the Yangtze to British trade will make me hasten my arrival at Shanghai; and as soon after my arrival there . . . I intend to go as far as Nanking in this ship [*Hartford, *drawing 15 feet] taking the* Dakotah *and* Saginaw *with me. At Nanking, I hope to come to an agreement with the Insurgent Chiefs to permit the free navigation of the river by American ships. . . . Otherwise they could not navigate the river in safety. It is the more important for us, as we have not the force, if it was our policy to station a man of war in each port as the British do. . . .*

This was to set a pattern. In 1866, Commodore H. H. Bell wrote the Secretary that the British had forty-five warships on the China Station, of which twenty-nine were gunboats, while the United States had only five. In 1871, the USS *Benicia* made a trip upriver to find every port held a British gunboat, but no other.

In preparation, Stribling wrote *Dakotah* at Shanghai to gather all information available and if possible, hire a pilot who knew something about the river. *Dakotah* sailors did not look forward to a summer trip on the Yangtze; even in *January,* her fireroom had temperatures of 160 degrees.

Stribling's three-ship squadron, *Hartford, Dakotah,* and *Saginaw,* got under way from Shanghai at the crack of dawn, 1 May 1861. *Hartford* grounded lightly the first morning, otherwise the trip was uneventful. Stribling left the *Hartford* a hundred miles above Nanking, shifting to the *Saginaw,* as he "did not think it proper to run the risk of getting so heavy a ship on shore."

The specifications of the *Saginaw,* first naval vessel built in California, suggest the hardiness with which mariners of that day braved the distant Pacific. Of 453 tons, she had two 20-foot paddle wheels, six feet wide, plus sail for economy or when the coal petered out. On deck was the odd assortment of ordnance then in vogue: one 150-pounder Dahlgren rifle, one 32-pounder, and two 24-pounders.

Stribling was acting minister to China at the time, in which capacity he appointed a Mr. O. D. Williams as the first U.S. consul (acting) at Hankow. Also in the party was the inevitable and indispensable missionary interpreter, D. B. McCartee, who must have struck the right note in his labors, as the governor general at Hankow reported to the emperor that "his expressions were compliant, so they were received with courtesy and everybody was pleased.[11] The governor general, Swisher went on, sent instructions to Yochow, Changsha, and Changte to treat the Americans courteously, and for all the merchants to carry on as if nothing out of the ordinary was happening, there being no cause for alarm, ". . . nor should any trouble be started."

Stribling got to Yochow, penetrated Tungting Lake a few miles, but never reached Changte. In fact, there is no record of any U.S. warship every visiting there. Returning to Hankow, it was proposed that three godown sites be selected on Hankow's outskirts, below the British location.

In his report to the Secretary, Stribling said:

The display of our Flag here on board a national Vessel for the first time, is worthy of being noted. The presence of three Steamers on the river at the same time,

I think has been of great service to our Countrymen residing on shore, and their Vessels upon the river.

He also ventured the opinion that "the Yangtze is singularly exempt from difficulties," continuing that anyone acquainted with navigation on American western rivers would have no trouble at all. Alas! One sparrow does not make a summer nor can one Yangtze trip make an expert, as any Old River Rat reading Stribling's optimistic view would be quick to perceive.

McLane and *Susquehanna* in 1854 had been the trailbreakers. They had founded the Yangtze Patrol. But Stribling was the first U.S. naval officer to show the flag, not only in "Woohoo," but one hundred and fifty miles beyond the farthest open port, Hankow. Unfortunately, all this fine effort was at that very moment being set aside for five years; while Stribling was on his Yangtze cruise, he could not know that just a month before, 12 April 1861, the Civil War had begun at Fort Sumter.

As soon as Stribling got news of the Civil War, although a southerner, he issued a stirring appeal to officers under his command to respect their oaths of allegiance. Whether this, or remoteness from the scene, or the sheer physical difficulty of getting away, or all combined, held the line it is difficult to say. But defections, resignations, and bitterness in the Far East ships was not a problem. Perhaps it was academic anyway, as by 1862, the steam sloop *Wyoming* was the only active ship remaining on the station.

The war's effect on American commerce was catastrophic. By August 1861, seventeen U.S. ships, all sail, from 1,800 to 180 tons, lay in Hongkong Harbor. They were unable to get cargo and afraid to venture out. Shippers were worrying about the rich field for enterprising holders of letters of marque from the the Southern Confederacy.

On 22 August 1861, Stribling was ready to leave Hongkong in the *Hartford,* headed home, when he read in a Hongkong paper a disquieting article titled "A Southern States Privateer," in which the U.S. naval storekeeper at Shanghai prominently figured. One could almost feel the glee with which the British viewed the collapse of their once bitter rival, the American merchant marine:

Well, we certainly must say that the disruption of the United States has soon assumed those calamitous characteristics for which civil wars are so celebrated. The U.S. naval storekeeper at Shanghai was a politician named "Judge" Cleary—we believe he earned his title from having been a magistrate in California. It seems from last advices from Shanghai that this man, in connection with "Colonel" Ward (the celebrated filibuster), Captain Allen . . . and Captain Lynch . . . purchased the schooner Neva, *equipped her from the USN stores, and intend to cruise off this coast as a privateer.*

The tale was soon exposed as pure invention, most probably to further damage American shipping. With the exception of the short cruise of the famous raider *Alabama* in those parts, privateers were no problem, simply because American commercial shipping had disappeared from the eastern seas.

The privateer hoax dissipated, *Hartford* sailed for home to make everlasting history at Mobile Bay. "Colonel" Ward, the "filibuster," when next heard of, had formed and led a highly effective mercenary force, "The Ever Victorious Army," instrumental in the downfall of the Taiping insurgents. As for those "young" River Rats who had had so precipitously to leave Shanghai's delights for the more pressing affair back home, they missed one of the outstanding local natural phenomena of all time: in early January of 1862, the city and environs lay paralyzed and silent under *three feet of snow!*

Concurrent with the American Civil War, China experienced another civil war, far bloodier and equally decisive as to the ultimate course of a great nation. This was the Taiping Rebellion, which rent the Yangtze Valley for a decade. Its final fiery apogee and extinguishment were dimmed for Americans by preoccupation with troubles at home, but the cacaphony lingered on.

Yangtze travelers who followed McLane might have noted a pinkish tinge to the river's yellow waters, and for good reason. During the decade ending in 1864, the blood of some twenty million Chinese drenched the Yangtze Valley, a mass carnage never exceeded in world history and only equalled by the Russian losses in World War II.

The ability of one man to create such havoc is familar in the careers of Napoleon, Hitler, and Stalin. Such a destructive despot was Hung Hsiu-ch'uan (Perfectly Accomplished), the self-styled "Younger Brother of Jesus" and founder of the Taipings.

Hung was a *Hakka*, a minority group which had immigrated to south China in the eighteenth century. Female members of that looked-down-upon group wore wide straw hats rimmed with a three-inch black curtain, a headpiece familiar to old China hands. The innocent trigger in the whole affair was the first Protestant missionary to China, the Reverend Robert Morrison, who arrived in Canton from Scotland in 1807. One of his biblical tracts fell into the willing hands of Hung in 1835. Like any other Chinese, Hung was not averse to receiving something for nothing, but he apparently put the tract aside without having been immediately swayed by its message.

In 1837 he took the government examinations which could open the door to success and riches for even the lowliest citizen, if he passed. But Hung failed. Physically falling to pieces, 24-year-old Hung spent forty days in a sort of trance, during which he had visions. Meeting the Heavenly Emperor in the 33rd Heaven in one of these, he was told that "You, too, are my son!" His troubled heart, he believed, had been cut out and replaced with a bold new one. As the Chinese were the first to use liver as an antidote for anemia, iron-bearing seaweed for thyroid disease, and ephedrine for colds, it is not surprising to discover that they also were first in heart transplants.

New heart or no, Hung failed the examinations again in 1843. That did it! He rediscovered Morrison's tract and was gratified to find it miraculously dovetailed with the pattern of his uplifting dreams. By 1847 he had gathered a band of fellow believers and by 1850 was in open rebellion against the local authorities. The new group named itself *Taiping Tien Kuo* (Heavenly Kingdom of Great Peace), probably as wildly inaccurate a forecast as has ever been perpetrated.

The *Taiping Tien Kuo* were, basically, "do-gooders," an odd mixture of Christian morality and the Chinese theory that man is inherently good. Their chief appeal to followers was the war to be waged on the rotten dynasty, corrupt officials, and ruthless landlords. Gambling, whoring, and the dissolute life in general were all prime targets. In essence, the Taiping creed was a blueprint for another "cleanup gang" one day to follow, the breeding ground for an even greater cataclysm which would rock the ancient kingdom a century later: the People's Republic of Mao Tze-tung.

Hung's new religio-political doctrines espoused common property, no class distinctions, land reform, examinations, abstinence from alcohol, tobacco, sex and opium, the introduction of the Julian calandar, and throwing out the classical form of literature and replacing it with the colloquial. Again, like the later communists, they were violently intolerant of dissent—to the death.

The Taipings completely rejected the passivity of Taoism, Confucianism, and Buddhism, but felt a close affinity for the Christian missionaries. The latter were first charmed, then repelled. Initially, it looked as though God had at last answered their fervent prayers that the heathen Chinee see the light. Then doubts began to creep in. Hung, now become the *Tien Wang* (Heavenly King), had never got around to being baptized. As the younger brother of Jesus, it really seemed superfluous.

The foreign *taipans* also initially were favorably inclined toward the new sect. For their part, anything would have been an improvement on the infuriating intransigeance and obfuscation of the Dragon Throne. Even Commodore Perry was disposed toward lending the Taipings military support. This, it very soon developed, they did not need. In 1853 they captured Nanking. By 1860 they had invested Shanghai and were turned back at the city walls, as they were again in 1862, by foreign gunfire. Their strategy and tactics had cut the Imperial armies to pieces: they hit at weak spots, avoided strong well-fortified positions, used "fifth columns," retreated when attacked, harried the enemy when he stood still and attacked when he retreated. Treason and desertion were punished by death. Chinese law required a confession for conviction and beheading; it is perhaps superfluous to add that the means used to extract confessions were ingenious and effective. A slightly comical band of fanatics had become a huge, brutal, ruthless, destructive horde.

As early as 1856, the Taipings suffered a severe internal crisis. "Heavenly King" Hung had begun to give himself over to meditation and active leadership was taken over by others. There were moral decay, factionalism, and excesses at the top. Any dangers of self-genocide through sexual abstinence no longer had to be feared. The old vows of abstinence—women, wine, and opium—had been replaced by downright debauchery in the Nanking palaces.

The once fierce bannermen of the Imperial Manchus, grown effete and useless as soldiers through centuries of peace, had to be replaced by provincial Chinese armies in the war against the Taipings. But it was like inviting in the ferrets to rid the henhouse of rats: the rats were vanquished but the ferrets became a greater scourge. The new armies were in fact the precursors of the "warlord" armies which would rack, rob, rape, and ravage China for another hundred years. From them would descend the ragtag half-bandit, half-parasite wretches who during the existence of the Yangtze Patrol would snipe at its gunboats.

The Taipings' downfall was as rapid as their bloody, flaming ascent. In 1864, Nanking, the Taiping capital from which they ruled a hundred million people in the Yangtze Valley, fell to the combined efforts of the new Chinese armies and the "Ever Victorious Army" of the British General Charles Gordon. This remarkable Chinese force, officered by Filipinos and Europeans, was organized by the American adventurer, "Colonel" Frederick Townsend Ward. Ward was killed in 1862 and his place was taken by an even more adventuristic Franco-American, Burgevine, who in turn was replaced by Gordon, then a major in the British Army.

Gordon, an officer of impeccable moral character and honesty (both near unique attributes in contemporary Chinese military circles), whipped his small force into a highly efficient fighting machine. Of its 130 foreign officers, 35 were killed and 73 wounded. In four years, the Ever Victorious Army had lived up to its name by capturing some 59 major objectives. At Nanking, the final one, sappers mined the 60-foot high city wall with 40,000 pounds of gunpowder. Through the resulting 150-foot breach, Gordon's troops charged for the *coup de grace*.

Awarded the yellow jacket and peacock feather by a grateful emperor, Gordon left China disgusted at the perfidious nature of its officialdom. Not long thereafter, this remarkable man met his death at the hands of the Mad Mahdi's Whirling Dervishes in the Sudan, where he commanded British forces.

The effect of the Taiping Rebellion, beside the horrendous loss in human life, was a paralyzing of trade and movement in the Yangtze Valley for ten years. In this, foreign trade suffered equally with Chinese. Whole cities, such as Chinkiang, were wiped from the map. Nanking, a city of a million, was reduced by half, and even in the mid-twentieth century great open spaces inside the walls remained where streets had once teemed with people.

The foreigners had chosen the side of the Imperials, in that as a result of the 1858 war, sweeping concessions had been wrung from Peking which made China in effect a European colony, with her finances in European hands. If the Taipings had succeeded, the foreigners reasoned, the terms would have been less satisfactory.

It so happened that the worst excesses of the Taipings had come during the American Civil War, when U.S. efforts in the Far East were at a nadir. Thus, by holding the British in check on the Yangtze during that period of enforced nonparticipation, perhaps the Taipings were American benefactors. By the time the American Civil War ended, the Taipings had disappeared from the Chinese scene. The Americans, made confident by the magnificent performance of their gunboats on the Mississippi, saw the Yangtze as just another slightly tricky but easily tamable stream.

1865-1898

The U.S. Comeback in China
Cruise of the Ashuelot
Szechwan Opens to Foreigners
American Decline on the Yangtze

Americans in the Far East, prior to the Civil War, played a mixed role. They had been far less truculent toward the Chinese than either the British or French, the other principal powers involved. But thanks to the brilliant diplomacy of the American envoy Cushing, they had gained "most favored nation" status, enjoying any privileges given others by treaty. This would have a profound effect on the lives of Americans in China for another century. Through "extraterritoriality" they would enjoy the same personal rights and guarantees as if they had been at their own firesides in the United States; Chinese law could not touch them anywhere in China.

Concessions and foreign settlements* had been established which were in practically every sense of the word a piece of the sovereign territory of the country concerned, and in which the controlling foreign powers retained the rights of policing and governing, delegated to a council of resident merchants.

There was a vast difference in the earlier American and British trading approach. The Americans were independent, with few government representatives in the Far East, and those generally noninterfering. The British merchants felt themselves representatives of their sovereign and under the control of their local authorities. This American independence, with its lack of coordination and control, contributed to the uneven tempo of American commerce. In 1844, the governor of Kwangtung, at Canton, memorialized the throne that American merchant ships were gathering like clouds, in numbers equal to those of England. Twenty years later, the American flag was a rare sight on the China coast.

The Civil War had enormously enlarged the U.S. Navy and broadened the popular interest in international affairs. These extensions of horizon had definite effects in China. With the coming of peace, the Navy shrank once more, but its officers had had

* A concession, such as at Hankow, Kuikiang, and Tientsin, was a foreign leasehold where land could not be subleased to Chinese and from which Chinese could be individually denied entry. In a settlement, such as at Shanghai, foreigners might lease land directly from native proprietors, who could as well hold properties for their own use.

a taste of salt and a good sniff of gunpowder. They had fought a great war and no longer felt themselves "younger brother" in the presence of Britain's ubiquitous John Bull.

After the Civil War the United States again devoted some thought and endeavor toward the Far East, where the Flowery Flag was by then a rare sight on a merchant ship and nearly as rare on a warship. The once great American merchant marine which, in a few decades, had experienced a meteoric rise, had even more rapidly declined.

By 1853 the American merchant marine was at its zenith. Sixty-six clippers were built in 1852. These faster ships won high freight rates—almost double that of their stodgy British competitors. But the decline was rapid, sped by a financial panic and depression in 1857. An overabundance of ships brought about a near collapse of freight rates to California. Insurance rates were lower in the new British steamers than in the sailing ships which made up the American merchant fleet. And British ships were subsidized, while American ships were not.

The real cause of the decline of the American merchant marine was a total redirection of national effort, based on the opening of the West, the discovery of gold in California in 1848, and the completion of the transcontinental railroad. Young men on the east coast no longer went to sea, but headed inland. In 1855, a total of 75.6 percent of American cargoes was carried in American bottoms. By 1865 the total had dropped to 27.7 percent, and by 1880 it was down to 17.4 percent.

In 1866, the *Wyoming* was reinforced by *Wachusett, Shenandoah, Hartford, Supply,* and *Relief*. Even more interesting were the sister ships *Monocacy* and *Ashuelot,* brand new double-ender paddle-wheel steamers whose design had profited from Civil War steam power development and inshore operating experience. They displaced 1,030 tons, were 255 feet long, but drew nine feet of water. Like most steamers of the time, they carried a full sail rig. Armament was ideal for the mission: four 8-inch smoothbores that could be loaded halfway to the muzzle with grape shot to mow down mobs, two 60-pounder rifles for long-range accuracy against ships or earthworks, two 24-pounder howitzers for close-in pirate junks, and two 20-pounder rifles for anything from firing ear-splitting salutes to massive pot shots at wild ducks.

These two ships would live out their lives in the Far East. When the Spanish-American War broke out, the *Monocacy* was too decrepit to get off the pile of ashes and coffee grounds on which she sat and had to be interned in China. She was sold in 1904 for US$8,000 which, said Admiral "Fighting Bob" Evans, was about $7,500 more than she was worth. The *Ashuelot* became the first U.S. warship to navigate the Middle Yangtze to Ichang. She was lost on Lamock Rocks, off Amoy, on 18 February 1883.

Maneuvering the *Monocacy* was a nightmare, due to the unfortunate propensity of her single huge crankshaft to stick on dead center. It took half a dozen sweating engineers to prize her over with crowbars. This brought about special difficulties with a new skipper who had been raised in sail and was not about to put up with the shenanigans of any damned *steam engine*. During his first formal call on the new skipper, the chief engineer explained the dead center problem, suggesting the wisdom of keeping the black gang informed well in advance of maneuvering plans.

"I have bell, you have engine!" replied the skipper, testily, making clear what in future he expected of steam.

Shortly thereafter, as the ship approached a bamboo dock, the skipper rang up "stop." Naturally the engine stopped on dead center. The black gang, with no idea of what was going on topside, saw no need for haste. As they leisurely broke out the crowbars, there was a second bell, "back full." Then she hit—and a belated reply to the back bell set the engine turning astern, which stopped her, with most of the dock sitting crazily atop the forecastle. Then came a frenzied jangle from the bridge for "stop." Stop now? On dead center again? With all that blasphemy blasting down the voice pipe? Nothing doing! The big paddles continued to churn back. Soon the *Monocacy* was out in the stream again—and so was most of the dock, still on her forecastle.

The ration inspection entry in *Monocacy*'s log at Hongkong in 1866 gave no hint of what culinary treats "Cooky" might concoct in the galley. From the storeship came "I bbl butter, 1 cask flour, ½ bbl dried apples, 1 bbl brans, ½ bbl molasses, 5 boxes soap, 11 bbls pork, 10 bbls beef, 1 bbl sugar, 10 cases pickels, 4 boxes coffee." There was no tea; apparently the Navy had already given that up in favor of its "jamoke"—strong, black coffee.

The daily issue of a half pint of watery grog had been abolished in 1862. Even before that, sometimes as many as two-thirds of a ship's company would relinquish their grog ration for a six-cent augmentation of their pay. This would be available for one glorious lost weekend on the next rare occasion when overnight liberty was granted. Most skippers in those days were "sundowners," who required liberty men to be back aboard before dark.

Daniel Noble Johnson, a ship's clerk in the *Delaware*, kept a journal which described the inevitable collision between sailors and rum when long liberty was granted:[11]

> *Part of the men who have been on liberty have returned aboard, but the majority of them in a condition that rendered it necessary to confine them, others were so beastly intoxicated as to render them totally insensible, and they were brought over the gangway and lain upon deck like so many logs of wood.*

Johnson's description of life on board ship suggests why an occasional blowing off of steam was not to be unexpected:

> *The men are confined exclusively in their lounges in the fore part of the ship. The luffs pace the weather side of the poop or quarterdeck at sea, as their exclusive provinces; the reefers as a matter of fact approximate whatever may be left of the official privileges of the ship, and exact their full honors, as well as a due share of cringing subservience from their inferiors in rank.*

Some of the more enlightened skippers endeavored to encourage amateur theatricals, fo'c'sle flautists, and squeeze box artists in an effort to alleviate the crushing boredom of the close confinement and beastly living conditions, the only other outlet being hard work. But such skippers were rare.

The merchant sailors in the Far East were a rougher lot than the men-of-warsmen.

Many had been in effect impressed by waterfront joints that first got the men drunk, then for a fat fee delivered them insensible aboard some ship about to sail.

Following the California gold rush in '49, the West Coast became a dumping ground for riffraff of the lowest sort who jumped ship in San Francisco, discovered that the streets were not paved with gold, and in many cases then made their way out to the Far East. Humphrey Marshall, U.S. commissioner to China, reported to Washington in 1853 that "there are now in this port [Shanghai] at least one hundred and fifty sailors ashore, men of all nations, who go into the Chinese city and drink and riot and brawl, daily and nightly. They presume to defy all law, because they have tried the jail and find they cannot be confined in it. . . ."

Commissioner W. B. Reed found that things were still rough all over in 1858, and informed Washington that ". . . there are consular courts in China to try American thieves and burglars and murderers but there is not a single jail where the thief or burglar may be confined." The French and British jails were inadequate for U.S. criminals, Reed added, and furthermore, he had no appropriation to pay for their keep. The hairiest part of all was that every vagabond Englishman, Irishman, and Scotsman plus anyone speaking what passed for the English language in those parts tried to claim U.S. citizenship, knowing full well he was safe from going to jail.

The naval officers, consuls, and *taipans* had their clubs, hotels, theatres, associations, and race meets. But what of the man in the bell bottom trousers and coat of navy blue? None of the better establishments would admit a sailor in uniform and sailors were forbidden to wear civilian clothes on foreign station. The men were in some respects in a jungle as soon as they set foot ashore in some of the largest cities in the world: Hongkong, Singapore, Tientsin, Shanghai, and Hankow. In 1889 J. W. Maclellan reported in the *North China Herald* that ". . . out of 10,000 Chinese houses [in the Shanghai foreign settlement] there were 688 houses of ill fame . . . not including opium shops."

Social diseases in the Far East were of an especially virulent character. Sometimes up to a quarter of a crew would be infected. At one time so many of the crew of the hospital and storeship *Idaho* were infected that they were unable to care for the patients. In 1869, Vice Admiral Keppel, commanding the British Far Eastern Fleet, proposed to the French and American Far East commanders that they cooperate with him in establishing medically inspected "facilities" in Japan, a prime source of infection, as had been done in Hongkong. To help maintain the quality of these establishments, and following the British lead, Rear Admiral Rowan, U.S. CinC, established a U.S. Navy first by instituting pre-shore-leave physical examinations of unmarried seamen under thirty-five years of age ". . . in private, so that the most sensitive would not be offended."

It may be appropriate to comment here on the apparent high degree of built-in immunity to physical ills in the early explorers which was not enjoyed by River Rats of a later day. The same bugs undoubtedly were proliferating in the Chinese filth when those first stalwarts beat their way along the rivers and across the passes, but their journals rarely reported illnesses. Yet even during the 1930s, boarding calls on warships returning to Shanghai or Canton from extended trips upriver were often received by chief petty officers because all the officers had been laid low by dysentery or related intestinal disasters. A French sloop lay at Hankow for weeks, unable to move downriver due to a mass outbreak of schistosomiasis picked up when her crew swam over the side in Tungting Lake. But it really should not have been surprising; many of these people lived as they would have in Europe.

In the river gunboats, flyscreens were exotic tinsel peculiar to Americans. Foreign gunboat officers would move the mess room table out on deck to escape the sizzling wardroom heat. There, flies by the hundreds would slog around up to their hips in the butter and troop over the pickles and anything else cool enough to bear the traffic. Their last previous stop would have been ashore, from whence they would stagger out to the ship under a burden of assorted bugs that would have delighted any microbiologist.

The thirsty sampan coolie simply dipped his cup in the river for water that would have a River Rat hotfooting it for sickbay for an internal and external disinfecting if he had had the misfortune to fall overboard. It is not surprising that the ordinary foreigner, lacking the natural resistance the Chinese had built up over the millenia, in the absence of strict precautions would soon be flat on his back, or at the very least, wearing a path to the toilet.

Mrs. J. F. Bishop, an English lady explorer of wide experience, solved the problem in an 1899 trip up the Yangtze and into Szechwan; she carried nothing but tea and curry powder. The former provided a sterile drink; the latter, doused in quantity on whatever the market offered, presumably smothered the germs as it disguised the taste. Judging from her description of restaurant food in darkest China, it was perhaps not unlike that often found seventy-five years later at roadside restaurants in the United States. "Ancient and fishlike smells abound," she wrote, "and strong odors of garlic, putrid mustard, frizzling pork, and the cooking of that most appetizing dish, fish in a state of decomposition, drift out of the crowded eating houses."[12]

Aside from the filth, disease, and degeneracy to which the men were exposed ashore, there were the potential horrors of wholesale contagion, which could immobilize a ship or even an entire fleet.

If sailormen occasionally were given to a bit of whoring, brawling, and boozing, it was not because the officers set them a visible example. "Usually they were an exceptionally fine set of men," wrote Tyler Dennett,[13] "sustaining a higher average, perhaps, than those of any other class. The naval officers were feared by the consuls whose delinquencies they reported, and not always welcomed by the civilian ministers who resented their lordly ways, but they had a fine regard for the honor of the flag under which they served and rarely disgraced it."

As for the consuls, Commodore Perry wrote:

Our country has no right to expect our consuls and commercial agents, many of whom were unfitted in every respect for their stations, either to represent or sustain the commercial interests of the nation so long as the system then existing was fol-

lowed. The fees at many of the places were our consular agents were accredited, it was notorious would scarce suffice to clothe them, and, accordingly, to eke out a scanty living, they were often obliged to resort to some sort of business, not of the most dignified character.[14]

Perry's views on consuls were shared by the Chinese. The governor of Kiangsu, one Chi-erh-hang-a, in a memorial to the throne, said that ". . . even barbarian chiefs know how to simulate loyalty and justice, and could make a reasonable judgement, whereas consuls, whose villainy has a hundred manifestations, are hard to reason with. . . ."[15]

Diplomats and dysentery were Perry's greatest burdens in China. Commissioner Marshall, the top U.S. diplomat in China, did his futile best to persuade Commodore Perry to forego his attempt to open Japan, making use of his ships instead as added protection at Shanghai and Amoy against the disorders then prevailing. Marshall's successor, R. M. McLane, an ambitious young politician, would be equally demanding on ships for protection and personal transportation.

For the edification of latter-day China sailors, most of whom learned to love Chefoo's Beach Cafe, Russian dancers, the Jew's Restaurant, Rose Creek, the Cloob 49, and sundry other delights, such was not always the case. Earlier-day Asiatic sailors detested the place, and even the old *Monocacy* felt their mood. Up through the China Sea to Chefoo, her engine wheezed like a weary, disconsolate man climbing uphill: "Cheeee—foooooo, Cheeeeee—foooooo!" But a month or so later, bound for the land of cherry blossoms and geisha, the old engine sang like a happy child skipping downhill: "*NA*ga-saki, *NA*ga-saki, *NA*ga-saki!"

In an interesting commentary on the *Monocacy,* "Fighting Bob" Evans wrote in his *Admiral's Log* that:

She had the peculiar distinction of being one of the few American ships that had ever been fired on by an enemy without returning the fire. When the Allied fleets attacked the forts at the mouth of the Peiho [Boxer Rebellion, 1900], she was fired on at her anchorage by a Chinese battery, and at least one shell passed through her rotten sides. The commanding officer did not return the fire, and we must assume that he was right, because the CinC, and afterward the government in Washington, approved his conduct. Such approval always makes the conduct of a naval officer right but in this case it did not add to our prestige either among foreigners or natives.

The new postwar CinC, Acting Rear Admiral H. H. Bell, had a rank introduced in the Civil War and commanded a force known as the Asiatic Squadron instead of the East India Squadron. A most energetic man, he very soon visited all the open treaty ports, including some where no U.S. man-of-war had ever been before, but failed to venture up the Yangtze. The Secretary had pointedly observed that unlike the great maritime nations of Europe, the Chinese government and people gave no encouragement or recognition to the rebels in the United States.

But forgetting British intransigeance, Bell soon was on good terms with his opposite number in the Royal Navy, Vice Admiral Keppel, cooperating in the control of pirates.* These gentry, Bell wrote Keppel, ". . . invariably boarded in calms, light weather, or at anchor, and overpowered by numbers, with stink pots, pistols, and muskets, and frequently in the very entrance of the ports open to trade."

In May 1866, cementing the friendship further, Admiral Bell moved the U.S. Navy headquarters ashore from silted-up Macao to Hongkong, rented godowns ashore for coal, supplies, and ammunition, and installed a Navy storekeeper.

The first postwar venture of the U.S. Navy up the Yangtze was made by Commander Townsend in *Wachusett* in August 1866. The ship visited Chinkiang and Hankow. Unfortunately, during the trip Townsend died of heat stroke. The *Wachusett* went up the river again in February and March of the following year, under the command of Commander R. W. Schufeldt, and visited "Chin Kiang, Kew Kiang, and Hankow, the open ports." In spite of rumors she would be fired on, Schufeldt anchored the ship at Nanking, where he and passenger McGovern called on the new, acting governor general under the pretext of delivering a message from the consul general at Shanghai. Schufeldt judged from "the anxiety of the officials at both Kiu Kiang and Nanking and from the great extent of the country devastated in a short time that the state of the rebellion was of a much greater extent and more successful than of late years." (This was the crushing of the vestiges of the Taipings by provincial troops loyal in a fashion to the emperor, and the beginning of the "warlord" armies which were to scourge China until the 1947 triumph of the communists.)

The American shipping people might have been spooky about privateers, but Yangtze pirates apparently didn't faze them. Schufeldt found "the immense freighting business on the Yangtze is almost exclusively an American trade—deserved to be fostered and protected by our government and if in the coming revision of foreign treaties, the American Minister were to succeed in opening the river to Ichan [*sic*]—400 miles of navigable water above Hankow—he would confer a great benefit upon the 'steamboat interest,' which in China must always be more or less of advantage to American enterprises and capital." It would be another twenty years before the Minister's and Schufeldt's dreams came to fruition.

With the exception of slight variations in terminology, Bell's 30 April 1867 report to the Secretary could very well be dated sixty years later: [*Wachusett*] "was in the Yangtze River, guarding the American steamboats and mercantile interests against the *Nanfi,* or robber bands, the Mandarins themselves soliciting Comdr. Schufeldt's assistance."

In 1867 a violent September typhoon came close to claiming one of Admiral Bell's ships. In those days, the mariner had only his barometer and the weather sense born of experience to tell him when to run for cover. During an average autumn on the China coast a ship could expect eight typhoons. The commanding officer of the *Monocacy,* Commander F. H. Carter, after his ship staggered back into Hongkong for repairs, described what hit him:

* As late as 1935, a Royal Navy office door in Hongkong carried on the tradition of the China coast. "Pirate Control Officer," it was marked.

Left Hongkong 7 September for Shanghai. The wind built up and the Monocacy was hove to under storm sails. At 7:45 A.M., the steering gear carried away and tiller on deck was resorted to, by which time it was blowing a hurricane.

At 1:30 P.M. the forestaysail blew away. The storm staysail was bent and set, but had to be hauled down to save it. At 3:30 P.M. the second and third cutters were lifted out, the davits of the latter being unshipped. Set the storm mainsail. The barometer at 28.44. Weather thick and gloomy and an uncommonly heavy sea running. The engines were kept turning slowly and the ship kept as well up to sea as possible.

*At about 5:00 P.M., the wind, which had hauled NNE to between E by N and SE, moderated and was quite light until 5:45 P.M., when it came out most noisily from S by E to S by W.**

At 5:50 P.M. the fourth cutter was carried away. We were taking in much water but the ship was bouyant and freed herself rapidly.

The fury of the storm from that hour until 9:00 P.M. exceeded in violence anything I had ever conceived and seemed to strike the vessel with the force of a heavy, solid body, and it was a matter of wonder that anything could resist its terrible force.

Just before 7:00 P.M. the launch was carried away, taking with it one of the davits. About the same time the fore and main topmasts went, and the fore-topsail yard was snapped in two at the slings. The storm mainsail was also blown overboard. . . .

At 7:30 the smokestack guys commenced giving and got up hawser and tackles to secure it. At 9:30 P.M. the smokestack went by the board, striking in falling the port fore braces and carrying away the foreyard. The steam soon went down. Increased after sail by hoisting as much of the head of the mainsail as would stand.

By 11:30 P.M. the typhoon had moderated. Steam raised and the engine worked slowly.

As the horizon commenced to lighten up about 3:00 A.M., the bone-weary skipper anchored his ship in thirteen fathoms of water, having been driven some forty miles by gales coming from all points of the compass. Daybreak revealed the craggy coast of China several miles to leeward. The escape from seas and rocks alike had been narrow. They had survived only through "the special Providence of the Supreme Ruler of the Universe." As a vital concomitant to this divine leg-up, however, the skipper did not fail to give credit to "the admirable bearing of the officers and crew of this vessel."

Fortunately, the *Monocacy* was new and tight. For another thirty-seven years she would beat her way along the China coast, in and around Japan and up the middle Yangtze.

If the weather was bad, at least the CinC could report that "health was good." In mid-July 1867, his flagship *Hartford,* with 30 officers and 350 men, had 38 on the

* They had experienced that rare and terrifying passage of the "eye," the very center of the storm, during which the barometer dropped to the phenomenally low level of 27.85 inches.

sick list—14 of them with "the Yangtze rapids," otherwise known as diarrhea or dysentery.

The Taipings and their leader, the "younger brother of Jesus" might be gone, but things were far from settled on the Yangtze in November 1868, when missionaries were roughed up at Nanking. The ubiquitous British, with a sloop and a gunboat, took matters in hand, and the affair seemed settled. But the Chinese backslid on their agreement. This time there was no equivocation. Up went line-of-battleship HMS *Rodney* with five supporting British warships. The Chinese immediately got the message; on 1 December 1868, the governor general of "the two Kiangs" (Kiangsu, Chekiang) warned the populace that they ". . . must not annoy religious establishments nor raise pretexts for disturbances, nor must they treat travelers with wanton disrespect. . . ." The populace was instructed to "obey with trembling." This was something less than the full treatment, the latter being, "Obey with *intense* trembling."

In such nuances and gradations the Chinese were magnificently adept at taking advantage of the ignorance and naïveté of the foreigner to make him "lose face," while imagining he had won a victory. A practice less subtle than damning by faint praise was the assignment of degrading ideographs to represent a foreigner's name as rendered phonetically. In 1834, Lord Napier, an unwelcome, bumptious British diplomat, was given the ideographs meaning "laboriously vile." Under such circumstances, it should have been apparent to the *fanquei* that his mission was hopeless from the start.

The issue of the Shanghai newspaper carrying the governor general's edict gave some additional interesting signs of the times. "New Patna opium" was 470 taels per chest (220 pounds), or roughly US$3.50 a pound. Malawa new, Malawa old, old Patna, and new Bernares were not yet available, hence unquoted. As a result of the Civil War, U.S. dollars were weak; it cost US$1.27 to buy one "Mex." This brought the Shanghai price of the best Rhine wine to the staggering equivalent of US$12.75 a case, a week's pay for a junior officer. By comparison, in 1932 an ensign on his magnificent monthly income of $143.28 could buy an excellent bottle of scotch whiskey in Shanghai for sixty cents.

Admiral Bell, true to form, died with his boots on—drowned 11 January 1868 when his boat swamped crossing the bar at Osaka. The squadron had been fortified by *Oneida, Iroquois, Unadilla, Aroostock, Maumee,* and a new flagship, *Piscataqua,* which had a variety of nicknames, most of them unprintable. The storeship *Idaho* predated by many decades World War II's floating logistics that freed the fleet from bases.

Bell's successor, Rear Admiral S. C. Rowan, apparently did not hold the same appeal for the British as his amiable, energetic predecessor. Perhaps a shooting affair in Hongkong in September 1867 had cooled relations generally. The officer of the deck of the USS *Supply* had wounded a Chinese boatman, resulting in a progressively heated exchange of letters with the British authorities over jurisdiction.

At any rate, Rowan was frozen out of a project he felt should be Anglo-American, which Vice Admiral Keppel, the British CinC, had earlier intimated. This was a survey of the Yangtze. "The Admiral has taken the matter in hand without inviting me to join," Rowan wrote the Secretary, adding he intended to remain aloof, ". . . as I know the Chinese are hostile to the work." There was a note of mild satisfaction in his concluding observation: "The Admiral's tender is hard aground above Hankow."

The whole thing had been set off by a letter from the Shanghai Chamber of Commerce in March 1869, apprising Rowan of a subject brought up in that august and powerful body concerning a revision of the Treaty of Tientsin. "The commercial importance of the rich province of Szechwan suggests the opening of its principal town, Chungking," wrote Chairman Porter. They had not sufficient information on the Hankow—Ichang stretch, but felt the class of steamer then employed on the lower river would do nicely. As for the upper river, the rapids and gorges, they knew nothing. Rowan was uninspired by all this, pointing out in his reply that in view of the reduction of U.S. forces on the station, no vessel was available to investigate. But he promised to ask the opinion of the Navy Department. As for Keppel, he was off like a shot, with so much enthusiasm that, as reported by Rowan, he didn't stop until he was high and dry on a mud bank.

Rowan might have used the double-enders, but clearly they were not his favorites: "*Ashuelot* and *Monocacy* are not fit for cruising, being helpless without steam," he wrote the Secretary. Besides, they were expensive to run; they burned 30 to 35 tons of coal a day of their 260-ton bunker capacity in order to make an all-out 11½ knots. To effectively use sail, a fourth of the paddle-wheel buckets had to be removed. Rowan suggested the ships be sold on station, as they were "of iron, light draft and speedy, and would sell well for Yangtze traffic."

The CinC took an equally dim view of another development: a recent friend become a possible threat. He reported that the Russians had taken possession of all Sakhalin Island, in the Sea of Okhotsk, and fortified the southern part. Their next grab, he thought, would be the Japanese island of Yezo [Hokkaido], "for the sake of the port of Hakodate, and she will take the north end of Nippon for the sake of the Bay of Awari [Awaji], and in twenty years Japan will be Russia, unless the foreign nations prevent her."

Things had been rough all over. On top of Keppel's indignity, the Russian bear's antics, needling by the Shanghai merchants, and the shortcomings of the double-enders, *Oneida* had been lost with nearly all hands in a Tokyo Bay collision on 29 January 1870. The *Monocacy* had just managed to get back on the line after having thirty percent of her crew down with smallpox, having neglected to vaccinate them.

If Rowan couldn't stay fate, he could at least put right some obviously ridiculous customs. For example, "setting the colors at nine o'clock in winter and beating the tattoo at 8:00 P.M." was something he felt could be changed. Only the Americans and the Russians followed this custom, he said, having borrowed it from England, where the sun didn't peep over the horizon before 8:00 A.M. Rowan was for 8:00 A.M. colors and 9:00 P.M. tattoo.

Rowan was willing to settle for "two seagoing vessels for the coast of Japan and two for the coast of China, with the flagship and a tender for the Commander in

Chief." He said nothing about Yangtze gunboats. As for *Ashuelot* and *Monocacy,* he felt they "are worthless as cruisers and are giving way, *Ashuelot* working like a basket."

American enterprise might have been dead on the high seas, but up the Yangtze it was going like a string of Chinese firecrackers. In November 1870, the wooden, screw-driven, heavily armed USS *Alaska* arrived at Hankow, reported by her skipper, Commander Blake, to have been the largest—2,400 tons—so far up the river. The new CinC, Rear Admiral John Rodgers, betrayed a touch of bellicosity slightly out of character for Americans in those parts, writing the Secretary that he thought Blake's visit "was important as showing to the Chinese that their cities are at the mercy of the foreign navies." Rodgers was even more pleased that "the Americans have from various causes almost monopolized the foreign trade and passenger traffic upon the Yang-tze Kiang, and here the American interests are greater than elsewhere in China."

Blake himself was enormously set up over the fine treatment he had received along the way. At Chinkiang the governor had even sent down his sedan chair to carry the skipper to his yamen, an honor usually reserved for the viceroy. The governor inquired of Blake whether he was a "war-man or a peace-man," to which Blake replied that friends gave him the character of peace-man, but when the occasion required, he would know how to take war measures.

Blake's upriver trip in *Alaska* was singularly unaffected by a violent antiforeign outbreak in Tientsin five months earlier, a phenomenon so typically Chinese of that period that it bears recounting.

It had been the practice of the French Catholic mission there to offer a small bounty for children given over to the orphanage, in the case of girl babies to save them from possible infanticide. Some of the proffered "orphans" had been kidnaped for the sake of the small reward. Others were delivered up in the last stages of some disease, to be baptized *in articulo mortis* to save a soul, then buried. It was totally beyond Chinese popular comprehension that this could be based on Christian charity. The real reason, they had been convinced by agitators, was that the eyeballs and other medicinal parts of the children were being ground up as something far more potent than pulverized tiger teeth or unicorn horn and sold at great profit in Europe. It was also a firm Chinese belief up the Yangtze that the vision of blue- or grey-eyed *fanquei* could penetrate several feet of rock, very useful in prospecting for hidden Chinese mineral treasures.

A trifle sometime sets the mob in action. A shot fired by an excited French consul surrounded by a menacing crowd triggered them on 21 June 1870. Soon seventeen French, some of them nuns, and four unlucky Russians taken to be French were brutally butchered. It was the opinion of Rear Admiral Rodgers that had the Franco-Prussian War not intervened, France would immediately have been at war with China, with major resulting alterations in the status quo concerning foreigners in general. However, as happened again thirty years later in the Boxer Rebellion, these disorders did not extend to the Yangtze, suggesting that China was indeed a dragon in fact, and like its ancestor the dinosaur, had one brain in its tiny head and another in its more important hip section, both brains feeble.

In January 1871, the 420-ton screw tug USS *Palos* arrived on the China station, the first U.S. warship to transit the new Suez Canal. Typical of the misfits the U.S. Navy consistently sent for river patrol duty in China, she soon drew the scornful disapprobation of the CinC, Rear Admiral T. A. Jenkins: "She burns a great quantity of coal, is slow, and draws too much water to go to many places which a gun-boat of her tonnage should be able to reach; neither her appearance nor her battery is calculated to produce respect for her."

On 4 October 1871, USS *Benicia*—coal bunkers filled and forty tons on deck—headed up the Yangtze. With a favorable tide and three boilers on the line at fifteen pounds pressure, the log recorded ". . . about six knots per hour. We steamed only from daylight to dark, using all sails when wind favored us, which was nearly all the way to Hankow."

At Hankow, ". . . the largest as well as the pleasantest treaty port on the river above Shanghai . . . very satisfactory visits . . . were exchanged with the *taotai*. . . ." The latter, a sort of mayor, was resplendent with an azure button atop his mandarin hat and a wild goose embroidered on his robe as badges of rank. A key official in the Chinese hierarchy, a *taotai* was just high enough to merit the equivalent of "excellency," standing 4a on a scale of 9.

There were 145 foreign residents at Hankow, and 45 at Kiukiang. As usually was the case in the smaller ports, an agent of the influential American firm of Russell and Company, Mr. S. C. Rose, was acting U.S. consul at Kiukiang.

As a sample of the continuing American interest in trade and shipping upriver, *Benicia*'s skipper found the U.S. consul at Hankow ". . . endeavoring to have the Yangtze opened to navigation some 300 miles further up [this seems to be his hobby] and no doubt will succeed if irrepressibility can accomplish it." Unfortunately irrepressibility alone could not turn the trick; it took another two decades, plus an assist from Japan's 1895 victory over China.

On the way downriver, "Nankin" proved to be a disappointment officially; the "Vice Roy Sin-Ko-Phau had left the morning of our arrival on a military inspection of the Province," a viceregal syndrome which had and would for long balk troublesome *fanquei*.

Benicia's skipper missed the viceroy but met a mandarin who ranked second to him, surprisingly enough a Scotsman, Dr. McCartney, former surgeon in the British Army. Assisted by four British workmen from the famed Woolwich Arsenal and three hundred Chinese, the versatile doctor was turning out "great guns," rifled field pieces, percussion caps for muskets, Hale's rockets, and "torpedoes" (mines) holding six hundred pounds of gunpowder.

The British were determinedly maintaining their grip on things other than the Nanking arsenal; in February 1872, Rear Admiral Rodgers pointed out to the Secretary that the British had upped their CinC to the rank of acting vice admiral to insure his seniority on the station.

In the last years of the Yangtze Patrol, before World War II, a River Rat enjoyed a breakfast of pancakes and eggs any style, per order, had his bunk made up and shoes shined by a Chinese "boy," and allowed the six Chinese boatmen carried by each gunboat to do all the dirty work on board. What was life like aboard *Alaska, Wachusett,* or *Hartford,* as they lay at Shanghai or worked their way up the lower reaches of the Yangtze in the mid-nineteenth century? Ship's Clerk Johnson of the *Delaware* described it in his journal as follows:

> *At the first appearance of day, the reveille is beaten and a gun fired to welcome the sun. All hands are then called to get up their holy-stones [large blocks of sandstone, used for cleaning decks] and sand, and from that time until 8 bells, the starboard watch is on the spar deck, and larboard watch on the main deck, and the steady cooks on the berthdeck, all busily employed hauling too and fro these immense stones, washing down the decks, drying them up, cleaning bright work, etc., etc. At 8 o'clock the decks are well dried, and as white as holy-stones and sand can make them. Breakfast is then piped, and the men receive their morning allowance of spirits of one-third of a half pint. At 2 bells or 9 o'clock, the hands are turned to, and the word passed for the uniform of the day, which in warm pleasant weather consists of white frocks, trowsers, and hats, in cool weather of blue cloth trowsers, white frocks and black hats, in uncomfortable weather comes blue jackets, blue flannel frocks and black hats, in which a man could double Cape Horn in the most unpleasant season with impunity. At 4 bells or 10 o'clock the drum beats to quarters and the different divisions are inspected by their respective officers, and wo' betide the unlucky wight, whose clothes are not in fit condition to pass around the Capstain, he receives his dozen with the well laid up colt of the nearest Boatswain's mate for "corking" while others were washing their clothes. As soon as the Divisions are reported to the executive officer who is pacing the Quarter deck, the drums sound the retreat, and the men knock about until 5 bells or 10:30 A.M. when the Boatswain and his mates call "All hands to muster," and every man makes his way quietly and orderly to the spar deck. It is a pleasant day and awnings are spread fore and aft, the Capstain is covered with a union Jack, seats are placed upon the poop for the officers; benches are also placed on the Quarter deck for the accommodation of the ship's company, this is a sure indication that we are to have a sermon, and prayers. The Chaplain makes his appearance, takes his place in the pulpit, and gives out the Psalm which is performed by our band, stationed upon the poop, for want of a proper organ. After the sermon is over, the benches are cleared away, and the crew sent aft to muster, this is the first Sunday in the month, and articles of war are read by the Captain's clerk, very much to the edification of the few persons who are within his effeminate voice. The crew are then mustered around the Capstain singly, and the petty officers inspected in the gangway. It is now hard on 8 bells or dinner time, and soon after dinner is piped accordingly and Grog served. At 2 bells or 1 P.M. the Hands are again turned to, and from that until sundown there is nothing transpires worthy of record. At sundown the drums again beat to quarters, and all hands are called to send down the top-gallant yards, after which come down all hammocks, and the day ends.*

China, in the early 1870s, was a composite scene of bewilderment, resignation, and incipient local conflict. The country was in effect a colony of the European powers, a humiliating condition for an ancient people whose culture was centuries older than anything Europe had to offer, and who, a mere hundred years earlier, had believed the Heavenly Kingdom to be the center of the world.

In 1862 France had gobbled up China's ward, southern Indo-China. Russia had completed the takeover of a large part of Chinese Turkestan (Sinkiang). Portugal, somnolent and feeble, had dealt the crowning indignity by confirming her *possession* of Macao, heretofore occupied only on sufferance and on the condition of her good behavior. A Britisher, Sir Robert Hart, inspector general of the Chinese Maritime Customs, dictated the five percent ad valorem duty which China might charge on imports, and held the purse strings of the country's chief and only reliable source of foreign currency. British forces, assisted by the French, had been instrumental in bringing about the 1864 collapse of the Taipings, who, at first the white hope and then the anathema of the foreigners, had strangled lower Yangtze trade for the previous ten years.

By the 1870s, American warships were a common sight between Shanghai and Hankow. On 19 January 1873, Rear Admiral T. A. Jenkins, the Commander in Chief, Asiatic Squadron, reported to the Secretary of the Navy concerning the November 1872 visit of the *Monocacy* to Hankow ". . . at the head of navigation." The *Palos* had been there in May 1873, for a five-day stay. The admiral himself, in the old Civil War flagship *Hartford,* had spent nine days in Hankow in October of that year, with visits at Chinkiang and Kiukiang en route.

As Admiral Jenkins sat watching the ponies at the Hankow Race Club, hearing British accents on every hand, he might have recollected that a scant twenty years before, the American merchant marine of five million tons—nearly all in sail—was about to surpass that of Great Britain. American tea clippers had once filled the roadstead of Hankow, but in 1873 not one was to be seen there. Russians dominated the tea trade, drank champagne exclusively, and preferred to live elsewhere than in the Imperial Russian concession, where they found the rules too confining.

Clearly, if Americans were not to lose out in the race for Chinese booty, it was high time they learned more about China—"look-see Ningpo more far," in the pidgin English of the godowns.

The opportunity came soon: on 15 April 1874, Jenkins reported to the Secretary of the Navy that the U.S. chargé d'affaires in Peking wanted a warship sent to Chinkiang to survey the channel. "The Chinese desire to close this highway. . . ," he warned.

The USS *Ashuelot,* under the command of Commander E. O. Matthews, was ordered to make the trip. After settling the "highway" business, she would proceed beyond Hankow, which had long been called "the head of navigation," and all the way to Ichang, where few foreign and no American man-of-war had ever been.

On 20 April, the "third-rate" *Ashuelot* paddled her way up the Whangpoo to Shanghai. There she commenced a three-month cruise, almost a thousand miles up the Yangtze to Ichang, gateway to the Upper River. After viewing the stupenduous rapids and gorges between Ichang and Kweifu, her skipper made some shrewed predictions on the feasibility and method of further upriver voyages. But another twenty years would pass before any U.S. ship, man-of-war or merchantman, would push up the mighty river beyond Ichang, up the roaring rapids and through the gorges to the end of navigation: Chungking.

On his return downriver, Commander Matthews submitted a report of the trip to Rear Admiral A. M. Pennock, commander of U.S. Naval Forces on the Asiatic Station. The opening paragraph of the report, dated 22 July 1874, established the routine for Yangtze sailors for seventy years to come:

> *On May the 3rd by order of Com-dr O. F. Stanton, senior officer present I landed 48 men and officers to protect the residents of Shanghai during a riot which was in progress.*

The outbreak had erupted in the French settlement, where attempts to put a new road through a long-abandoned cemetery had outraged ancient Chinese custom. Any foreign desecration of a Chinese graveyard would disturb the bones of venerated ancestors—a sin as heinous as disturbing the unseen, ever-present *feng-shui.** The mob reacted promptly and predictably, and by the time help had arrived—landing parties from two British and one French ship and the USS *Yantic* and *Ashuelot*—fires were raging and an attack on the municipal buildings was imminent. The American landing parties, in cooperation with their Anglo-French allies, had matters under control in the course of the day without having to fire a shot from the Gatling gun they dragged with them.

The riot, completely incidental to *Ashuelot*'s visit, cooled off in three days. The real reason for calling at Shanghai was to enlist more men to fill the ship's complement. Those were the days when desertions were common, disabling illnesses were frequent, and sailors convicted by courts-martial and wearing irons were being returned to the United States in every ship that sailed for home. As a result, ships were always short-handed, and many foreigners were enlisted.

The *Monocacy,* for example, had left the United States on 1 September 1866. Since then, she had had few American replacements and, judging by the records, she had few Americans left. "Send more Marines!" was, in effect, the appeal in a letter Admiral Jenkins had sent the Secretary in September 1873. "On this station," he wrote, "this deficiency [of Marines] is a very serious inconvenience owing to the class of men with which it is unfortunately necessary to man some of the vessels." Almost a year later, Commander Kautz, of the *Monocacy,* was still trying to requisition more Marines, "owing to the fact that the crew of this vessel has been shipped on this station, there are but few Americans on board, and there is therefore a great necessity for Marines."

By 1878 the CinC had become so exercised over the issue that he complained to Washington that the practice of recruiting all over the world had given the Navy a

* Spirits of wind and water, who dictated placement of buildings and temples, or almost any human activity that might be offensive to them.

mixed bag of seamen: waterfront riffraff, in many cases with no useful knowledge of the English language, and with a manner of living, morals, and principles woefully inferior to and at odds with what he considered American standards.

The results of the on-the-spot recruiting policy finally touched U.S. diplomats abroad. The U.S. consul general at Shanghai complained in a March 1871 letter to CinC Rear Admiral John Rodgers that he faced an intolerable burden in having to care for destitute ex-U.S. men-of-warsmen. Admiral Rodgers explained that only men shipped abroad were discharged abroad. If they were American citizens, the consul would have had the responsibility in any case—shipped or not. If they were foreigners, they remained foreigners and no responsibility whatever accrued to the consul. That ended that.

On 8 May 1874, with the Yangtze swelling to its summer crest and the *Ashuelot* presumably up to strength with Shanghai waterfront riffraff, she set off on a round of generally amicable visits with upriver dignitaries, surveying as she went.

At Chinkiang, by then fairly familiar to U.S. warships, *Ashuelot* found the British merchants and foreign missionaries well dug in and all traces of the Taipings' destruction papered over.

There the ship would have taken on fresh provisions, possibly from Middleton and Company, which had supplied an earlier American gunboat with the following:

17½	lb mutton	$1.72
12	lb roast beef	.84
4	lb bacon	1.57
3	lb vegetables	.55
2	tins peaches; 50 eggs	.91
6	bread	.60
4	fowls	1.00
2	tins butter	2.50
8	lbs white sugar	1.00
16½	lbs corned beef	1.34
2	tins tomatoes	1.00

Prices were in Mex dollars, at that time equal to US$1.25.

It is interesting to note the high cost of imported items, such as butter, tinned goods and sugar. No soap was listed, but an item for five bottles of perfume would imply that aboard ship then, as later, fresh water for bathing was sometimes at a premium.

Judging from another bill tendered at the same time, fresh water was not too often used for drinking, either. Jack Ayong and Company supplied the following potables to the account of a skipper who will be nameless:

1	cask ale	$15.00
1	dozen claret	4.00
1	" portwine	11.00
1	" quarts champagne	14.00
1	" ales	1.50
1	bottle best bitters	2.00

The order was repeated twice in the same month, plus a half-dozen each tumblers, liqueur, sherry, claret, hock, and champagne glasses. The skipper had either a clumsy messboy washing dishes or more guests than expected; he repeated the glassware order almost immediately. All this liquid cheer must have led to a desire for accompaniment by sweet music, as a subsequent order included a concertina, tambourine, triangle, and "set of bones." Remorse or exhaustion must have set in about this time, as the final entry on the month's bill was for a cask of soda water.

In the interests of fairness to well-cellared skippers of those days, it must be remembered that they were called on to do much entertaining of an official and semiofficial nature. In most cases, expenditures for such official entertainment were accompanied by very little official reimbursement—a situation which has changed but little over the century which followed. Commodore Perry, while on the diplomatic mission which "opened the door" to Japan, complained that he made a severe dent in his supply of drinkables keeping his guests in the state of inebriation which they appeared to find most pleasurable. Unlike the Japanese, the Chinese drunk was an extremely rare sight. Their infrequent tippling was reserved for the privacy of the home, where their faces could turn beet-red without public embarrassment.

At Chinkiang, *Ashuelot*'s Captain Matthews called on the *taotai,* an official with rank equivalent to "excellency," who wore an azure button on his mandarin cap and a wild goose embroidered on his robe. His name was recorded as Lee-Chang-Whoa. The Honorable Whoa volunteered to send a runner to Nanking with the news that the Americans were coming, so that the appropriate protocol could be set up for these exotic visitors from beyond the seas. At *Ashuelot*'s three and a half knots over the ground against the spring current, any medium slow runner could have taken a tea break or two and still have beaten the ship. Really important communications could be carried two hundred miles a day.

Although a major city, Nanking (Southern Capital) was less frequently visited by U.S. warships than smelly Chinkiang or burgeoning Hankow, in that it had not been opened as a treaty port until 1899. It was the seat of the *tsung-tu,* or governor-general of the provinces of Kiangsu, Kiangsi, and Anhwei, Lee Ching-shee. Lee, resplendent with a top-rank ruby button on his hat and a white crane emblazoned on the "mandarin square" of his gown, frequently had acted as Imperial foreign minister after the decline of Canton and the rise of Shanghai.

His experience in foreign affairs extended to knowledge of Formosa, named Islha Formosa (Beautiful Island) by the Portuguese who, with the Dutch and British, had been in and out of that island for centuries.* But the pickings had been too slim, international rivalries too sharp, an occasional Chinese administration there too tough, and the natives too savage to tempt anyone to remain there permanently. The Japanese were different, they had already pinched off some of the fringe on China's coattail, the "Loo Choo" (Ryukyus), and obviously meant to keep them. Now, the viceroy thought, the Japanese had chartered a steamer from their new friends, the Americans, to carry their troops to Formosa.

"I was careful to inform him," Matthews reported of his conversation with the viceroy's secretary, "that the American steamer which had been chartered was permitted to carry troops in the belief that the Chinese government did not object to the movement. As soon as it was found out that they did object, the permission was withdrawn."

The *Ashuelot* stopped briefly at Kiukiang, where Matthews was unable to see the *taotai,* who was reported as having been indisposed for some time. Anyone less naive and more experienced in Oriental dissemblance would have known the term deserved quotation marks.

From Kiukiang and its "indisposed" *taotai,* the *Ashuelot* pushed on to Hankow, even then a sprawling complex of three cities of over a million population and the hub of central China politics and commerce. There, the Americans exchanged calls with the inevitable *taotai,* plus far faster company in the form of "Lee-Chung-Ling, viceroy of Honan and Hupeh." Lee was none other than the brother of the famous Li Hung-chang, viceroy of Chihli, Guardian of the Throne and one of the foremost generals and statesmen of China for half a century. Consequently, what *Little* Brother had to say to Matthews should have carried some weight. It was, in effect a warning, in Matthews' words, of "possible incivility on the part of the inhabitants who, totally unaccustomed as they were to foreigners, might cause trouble by their rudeness."

Despite the viceroy's dim view of Matthews' prospects, *Ashuelot* chugged on upriver, where a day beyond Hankow she was met not by suspicious Chinese but by extremely friendly and hospitable Russians, who took the skipper and a party of officers thirty-five miles inland for a briefing on tea culture, a delicacy become a necessity which the Russians had been importing by coolie back and camel caravan to Novgorod and Moscow even before the time of Peter the Great.

Beyond Tungting Lake, the river "narrowed to half a mile," and the going was far less easy. With his clumsy, low-powered paddle-wheeler, Matthews complained about "short bends and strong chow-chow water which is so powerful as, from time to time, to take control of our vessel out of our hands and force us to back to prevent striking the bank." At one point the ship was going ahead 7¾ knots by log but moving astern relative to bearings ashore. She ran aground several times, once to the point of bending the port paddles against the side so that "four inches had to be sawed away from three buckets."

If the Americans found the river and the Chinese a new, fascinating experience,

* For a year (1857) the American flag flew over Formosa at Takow, through the efforts of American merchant Gideon Nye, trying to recoup his Hongkong bankruptcy. Commodore Perry strongly supported American occupation of Formosa as a coaling station, backed by U.S. Commissioner to China Dr. Peter Parker, who felt the United States should take over the island in retribution for Chinese intransigeance in fulfilling treaty obligations.

the sentiments were mutual: "During the whole passage the river bank was lined with people as far ahead as one could see. The inhabitants came running in every direction to gaze at the ship as she passed by."

Reaching Ichang at last, the Americans found the city to be more peaceful and friendly than it would become later, during the long existence of the Yangtze Patrol. The *taotai* was the very model of a good host; he proclaimed the arrival of the foreigners, was keenly interested in the ship, and when the captain announced his intention of going through the gorges, offered to supply chair-bearers, and to send runners ahead to announce the peaceful intentions of the visitors.

It is appropriate here to note the views of Tseng Kuo-fan, viceroy of Liang-chiang, as he expressed himself to the emperor in an 1861 memorial: "Of all the western barbarians, the English are the most crafty, the French next; the Russians are stronger than either the English or French and are always struggling with them. The Americans are of pure-minded and honest disposition and long recognized as respectful and compliant toward China."[16]

Notwithstanding the favorable views of the Honorable Tseng, six years and a thousand miles removed from the gorges, the Americans received a mixed reception. *Ashuelot* carried the first American naval party ever to view the gorges, and the account of their adventures and impressions, as written by Commander Matthews, is worth reading:

The party leaving Ichang in chairs found after about four miles, that it must work its passage—must make the trip virtually on foot; and from that time forward the going was a toilsome march of eight days over mountains and through gorges in boats; passing through a highly cultivated country, part of which is very rich, and meeting with large quantities of rice, tobacco, new potatoes, corn, beans, pumpkins, cotton and a variety of fruits, some strange, some familiar.

The scenery is very grand being vast but not grotesque. The natives are entirely good natured and well disposed, courteously returning a salutation, and refusing pay for the tea drank at their doors.

Reaching Kweifu on the morning of Saturday, 4 July [1874], we stopped on board of a junk while securing another for the return downriver. As our clothes were not in a proper condition for calling upon officials, I sent cards to the city saying that I meant to walk through the city that afternoon, with a party of officers, and would be obliged by the despatch of a couple of yamen runners. The men were immediately sent. Arriving at the gate, however, we found an officer stationed before it who said he had orders not to allow any foreigners to enter. Answer was made that we intended to pass in; and on seeing the yamen runners he stepped back. In passing before the French mission we were invited by a Christian convert to enter and see the Superior. We were received by Mon's J. Pous the French Missionary present. He told us he had had unofficial notice not to admit any foreigners. We also heard in several places that word had been sent to the inns and tea houses forbidding them to entertain any strangers from abroad. This would seem to show the spirit of the authorities in carrying out treaties.

At Kweifu I found a large number of junks loaded with foreign goods and having transit passes for Chungking and other places above on the river. These are detained by the Customs Mandarin in violation of the treaty clause regarding transit passes.

I was told that the Mandarins had even gone so far as to take some of the leading boatmen and bamboo them for bringing foreign goods here. I was also told that some of these boats had been detained since April.

On the morning of the 5th July I left Kweifu to pass down the river in a boat. The gorges are grand in point of scenery and interesting as a thoroughfare.

The width is in some places not more than a cable's length [600 feet] and such a compression of so large a stream produces chow-chow water and which sometimes threaten a boat rather seriously.

Having passed through the gorges I am satisfied that no ordinary vessel would be able to thread them; but believe that a steamer specially constructed, with two independent engines, capable of driving her 14 knots an hour, at least, and powerful steering apparatus, would find but little difficulty in getting up or down.

As Commander Matthews arrived safely at Ichang, his description of the terrors and pitfalls of the gorges and rapids were more muted than most accounts by those who traveled the river before or since.

The return trip to Shanghai was uneventful. The ship rammed a bank below Hankow, unshipped her bowsprit and carried away the foretopmast, but on the Yangtze such things were to be expected.

In that first adventuresome voyage up the turbulent Yangtze between 8 May and 21 July 1874, Commander Matthews and his small side-wheel command had added another chapter to the history of the U.S. Navy. The *Ashuelot* had established a thousand-mile-long route between Shanghai and Ichang that would be followed by countless American ships, navy and merchant, for nearly three-quarters of a century.

Ashuelot's trail blazing and the continuing interest of the merchants in going still farther into the hinterland to sell and buy arouses speculation as to what lay above the gorges which so attracted this keen attention. The province of Szechwan, so huge, so rich, so populous and at that time so independent, merits a special explanation. Szechwan was the goal of the British from the east via the Yangtze and the French from the south via any way they could get there. The Yangtze route has been well described. The southern route, and how the two rivals fared there, merits equal accounting.

After the Yangtze was opened to foreign trade in 1859 at Chinkiang, Britain considered herself, by right of conquest in the several Chinese wars, to be the senior Yangtze foreign power. This *de facto* situation became *de jure* following China's defeat by Japan in 1895, after which all China came up for grabs. Foreign "spheres of influence," which tended to metamorphose into "protectorates," then into colonies—proliferated. Japan took Korea. Russia occupied Manchuria and Port Arthur. On the pretext of a couple of murdered German missionaries, the Kaiser seized Shantung and its magnificent harbor of Tsingtao. France moved into Yunnan and "leased" Kwangchow Bay.

But the real plum fell to Britain; China agreed never to alienate the Yangtze Valley to any other power. The port of Wei-hai-wei, in northern Shantung, was an extra dividend—a salubrious summer resort for the fleet, plus a convenient spot to keep an eye on the Russians in Port Arthur, a hundred miles away. On 9 May 1899, the Under Secretary of State for Foreign Affairs, speaking in the House of Commons, defined the British sphere of influence in the Yangtze Valley as all the provinces adjoining the Yangtze plus Honan and Chekiang—an area of about 650,000 square miles with a population of from 170 to 180 million.

The only American assays toward ground rights in China were made in December 1900, again a year later, and once again in May 1902, when Secretary of State John Hay, under pressure from the U.S. Navy, sought a coaling station at Samsah Inlet, north of Foochow. But the Japanese, with newly acquired "inalienable rights" in Fukien, and in occupation of spoils-of-war Taiwan, just across the straits, effectively blocked *that.* Hay's efforts were not made public until twenty-four years after the event.

Actually, Americans were then almost wholly preoccupied with the Spanish-American War. Secretary of State John Sherman's mediocrity was a strong minus factor. But John Hay, then U.S. ambassador to London, had a more objective view. On becoming Secretary of State in the summer of 1898, he, along with the missionary element, pushed the idea of the "open door." The British, not adverse to freezing the status quo once they had the proverbial lion's share, were favorably inclined. In essence, it was that the foreigners were not to tinker unilaterally with the Chinese Maritime Customs, charge discriminatory rail rates through their several bailiwicks, or interfere with the administration of treaty ports, and on 6 September 1899, notes to this effect were sent by Secretary Hay to the British, Germans, and Russians.

But no foreign power ever succeeded in carving out a "colony" in vast, rich, populous Szechwan, where the Chinese central government itself had sometimes found difficulty in maintaining even a tenuous control.

Szechwan ("Four Rivers"—Min, Lu, Fu, Kialing) from the time of the very first explorations by foreigners has excited wonder and envy. A marvelously mild, equable climate and plenty of water allow continuous cropping. Timber, minerals, and what Marco Polo described as "rocks which burned like wood"—coal of mediocre quality—were all available in quantities. For centuries, salt brine has been raised from hand-driven wells 3,000 feet deep, making the region self-sustaining in that essential ingredient of human life. Since the year 100 B.C., the Min and Lu rivers have been split into many channels so that the whole great Chengtu Plain is abundantly and scientifically irrigated, turning it into one of the most densely populated, most productive agricultural areas in the world.

This central plain, the Red Basin, was so christened by the German explorer von Richthofen in 1870 for its heavy red sedimentation laid down in the early and middle Triassic Age, when southwest China, the Yangtze Valley, and Szechwan lay below sea level. Later, when, the land rose, the earth had split to the east, allowing the Yangtze to carve its bed through the mighty gorges of the upper river.

Frequently, as men pushed up the river, Chungking was called "the end of the road." In respect to its status as a treaty port on the river, this was so. Chungking was indeed the last one upriver. But merchant ships and occasionally warships as well, went beyond Chungking (*Pahsien* to the Chinese) to Suifu *(Ipin),* 238 miles farther upstream. There the Ta Ho, the Great River which Americans call the Yangtze, splits into the Min and the Kin'sha Kiang, the "River of Golden Sand." The Min squiggles off northward, toward Szechwan's capital, Chengtu. Only a short distance beyond Suifu, the Kin becomes an unruly series of rapids from then on to its source in the sky-high tablelands of Tibet, untamed and unnavigable even for sampans. This discovery was a sad blow to the French, as the Kin dips far southward in its course and would have made a magnificent outlet from their "colony" of Yunnan, straight to the sea at Shanghai.

By treaty, foreign businessmen could not own or even lease property in a non-treaty port. But this restriction did not apply to missionaries. These privileged parties were permitted (by treaty) to live, own, and operate anywhere in China that the inhabitants would suffer Christian jossmen to remain in competition with their own regiments of demons and gods. Thus, under the aegis of missionary protection, the international gunboats occasionally left Chungking behind and made their unfamiliar way upriver.

Suifu had been visited in 1861 by an English mission under Lieutenant Colonel Sarel. The Frenchman Francis Garnier had passed there in 1868, coming up overland from France's new conquest, south Indo-China, in the hopes of finding an outlet through the back door for Yunnan. In the next decade, Szechwan was literally criss-crossed by European explorers hoping to discover the most inviting area on which to stake out a claim.

In early 1875 there occurred an incident which added materially in its consequences to the opening up of Szechwan. An Englishman, Augustus Raymond Margary, exploring interior China, managed to get himself bushwhacked and murdered in Yunnan. An accurate account of what happened was impossible to come by, but speculation put the blame on the Oriental concern with "face." Margary, bored by the jouncing of his sedan chair and desirous of getting his daily exercise, was walking at the rear of his safari. Meanwhile, he had put his dog in the sedan chair, to spare the pampered animal's feet. On entering the domain of a rather important magistrate, the procession was greeted by this dignitary who, assuming that the impressive chair with four bearers and drawn curtains contained the august personage of the expedition's leader, bowed low as it passed. When the magistrate saw Margary afoot and learned that he had just made obeisances to a dog, there clearly were only two options: either take an overdose of opium and join his ancestors, or see that—the least objectionable from his viewpoint—Margary joined his.

Taken in his own favor, the magistrate's decision, if indeed the incident was as described, led in 1876 to the Chefoo Convention. Primarily, the agreement settled the Margary affair. A few coolies—the first within reach, most probably—were rounded up and beheaded. It also did a great deal more, such as controlling *likin* (local transit taxes on interior commerce), dealt with many sticky questions of trade, set up the protocol for sending Chinese diplomats abroad and for foreign diplomats in China, and opened Wuhu and Ichang on the Yangtze, plus Wenchow and Pakhoi elsewhere. But most importantly for Szechwan, it opened up Chungking to foreign trade, with the proviso it must first be reached by steamers. Indeed, if there was no statue erected somewhere in Margary's honor, there should have been. Or perhaps more appropriately, it should have been in memory of Margary's pooch.

The door thus opened a crack, the ubiquitous British soon had a foot in it, steamers or no. Their explorers—Baber (1878), Parker (1881), Hosie (1884), and Litton —were all around the province, measuring, probing, calculating, questioning, estimating, surveying, and recording in detail what they saw and found.

The Britisher, Archibald Little, great pioneer of the upper river, arrived in Chungking in 1883, fifty-nine days out of Shanghai. The twenty-one days spent sharing an upper river junk with assorted vermin and incessantly garrulous Chinese strongly reinforced Little's premise that what the Yangtze needed was not so much a good five-cent cigar to replace the foul stench of opium that filled his junk, but a reliable steamship line up the rapids to Chungking.

In 1889, Little built the small steamer *Kuling*. The Chinese, dumbfounded that somebody actually was proposing to accomplish the impossible, thus forcing them to honor their agreement to open Chungking when a steamer reached it, managed on one pretext or another to stop the *Kuling* at Ichang, where she eventually was sold to Chinese interests and ran on the river for two decades thereafter.

Nothing daunted, the indefatigable Little then built the steam launch *Leechuan,* too small to alarm even the junkmen as being competitive. She reached Chungking 8 March 1889, three weeks out of Ichang, the first fire-belching river craft that port had ever seen.

The nineteenth century brought wondrous things to the middle and lower Yangtze. But Szechwan remained locked behind the gorges until events of the twentieth century would pry open the ancient gates. For nearly a half century after that, the Yangtze Patrol and their foreign colleagues would fight to keep them open.

It would be repetitious to record in detail all the visits of U.S. warships up the Yangtze during the quiet years following the Chefoo Convention. The U.S. Navy maintained no such station ship system as did the British; ships were unsuitable and so few that generally only one a year was sent up the river. *Swatara* was a good example of a poor gunboat: 16 officers, 174 men, and 26 Marines manned one 8-inch, one 60-pounder, six 9-inch, two 20-pounders, one 12-pounder, one 3-inch howitzer (for landing) and a .45 caliber Gatling rapid fire gun. One marvels that the gunnery officer and his powder monkeys kept their sanity, attempting to supply that galaxy of weapons. But if the artillery failed, they could always close and board, for which purpose they were supplied with 83 cutlasses, 82 battle axes, 83 single-sticks, and 28 boarding pikes.

The focus of interest was on Japan, rather than China. Japanese ports were easier for the clumsy ships of that day to enter. Excellent coal was available at

Nagasaki, which the Navy used as a logistic base. Health conditions and climate were far superior in Japan. In 1872 the U.S. established at Yokohama a naval hospital that endured until after World War I. Many Japanese attended the U.S. Naval Academy, engendering a number of close personal friendships among the two navies, a situation that never was remotely approached in China. It is clear from the correspondence of the times and the operating schedules of ships that there was little question as to the Navy's preference. The land of cherry blossoms, geisha, and orderly people was the choice on every count.

After 1875, American commerce and naval power continued their steep decline. The report of the Secretary of the Navy for 1877 noted that "of the exports from China, more than three times as much go to Great Britain as come to the U.S., and with imports, the difference is still greater. The superiority which both England and France have obtained over us with reference to this vast trade may have been attributed in a large degree to the fact that each of these governments pays annually to its steamship lines between $4,000,000 and $5,000,000 in subsidies." By 1880, only 17.4 percent of American worldwide commerce went in American bottoms.

During this period there was at least one cruise up the Yangtze worth mentioning —that of the old side-wheeler *Monocacy*. "Two-side walkee," the Chinese called her. She was commanded at the time by Commander "Joe" Fyffe, about whom many tales, some perhaps less apochryphal than others, abound.

An entry in *Monocacy*'s log for 1 March 1877, when she was alongside a Shanghai wharf, is typical: "From Meridian to 4:00 P.M. By order of the commanding officer sent on shore a reward of one cent for the apprehension and delivery on board of Harry Johnson. Received on board in Navigation Department 8 dozen lamp-wicks and 15 chimneys."

The sequel to that one-cent reward was described some forty years later.[17] Johnson was something less than a sterling character, who was habitually absent without leave. Outraged by the one-cent reward notice, he appeared on the dock in short order and under his own power, only to be met and told by the executive officer, Mr. Clark, that he would not be recognized unless brought back in custody of the police. After trying unsuccessfully for some time to convince Clark he was worth more than a penny reward, Johnson finally accepted the inevitable and came aboard to collect his belongings and back pay before disappearing into the depths of Shanghai. He then turned to make his farewells. "Good-bye, me auld shipmates!" he cried out. "God bless you all! God bless you, Captain. And you, sir!" [To the officer of the deck.] "And God bless you too, Mr. Clark, to a sartin extint!"

The *Monocacy,* under way upriver the same year with her paddle wheels making twelve turns a minute on twenty pounds of steam, was a victim of the lower Yangtze's disconcerting tidal effects. First she stopped at Chinkiang, then Nanking, which was left behind so late in the afternoon that only about twenty-five miles had been covered before darkness forced the ship to anchor.

The destination for the next day was Poyang Lake, and an early departure was set. Fyffe was not feeling well and told Clark to get the ship under way; he would be up later. At the first faint sign of day Clark had the ship clunking merrily along. As the sun popped over the horizon, Clark climbed up on the paddle box to have a look. There ahead of the ship a large city appeared where none showed on the chart. To Clark's horror, it was no mirage; the city was obviously Nanking which they had left the afternoon before. The tide had come up the river during the night and swung the ship around; they had failed to check the compass in the morning darkness, and they were going the wrong way—down the river.

The problem now was to find room enough to turn around without having to back down on the paddle wheels, as that commotion would have brought Fyffe bounding topside in an instant. Clark managed the turn without incident, but Nanking was still in sight astern when Fyffe finally hauled himself sleepily to the bridge. It took all of Clark's ingenuity to keep the skipper occupied until Nanking had sunk out of sight astern, even to the desperate point of encouraging the "Old Man" to repeat for the hundredth time some of his favorite sea stories.

Naturally enough, after losing almost fifty miles, they never made Poyang Lake by darkness, which considerably shook Fyffe's confidence in the charts, the navigator, and *Monocacy*'s ability to eat up distance.

This cruise was for the specific and epic purpose of establishing the first U.S. consulate at Ichang. The city had just been opened as a result of negotiations following the affair of Margary and his footsore dog.

Racing to plant the British flag first, HMS *Kestrel* was *Monocacy*'s "honorable opponent." Well, mostly honorable, that is. For once, the chivalry of the Yangtze was in abeyance. There was no stopping *this* time to render the aid which inevitably would have been the case in normal circumstances. First, *Kestrel* would run aground. As she frantically laid out kedges to drag herself off, *Monocacy* would pass her with a cheer. Then *Monocacy* would pile up on a sandbank, by which time *Kestrel* would have freed herself and soon come charging by, her men waving hats and cheering lustily at *Monocacy*'s plight.

In this game of nautical leapfrog, *Monocacy* won the final jump. Having departed Hankow on 15 March, she dropped anchor at Ichang on a date that many hundreds of Old River Rats yet to come might consider singularly appropriate for arrival in such a backwash: April Fool's Day, 1877. Five days later, after the inevitable tortuous negotiations with the local Chinese authorities, seven guns banged out as commercial agent Bryant hoisted the U.S. flag. They had no sooner got back aboard ship before the Chinese hauled it down again. But this small contretemps was soon straightened out and to a salute of three guns, the standard parsimonious Chinese number, Ichang's *taotai* viewed the wonders of fire-belching, "two-side walkee" *Monocacy*.

The same year *Monocacy* also "opened up" for the United States the important commercial city of Wuhu. Consul Colby used *Monocacy* as his headquarters there until he could plant a flagpole and establish himself ashore.

By the last quarter of the nineteenth century, the two great American mercantile houses, Olyphant and Company, which eschewed the opium trade, and Russell and Company, which didn't, had gone bankrupt without replacement. But if the merchants were backward, the Navy was more so. In 1886, at the age of seventy-three, superannuated David D. Porter was still on duty as Admiral of the Navy, the top uniformed office, and strong for continuing the practice of rigging new warships for sail. "From the amount of fuel burned by our ships one would suppose we owned coal mines all over the world," the Secretary's *Annual Report* quoted him as pontificating. Some of the more progressive younger officers could see themselves being shot to pieces in action before they managed to clear the decks, furl those confounded sails, and hoist the smoke pipe. But Porter had that one figured out, too, in a vein which scarcely matched his Civil War aggressiveness: "A ship may sail her whole lifetime and never be fired on," he continued in justification. "She can use her sails to go all over the world without touching her coal."

If there was some doubt about design wisdom, there was even more about crews. By 1889, conditions were no better than when *Ashuelot* shipped her replacements from the ranks of Shanghai waterfront riffraff a decade and a half earlier. The Secretary's *Annual Report* stated: "The crews of our naval vessels are in large part composed of foreigners, or of men whose nationality is uncertain, and who are ready to serve any government that will pay them. It cannot be expected that crews so composed will be a safe reliance for the country if their services should be needed in war." It was indeed felt in Europe that the Spanish would have no problem with the U.S. Navy in 1898, as a result of such polyglot crews. But to the astonishment of many, including some of the doubters, the "riffraff of the foreign waterfronts" behaved as magnificently as they had over a century earlier in the *Bonhomme Richard* when John Paul Jones whipped HMS *Serapis*.

In the last decade of the nineteenth century, the Yangtze seldom saw more than one U.S. man-of-war a year, although in some cases the "station ship" routine of a later day was being approached. The *Monocacy* was upriver at Chinkiang, Nanking, Wuhu, Hankow, and Kiukiang from early January to July 1890. The *Palos* did the same from mid-February to mid-May, spending one whole fabulous month in Chinkiang. That trip was close to being her swan song. In January 1893, she was sold at auction in Nagasaki for 7,000 silver yen ($3,500), not bad for an old wooden bucket that had cost only $128,000 three decades earlier.

That same year, 1893, talk started in Washington about a ship *really* suitable for river use: steel hull, eight 4-inch, six 6-pounders, two 1-pounders (ducks and saluting!), and two of those famous early-day "pom-poms," Gatling guns. The ship should be 250 feet long, have a 40-foot beam, and draw 9 feet of water. The odds are against anyone's ever deliberately designing a less suitable Yangtze gunboat, although the French made a good try in 1931 with the *Francis Garnier*. The foregoing specifications resulted in the USS *Helena* and *Wilmington,* commissioned in 1897. They were poor sea boats, but did have a cutaway underbody aft, which gave them a short turning radius useful in river service.

After a 4,600-mile shakedown trip to the Amazon River and as far upstream as she could float—Peru—*Wilmington* arrived in China in May 1901. *Helena* had preceded her by six months. Both of them had long and generally useful lives.

For "YangPat," 1895 brought two developments of considerable interest. The most important was the first of the riverine revolutions in which the gunboats could play a role. The second was the alarming *drang nach osten,* the pressure eastward of the Germans, whose merchant fleet in the China trade was rapidly approaching the British in tonnage. Prussian engineers had constructed the Shanghai-Nanking Railway. The German gunboats SMS *Voerwaerts* and *Vaterland* were causing the British acute anxiety by their activity around Poyang Lake, where the resources were hitherto untapped by foreigners. In 1898 the German seizure of Shantung led the Kaiser to piously state that "Providence has wished that the necessity of avenging the massacre of our missionaries should bring us to the acquisition of a commercial port of the first order."

As to the role of the U.S. gunboats, in August 1895 the Secretary of State, following the paternalistic pattern, lectured the Secretary of the Navy on China. As if this were all news to the Navy Department, the Secretary of State wrote that "the current reports indicate that the entire basin of the Yang Tze, from the sea to the inland province of Szechwan, is permeated by an intense antiforeign sentiment, and outbreaks . . . are more than likely to occur." He added that the visit of USS *Peterel* in June to Wuhu and Hankow had produced some "temporary good" and ventured the suggestion that some craft of a like draft be held in readiness to again ascend the river.

The diplomats on the spot did less pussyfooting around. The American consul general in Shanghai, Jernigan, was all for a shoot-out. "Urge Admiral land marines and shoot the murderers of our ladies!" he stormed, as he described how the missionaries' houses had been stealthily surrounded and the sleeping ladies and children done in with spears in the most dastardly manner.

Evidently all this hue and cry brought results, as the 2,000-ton cruiser *Detroit* very shortly moored in the Whangpoo, called in her log the "Woosung River." She must have had disciplinary troubles, sitting it out in the Paris of the East, as every now and then she would steam off to Woosung bar for a week or ten days to clear the docket of mast cases and courts. The going reward for an overleave sailor's return was US$10, although one gunner's mate, apparently of less value, brought only "ten Mexicans." Smoking out of hours cost one sailor "one week on the poop after supper until hammocks." Some less fortunate culprits wound up in double irons.

Following the State Department's mild suggestion, the *Detroit* made a trip to Hankow, in April 1896. Preparing for the big adventure upriver, the little ship sharpened her teeth at the Yangtze's entrance by firing fifty-four slugs from her nine 5-inch guns and eighty-four from her six 6-pounders, half at anchor and half under way at 1,500 yards. If her four torpedo tubes launched any fish, no mention was made of it.

Setting clocks back or forward seven minutes at a time to maintain a near uniform state of daylight, *Detroit*'s trip up and down was uneventful. If incipient Chinese troublemakers were properly impressed, that was all to the good. One can be less sure that the foreign naval powers stood in awe; U.S. naval tonnage on the station totaled 18,000, but the British had 59,000, the Russians nearly the same,

France 28,000, and Germany 23,000. But though the U.S. Navy had fewer ships on the Asiatic Station in 1896, two years later it accomplished what the others would have dearly loved: the acquisition of the Philippine Islands.

1898-1918

The Upper River Opens to Trade
Post Spanish-American War Expansion
A British View of the Yangtze
The 1911 Chinese Revolution
The Great War of 1914–18

On 7 May 1900, almost as if celebrating the new century, the British middle-river gunboats *Woodlark* and *Woodcock* arrived at Chungking. They were thirty-one days out of Ichang, much of the delay due to *Woodlark's* grounding en route, which required twenty days spent repairing damage on the scene. In June of the same year, the 180-foot paddle steamer *Pioneer* made the trip in a more realistic seven days. But her 60-foot beam over the paddle boxes made her a poor candidate for permanent use in those parts. She was commandeered very shortly thereafter by the British Navy, to be prepared for evacuation of foreign refugees from Chungking, in case the Boxer troubles spilled over into Szechwan. The expected disorders never materialized, but *Pioneer* was retained by the Royal Navy, renamed HMS *Kinsha*, and served handily for many years, often as flagship of the British Yangtze Flotilla.

As a result of Baron von Richthofen's 1872–73 explorations in Szechwan, the Germans had become well aware of the resources available in the province. But their joss was not powerful enough against the River Dragon; on her maiden voyage their *Sui Hsiang* was wrecked—on Christmas Day of 1900—in the Kung Ling T'an, thirty-three miles above Ichang. Loss of life was heavy; the skipper's body was found down-river, four days later, well chewed by the Dragon's offspring—or perhaps crabs.

The French, meanwhile, had not been lying doggo. As far back as 1863, their gun-boat *Kien Chan* had penetrated upriver to Hankow. But it was not until the turn of the century that they really became active on the Yangtze. On 2 October 1901, the brand new built-in-France gunboat *Olry* (140 feet, twin screws, 11 knots, six 37-mm cannon, and twenty men) left Shanghai and on 13 November reached Chungking. It had been a bitter struggle. To claw up the rapids, the French had stripped *Olry* of guns, ammunition, and everything else movable, including most of her coal. Five percent lubricating oil was mixed with the remaining fuel to improve its heating properties. The movables had been transported upriver in junks, to be reinstalled above the rapids.

Their smartest move had been to hire the veteran British river expert, Captain Cornell Plant, who had learned his trade on the Euphrates River. Plant served France well and faithfully for a decade, until the penny-pinching French ministry of marine decided his salary of fifteen hundred francs per annum was too great a drain.

The real French contribution to navigation on the upper Yangtze was their tremendous effort in charting and exploring the waterways. It was, of course, not a matter of abstract philanthropy; by 1900, Yunnan had been well secured and by February 1910, the 520-mile railway from Haiphong to Yunnanfu was in operation. It was strongly hoped that the railway could be continued to Suifu, about three hundred miles cross country to the north, at the headwaters of Yangtze navigation.

The British, naturally enough, viewed these proceedings with a jaundiced eye. A French account complained that everybody in Chungking was cooperative except the British chief of Chinese Maritime Customs, who "looked maladroitly for ways to create difficulties."

The *Olry,* nicknamed "Soapbox" by her British rival HMS *Woodcock,* wasted no time steaming up to Suifu to establish the French Navy ashore. As Suifu was a non-treaty port, they could not purchase land there. But the missionaries could. "This subterfuge fooled nobody," wrote Captain G. Carsalade du Pont,[18] "but the Chinese authorities, without losing face, simply ignored it. Soon the French Navy's soccer field, barracks, and vegetable farm were in full operation."

A gunboat bound for Kiating had to steam up the narrow, crystal-clear, turbulent Min, where the current sometimes ran at eight to ten knots and a sudden drop in water level could leave a ship marooned 1,700 miles from Shanghai until high water the following year. The only recorded U.S. gunboat voyages up the river were made by the *Monocacy* in 1914, 1916, and 1924, and the *Palos* in 1920.[19] *Palos* made the trip with a frightening six inches of water under her keel all too often.

But to the British and French of 1902, the stakes were high. *Olry* had no sooner got her new Suifu digs in order than she set off on a cruise up the Min. *Woodcock* had already gotten a lap ahead by exploring fifty miles up the Kin to Pingshan, the head of Yangtze navigation.

On 11 July 1902, *Woodcock* kept the British lead by getting under way from Suifu an hour ahead of the *Olry*. It took the "Soapbox" twelve days to make those ninety tortuous miles to Kiating. But the Frenchman's bag of tricks was not yet empty. The little 60-foot steam junk *Ta Kiang,* described by Captain du Pont as ". . . of uncertain stability" and mounting a 37-mm cannon, joined the two gunboats at Kiating. There, *Olry*'s skipper, Lieutenant Hourst, "shifted his flag" to *Ta Kiang*. Loading her with rifles, ammunition, food, a bag of guncotton and three French seamen, he set out for Szechwan's capital of Chengtu, where disorders stemming from the Boxer Rebellion were threatening foreigners. *Woodcock,* upriver as far as she could go, had to be content to see the French unilaterally take on the defense of European dignity and safety in Chengtu, where both British and French consuls general were in residence.

Lieutenant Hourst's flambouyant little mission set off grave repercussions. The French home government, recently taken over by the Radical-Socialist party, was not only antimilitary and anticlerical, but as a result of the 1901 visit of the new King Edward VII, had engaged in a warm rapprochement with their centuries-old enemies and rivals, the British. What the devil were these cocky naval officers and their cronies, the missionary padres, up to, they wanted to know back in the Navy Ministry. A telegram from the Minister of Marine went straight to the point, storming that Hourst "... with six cannon and sixty men pretended to police Szechwan, become the help or menace of China, determine French policy, and bar the route to other European powers." It was not the proper role for a French naval lieutenant, they sternly admonished. How those three seamen became sixty, no one explained.

Hourst's superior, Admiral Maréschal, and the Navy Minister forthwith launched themselves into an exchange of vitriolic telegrams that delighted the readers of the public press in Paris and Shanghai. "I have the right and the duty to get on with the exploration of the Kinsha," shot back Maréschal, adding as a final insult, "If your cabinet offices would keep themselves informed of what was going on, they would perceive this." It is hardly necessary to add that Admiral Maréschal remained the incumbent Far East French CinC only a short while longer.

To settle any question as to what all this had to do with the Yangtze Patrol, it should be explained that without the French efforts at surveying, both gunboats and commercial shipping would have been heavy losers. By January 1903, two French naval surveying parties had completed a detailed charting of the Yangtze between Ichang and Chungking, having spent close to two years in the project. The Chungking-Suifu section had been done, too, as well as a rough survey of the Min to Kiating, with a detailed re-survey of the latter in 1912–13. In 1914, another party charted the Kialing as far as Hochow, fifty miles up. The British, in charge of the Upper River Inspectorate, had used all this information, in some cases without indicating the sources. As for the Americans, the urge to explore and survey was conspicuous by its absence. The sole known exception was Lieutenant Commander T. G. W. Settle, who as commanding officer of the *Palos,* collaborated in charting the Lungmenhao area of Chungking Harbor in 1934.

The greatest French exploratory feat of all was in a sense negative; it proved that the Kinsha was *not* navigable. In 1910 Commander Audemard determined once and for all to establish what the Kin offered as a path of commerce. He set out from Likiang which, approximately a thousand miles above Chungking, perches at an altitude of 6,240 feet. By sampan, his party drifted—or rather more appropriately—chuted downstream approximately 720 miles to Suifu, altitude 1,520 feet. Afloat actually only 102 hours, he spent the rest of the time walking around cataracts while the boats were let down by ropes. The average current was found to be 6.2 knots. There had been 410 rapids en route: 21 extremely violent, 60 very violent, 66 violent, and 263 merely ordinary. Captain du Pont put it succinctly: "On sait maintenant, grace à lui, tout ce qu'il importe de connaître sur le Kinchiakiang, et nottament qu'aucun service de vapeur ne pourra y fonctionner." In a word, "after *that,* we know all—no steamboats."

In 1903, an Australian, R. Logan Jack, wrote that Chungking already had commenced to take on some of the cosy small-town atmosphere that the post-1920 American gunboaters found there. The U.S., he said, was represented by Consul Smithers, with a British colleague named Fraser and a French one, Monsieur Bons d'Anty. The Standard Oil man, Hancock, was prepared to furnish oil for the lamps of Chungking, while seven missionaries were on hand to concurrently illuminate souls. Also present, to round out the bridge foursomes and read ancient copies of the *London Times,* was a strong contingent of British military: Captain Watts-Jones, Royal Engineers; Captain Watson, HMS *Woodcock;* Captain Hillman, HMS *Woodlark;* Dr. Burmeister of the Royal Navy, from HMS *Esk,* a cruiser then at Hankow; and *Esk*'s skipper, Captain Bingham. It would be another eleven years before those durable twins, *Monocacy* and *Palos,* would first show American warship colors at "the end of the line."[20]

At the dawn of the twentieth century, the Asiatic Fleet resembled nothing so much as a fat mother hen and a great flock of tiny chicks. Clustered around Rear Admiral George Remy's big armored cruiser flagship *Brooklyn* were twenty-four gunboats, all but six of them spoils of war or purchases stemming from the Spanish-American War. Capable of guarding the henhouse but unfit for high seas cruising were two short-legged bulldogs, the heavily armored and heavily gunned 4,000-ton monitors *Monterey* and *Monadnock.*

The new China gunboats were *Elcano, Villalobos, Samar, General Alava, Quiros, Pampanga, Pompey,* and *Callao.* Several of them were already fifteen years old. They would fortify American ship count on the Yangtze, but add little to prestige. They were even less suited to Yangtze river operations than *Helena* and *Wilmington* because of deep draft and low power.

The majority of the newly acquired gunboats were busy chasing insurrectos in the Philippines, an activity on which most U.S. naval interest in the Far East was concentrated through 1900 and into 1901. By mid-1901, Manila was still the center of operations for the fleet, but Aguinaldo, the insurrecto chief, had surrendered and the "Little Brown Brothers" were no longer a problem.

Thus, while the new CinC, Rear Admiral Frederick Rodgers, concurrently Southern Division Commander, personally hovered around the islands comprising that new slice of American empire, he felt free to allow the Northern Division, under Rear Admiral Louis Kempff, to range the China coast and visit Japan, as of old.

The *Monadnock* lay at Shanghai, where people were disturbed about the flooded condition of the Yangtze. Collections were being taken up for refugees—one dollar (Mex) would keep one person one hundred days. There already were two hundred Americans at Nanking. Three thousand foreign troops guarded Shanghai: one thousand British Indians, one thousand Germans, eight hundred French, and two hundred Japanese sailors. But no Americans. There was no cooperation between units. Everyone for himself. In the Whangpoo lay the foreign gunboats—two British, two French, one German, one Russian, one Japanese. But, again, no American.

In his summary of the preceding year, the Secretary of the Navy, in 1902, explained that "the existing conditions in the Yangtze Valley have not required an American vessel to be stationed at any point on the river." But *Helena* and *Wilmington* had been standing by on the coast, ready to charge in and avenge any affront to the American flag, provided the water at the scene of the indignity was deep enough.

In April 1902, fiery Admiral Robley D. ("Fighting Bob") Evans took over as Commander, Northern Division. His preeminent interest was China. Very soon he made a visit to Peking to meet the "Old Dragon," the empress dowager and real ruler of China, and her puppet son, the Emperor Kuang Hsu. The gunboats *Helena, Wilmington, Vicksburg, Monocacy, Elcano, Callao,* and *Villalobos* were taken under Evans' personal wing as part of the Northern Division.

Anxious to see all his new domain, Evans transferred his flag to the *Helena,* which was nicknamed the "jam factory" by the British because of her lofty stack. Crowded by the admiral and his staff, and the band, the ship set out for Ichang on 27 September 1902. Evans found the Chinese population dense and friendly, but could not claim the same charm for the dogs and cattle. "The former," he wrote,[21] "would bite on the least provocation; the latter, the carabao, a clumsy sort of ox with enormous horns, would attack us simply because we were not Chinese."

At Ichang, Evans learned that *Helena* was the largest ship (1,400 tons) so far to visit the place. He also discovered that following the usual pattern, the missionaries were confusing the Chinese as to which of the competing Christian sects provided the straightest road to Heaven. "The English Archbishop," he wrote, "rather complained of the advantage the French Bishop had by reason of the fact that his government, through diplomatic channels, had secured the right for a native of his faith to communicate *directly* with the Tao-Tai, or governor. In case of a land dispute . . . the French convert . . . could have a claim settled in his favor before the English Catholic could get into court, so to speak."

Evans also noted with interest the purely personal variations in approach: "The British Catholic Archbishop was in the orthodox dress of his order, while the Frenchman was in Chinese dress,* his hair in a long pigtail down his back. It was a great game these two shrewd men were playing, attempting to reform and 'civilize' a people who were highly civilized a thousand years before the nations they represented knew the meaning of the word, when they were, in fact, roaming wild in the woods."

Evans found a great many junks in the middle river flying American colors, a little matter that gave them a great advantage in reduced liability to customs duties. "Upon inquiry of our consul at Nanking—who was clean and honest, one of the best in the service . . . I was told that the vessels came from Shanghai and had secured

* It was a fact that even when Commander Matthews took the *Ashuelot* up the river in 1874, French missionaries had lived in Szechwan for several hundred years prior to that time, dressing in Chinese clothes, wearing either real or false queues, and otherwise appearing wholly indigenous.

papers from the consulate at that port." In typical "Fighting Bob" style, he requested the consul to notify the vessels "... that I would start an investigation ... and if found any not bona-fide American citizens, I would have them severely punished. Then most of the flags came down!" Evans added that the consul at Shanghai was later removed for various other offenses, "... and this consul was not the only one who came to grief in the same way." His views on consuls closely paralleled those of his illustrious predecessor, Perry.

Along with his other observations, Evans made note on how to drink tea in China. The tea-drinking ceremony in Japan is a complicated ritual, stylized and rigid in its execution. But the major burden falls on the preparer of the brew. In China, tea drinking is almost equally stylized, with the difference that the *drinker* carries prime responsibility for doing the right thing. Many an ignorant foreigner has perpetrated crudities which to a Chinese were as revolting as their custom of spitting on the dining room floor would be to an American. To prevent any such *gaffe* on the part of Evans and his aides during a call on Viceroy Cheng Chi-tung, at Wuchang, the viceroy himself set his guest straight, and Evans recorded the explanation:

Tea must be strong and without sugar or other condiment. When all have been supplied, the host, followed by his guests, raises the cup to his lips with both hands, tips the cover a bit with the forefingers, drinks about half the contents, and replaces the cup on the table. When the guest wishes to leave, he raises his cup as the host has done, drinks what remains in the cup, and takes his departure. If the host takes a second draught of tea without waiting for his guest to do so, it is a signal that he wishes to terminate the interview, and so it may be considered an insult.

A call farther downriver, on the acting viceroy at Nanking, was something short of ideal. The weather was hot and sticky, the Americans were gussied up to the nines in special full dress with choker collar, fore-and-aft cocked hats, swords and gloves, and everybody was sweating like a horse.

The principal object of the call was to lay a wreath on the coffin of the late viceroy, who clearly had gotten a distinctly spotty embalming job. The next item on the agenda was a Chinese banquet in an adjoining room, amidst swarms of flies swimming in air made heavy by the odors from the next room.

On 29 October 1902, Evans relieved Rodgers as CinC, and less than a week later took advantage of his total power to put out General Order No. 4. He had been heavily exposed by then to the formality of British wardrooms, which must have impressed him with their dress, the "bum freezers" (mess jackets) that the Royal Navy invariably shifted into for dinner aboard ship. The order shook the fleet and gladdened the hearts of the tailors in one fell swoop: "The uniform for dinner for all commissioned officers, except ... small gunboats ... with one or two officers, shall be ... either evening dress or mess jackets."

Admiral Evans repeatedly mentioned cholera and occasionally, typhus as being endemic on the Asiatic Station. He constantly was concerned about finding disease-free liberty ports for his ships. "Chefoo is the summer resort for North China," he wrote. "The city is filthy and Asiatic cholera can always be found there . . . but all Chinese cities are filthy, and all of them have cholera or something equally bad." He even tried out Vladivostok, where "the city was a lake of mud . . . saloons everywhere, large buildings, well-lighted, attractive by reason of women and wine, and full to overflowing, day and night. Champagne and vodka literally flowed over the door-sills." He cut his visit short.

In fact, by 1903, Evans had got to the point where he felt the only solution was the reintroduction of grog on board, after a hiatus of half a century. He strongly recommended the sale of beer and light wine to the enlisted men in the ships, where some of the less discriminating had died during the year by drinking shellac. He concluded that the principal cause of disciplinary problems was drinking ashore and smuggling liquor aboard. It might have been simpler to improve the drinking quality of the shellac, as nothing came of his recommendations.

Actually, ten years later, a retrogressive step was made: Secretary of the Navy Josephus Daniels, who strongly pressed for the wearing of pajamas by enlisted men and advocated that there be an officer at each enlisted mess table to suppress foul language, also did away with liquor in officers' messes afloat. The Navy's wardrooms have been dry ever since, although the pajama and clean language programs failed to prosper. Very few naval officers of the nuclear age would understand that a broken match carried on a tray by the captain's steward meant that the recipient was expected in the cabin for a predinner drink.

Prior to the prohibition edict, Navy wine messes ran like the present cigar and wardroom messes. One bought a share to join and paid at the end of the month for what he had drunk, calculated from his little pile of chits.

In 1913 Shanghai, a case of Haig & Haig cost the equivalent of US$10. Pre-World War I America looked down on gin as fit only for the dregs of society, but in China, where Gordon's best gin cost US$3 a case, officers soon took their cue from the British and became devotees of the prelunch pink gin or antiscorbutic gimlet.

It was understandably a source of wonder to most foreigners to discover that U.S. warships were "dry." After all, such proclivities scarcely were suggested by the habits of American men-o-warsmen ashore, sailor or officer. Governor Hung Ho-chien, of Honan Province, reflected this mild disdain at U.S. alcoholic ambivalence following a call on Rear Admiral Yancey Williams, ComYangPat, aboard his flagship at Changsha in 1932. Williams explained that his stay in town was shorter than expected, because of his fear of being trapped there by falling water.

"All I got on board his ship," Hung ruefully remarked later, "was water."

Evans wrote that he had once passed close to the British Mediterranean Fleet and counted twenty-four battleships and two divisions of cruisers. "We watched them with longing eyes," he said, "and wondered if we might ever see such a fleet under our own dear flag."

He must have felt a little like that when up the Yangtze. Although the *Susquehanna*'s trip to Wuhu in 1854 had marked the birth of the Yangtze Patrol, he saw no American gunboats on station, although there were plenty of British craft.

It was only after the Spanish-American War that suitable warships were available for the river and not until 1903 was there a "Yangtze Patrol" *de facto,* if not in name. That year the *Elcano, Villalobos,* and *Pompey* arrived in Shanghai and were placed under the general operational control of *Monadnock,* the station ship, whose three-stripe skipper was the senior officer of the lot.

The *Monadnock* was a remarkable old 4,000-ton monitor, built in 1864 and rebuilt in 1883 with "only the nameplate remaining original." She was station ship at Shanghai the winter of 1902–1903, spent the summer at Chefoo, shot her 10-inch guns at target practice in Nimrod Sound in March of 1904, and was back for a hot 1904 summer in Shanghai. In November, as a swan song, she visited Nanking, then after a gay Christmas in Shanghai, steamed down the Whangpoo for the last time on 15 January 1905.

In the first quarter of 1903, "Fighting Bob" had something more to worry about than bum freezers and bad embalming: *Villalobos* and *Elcano* had been busily sniffing around middle and lower Yangtze ports in furtherance of his desire to show the flag and find out what was doing in out-of-the-way places. In the course of this, *Villalobos* eased into Poyang Lake, hitherto a sort of never-never land, principally because of its shallow water and the nine-foot draft of most of the U.S. gunboats. The U.S. consul at Kiukiang had received a letter of protest from the *taotai,* who put forth the argument that the inhabitants of Poyang Lake district were "bad men." The consul read his letter to the commanding officer of *Villalobos* and appeared to agree that the *taotai* had something there.

This was not the sort of mishmash that Evans was accustomed to stand still for. He approved the action of *Villalobos*'s skipper and instructed him to inform the complaining *taotai* (and to hell with the consul, presumably) that the Chinese objections would not receive consideration. Furthermore, he could tell any other Chinese official who might raise objections that ". . . our gunboats are always amply provided for dealing with 'bad men,' and if there should be any indication . . . other than proper respect to American life and property on the part of these men, that they will be dealt with immediately, and that the gunboats will, without further instructions, administer severe and lasting punishment . . . and if these officials fail to provide proper protection it will be taken in hand by our gunboats, and that our gunboats will continue to navigate Poyang Lake and elsewhere wherever Americans may be."

Admiral Evans was greatly surprised when Minister Conger in Peking sided with the consul, and asked Evans for his authority. This Evans gave, as based on the broad principle that wherever the Chinese government allowed American citizens to reside, he claimed the right to send proper force to defend them in case of necessity.

In a letter to Minister Conger, copy to Evans, Secretary of State Hay stated that Evans was absolutely and entirely correct in his position and expressed surprise that Conger had differed. The gunboat action was based on the "most favored nation" clause in all the Sino-foreign treaties, which in effect said that whatever the other foreign gunboats did—such as sail Poyang Lake—so could and would the Americans. And that was *that!*

It was about this time that the only large American concession in the history of Sino-American relations was in the news. The United States had obtained the right to build the Canton-Hankow Railway, the concession involving a financing equivalent to US$40,000,000 and the railway to revert to China in eighty years on payment of the market value of the stock.

The project interested Evans because he would have American engineers and workmen deep in the interior of south-central China, with the only access being rivers three feet or less deep whereas his gunboats drew twice that and more. Operations had already begun and clashes had occurred over the engineers' natural desire to lay a straight track versus the natives' insistence they zig and zag around burial plots. Evans wanted gunboats that would float on wet grass. Or at least, with not more than eighteen inches draft.

But nothing came of the wet grass gunboats *or* the railway. In 1905 the Chinese for the first and last time in the period of mutual relations with the United States, staged a violent boycott of things American in reaction to the abusive attitude toward Chinese immigrants on the West Coast, and the concession was canceled. The Chinese finally completed the railway in September 1936.

In March of 1904, Evans was relieved by Rear Admiral P. H. Cooper. In July Cooper was relieved by Rear Admiral Yates Stirling. The gunboats still remained part of the Battleship Division, which included the *Wisconsin* and *Oregon*. Although the Secretary's *Annual Report* said that *Monadnock's* commanding officer "had general charge of the gunboats operating in the Yangtze River," they actually were playing it mostly by ear, the rest of the time operating according to the CinC's instructions via Chinese telegraph.

During the Civil War, Russian naval squadrons were present in San Francisco and New York, silent but clear reminders of which side Russia would be on if Britain came out in open military support of the Confederacy. Following Perry's 1854 opening of Japan, he and the Russian CinC, Admiral Count Poutiatin, were on the best of terms. Russian and American sailors of that day were buried side by side in the little foreign cemetery at Shimoda, Perry's "home port" in Japan. In 1858, following the Chinese defeat at Taku, the negotiators, British Lord Elgin and French Baron Gros, rode upriver in a British ship flying their two flags. Admiral Poutiatin, the Russian envoy, and American Minister Reed, went up in the Russian steamer *Amerika,* flying Russian and American colors. Indeed, it was this Russian-American "togetherness" that strongly influenced the British change toward a closer cooperation with the Americans and acceptance of them as near equals.

This eagle-and-bear cameraderie, albeit based on the practicalities of world power balance, remained steadfast over the years. Thus, in 1904, following the disastrous Russian defeat at Tsushima by the Japanese, Admiral Yates Stirling, CinC Asiatic, had an opportunity to make a small repayment in kind.

A few Russian survivors had managed to steam off in several directions. One destroyer interned under Chinese protection at Chefoo. Two Japanese destroyers, over the anguished but powerless protests of the Chinese commodore, took the Russian destroyer after a short scuffle with the crew and towed her to sea. The cruiser

Askold, badly damaged, fled to Shanghai. Under international law, she was eligible to receive repairs sufficient to ready her for sea, plus fuel to take her to the next friendly port. Three Japanese armored cruisers yapping at her heels in hot pursuit soon anchored in the mouth of the Yangtze.

That night, Lieutenant Yates Stirling, Jr., the Admiral's aide and son, sat next to the Japanese consul general at the *taotai* of Shanghai's banquet. Explaining to young Stirling that the Japanese squadron was commanded by Admiral Uriu, a graduate of the U.S. Naval Academy, the consul general intimated that the American course was obvious—stand clear. The British, traditional protectors of the Yangtze peace, wanted no part of *this;* the Japanese were their allies and the Russians their traditional bogy man.

Off Woosung sat the spanking new American battleships *Wisconsin* (Stirling's flagship) and *Oregon;* in the Whangpoo were the powerful monitor *Monadnock,* a U.S. collier, and five destroyers. The Standard Oil compound was cheek by jowl with the drydock in which lay fugitive *Askold,* licking her wounds. It would have been a real shame, figured Stirling, if some unfortunate incident occurred which might punch a hole in one of Standard's tanks and fill the river with flaming oil, or set off the excitable millions of Shanghai.

With that possibility in mind, Admiral Stirling moored the *Monadnock* off the entrance to *Askold*'s drydock. To make the cheese doubly binding, he moored the collier and three destroyers directly across the drydock's gate. The two battleships remained off Woosung, cleared for action.

About noon, a small Japanese destroyer appeared, headed for the Whangpoo entrance. "Follow her!" Stirling commanded Lieutenant George Williams, commanding the destroyer *Bainbridge.* Williams did even better; at four bells and a jingle, the *Bainbridge* boiled past the Japanese destroyer just as she was crossing the Woosung bar, coming very near to rolling her over on her beam ends. "So sorry! *Bainbridge* was simply in a hurry," explained Lieutenant Stirling to the infuriated Japanese skipper later, after he had anchored at Woosung to sweep up the broken crockery.

The Russian admiral soon called aboard *Wisconsin* to express his gratitude, bussing the astonished Stirling on both cheeks and giving him a proper Slavic bear hug of affectionate greeting. "After all," explained the embarrassed American, "it just happened that I was giving an annual inspection at this time, and it is of course customary to have all ships cleared for action."

Stirling took the *Wisconsin* upriver to Nanking in October of 1904, continuing on to Hankow in *General Alava.* He had already (in August) restated policy for *Elcano* and *Villalobos,* now the "Yangtze Patrol," leaving them a nearly blank check. *Monadnock* was still to have "general charge," but in view of lack of communications, the CinC left to each skipper's "discretion and judgment the cruising of the vessel under your command and you are authorized to visit any port of the Yangtze or its tributaries which you may be able to reach and at any time when you may consider a visit desirable to be made." Obviously, nobody got around to having the governor at Nanchang countersign and approve this. The *Villalobos* had been back to Poyang Lake, Nanchang, and the Kan River in May of 1904 and had met the same Chinese howls of protest as in the previous year. It was a pity that Nanchang could not have been on the list of ports frequently visited. Perhaps nowhere other than Peking was there such a store of curios and works of art; the city's twenty-two miles of walls had never been scaled by an enemy during its nine hundred years of existence. It had never been looted. In Nanchang the exotic things one associated with old Cathay—porcelain, tea, grass cloth, indigo, camphorwood chests, wood carvings, and rice paper—were readily available.

During this period, China Station CinC's arrived and departed like people in a revolving door. Rear Admiral Cooper, who relieved Evans, held the command three months. Stirling was headed back stateside in eight months. His successor, Rear Admiral W. M. Folger, held the command all of a week. None had had the time nor inclination to alter the zany organizational setup of the Fleet. But in 1906, events took a drastic turn. Apparently it was felt the only way to put things back in battery was to follow Lincoln's procedure in getting his militia company through a fence: fall out and fall in again on the other side. Thus, the Asiatic Fleet was done away with in April 1907, and consolidated with the Pacific Fleet. The former proud Asiatic Fleet that had recently won the Battle of Manila had become the Third Squadron, Pacific Fleet. Its Fifth Division contained the auxiliary tender *Rainbow,* the light cruiser *Concord,* and the "jam factories," *Helena* and *Wilmington*. The gunboats, at long last divorced from the departed battleships, had become the Sixth Division. The *Callao, Elcano, Quiros, Villalobos, Paragua, Pampanga, Panay,* and *Arayat* were scattered from the Philippines to Canton to Ichang. Like Chinese urchins, they had "no mamma, no poppa," but, finally, a unit designation of their very own.

Under Rear Admiral J. N. Hemphill, Commander Third Squadron, Pacific Fleet, there appeared in 1908 for the first time the designation—Second Division—that would cover the Yangtze gunboats until 1919. The Secretary's *Annual Report* for 1908 stated three of them were on the river, which strongly suggests he was not reading the papers.

A revolution had broken out in the Anking area in November 1908 and wives and daughters of missionaries had taken refuge aboard the British cruiser *Flora*. American Consul J. C. McNally advised that the missionaries "deplore the absence of our gunboats from their regularly assigned stations on the Yangtze." He added that no American warship of any size had been either on the Yangtze or at Shanghai since about the middle of September. "Who should I ask for a gunboat?" he plaintively if somewhat ungrammatically inquired. He usually asked the station ship at Shanghai, but now there was none, he said.

Minister Rockhill in Peking gingerly picked up the ball, to "venture to express the hope [to Secretary of State Root] that the Navy Department may see its way clear to return to service on the Yangtze River the two or three small gunboats which have been kept there for some years past. . . ." Bishop Roots got in a sly dig, suggesting that "American residents would appreciate the presence of American ships so that they might not be so wholly dependent upon the courtesy of our British cousins." There ensued a fast exchange of letters as the Secretary of State informed the Secretary of the Navy of the facts of life in China, and the latter rather lamely replied as to scarcity of forces.

The "Yangtze Patrol" was a body without a head—the stepchild of an orphan. By 1910, nobody could stand this absurd Pacific arrangement any longer. With radio still a horizon-limited toy, with mail thirty days en route from Manila to San Francisco, it was clear the whole thing was unworkable. To the relief of most people, the Asiatic Fleet was reestablished that year under Rear Admiral John Hubbard. The way events were developing on the other side of the world, it was not a moment too soon.

Our British cousins, as noted earlier, always carried a notebook and sharpened pencils, and their observations of the many facets of the Chinese scene were well done. Lieutenant Douglas Claris, Royal Navy, who served in His Majesty's 1,200-ton gunboat *Thistle* in 1911, filled his diary with a colorful account of life in Hankow and along the river. His notes provide an insight into China second only to being ordered to duty under ComYangPat.

"Ships always moor at Hankow," he wrote, "and the ground is good for holding. In the winter you have to sight your anchors once a week, but in the summer every three or four days and even then one sometimes has great trouble to break the anchors out of the mud which has collected on top of them." He then added the inevitable British remarks on shooting, sorrowfully noting that snipe were scarce thereabouts.

The climate in Hankow in 1911 was no better or worse than the last River Rat found it thirty years later. "On board HMS *Thistle* from July 22nd till August 3rd the temperature never went below 100° at night and during the day went up to about 103° in the shade." A familiar refrain followed: "The mosquitoes and flies swarm!"

The rest of the year it was cooler—in winter, almost frosty, and then people made out better. "One can obtain plenty of exercise at Hankow," Lieutenant Claris wrote, raising again the British preoccupation with "keeping fit." Riding was cheap, he reported, ". . . and the ponies go very well; in fact they go too well sometimes."

The Recreation Club had a race track, twelve tennis courts, year-around nine-hole golf course "in very good condition" and naval officers paid no subscription. The local cricket team could defeat anything the gunboats had to offer, although the latter sometimes put up an even go at football.

The River Rat of the mid-1930s might almost feel himself at home as Claris described the Hankow (town) club of 1911: "Very well run, with three billiard rooms, bowling alley, splendid library and bar." He carefully underlined "bar," and added a warning: "Keep away from it from 6:45 P.M. during the winter, especially Saturday nights. The members are a heavy drinking set and although good fellows, meet them at the 'race club,' not at 'the club.' " They played bridge at the club, too, which brought another discreet word of caution from Claris: ". . . it is just as well only to play with people you know really well, as some of the members . . . (no proof)."

Claris described the temper of the natives, which was not what one would loosely call chummy toward Europeans. "Antiforeign feeling is very strong," he wrote, adding that any excuse to form a crowd—fire, festival, or simply to gawk at something interesting—could, with the greatest ease, break out into a riot wherein foreign property, and lives as well, became threatened. "The scheme of defense [of the foreign concessions] is very complicated as all the big nations have concessions except America, and the various volunteer corps will not unite, but all want to defend their various properties."

Concessions or not, the Americans were there just the same. "All night leave is given here," Claris recorded, adding that "a certain amount of trouble often occurs ashore, as it often does in all the big ports on the river, between the British bluejackets and the American dittos. The Yankee has much too much money and buys up everything and our men try to clear them out of some of the bars and the result is usually a few broken heads and one or two people being locked up for the night in one of the foreign jails."

This obviously was not unique to Hankow; in describing life at Nanking later in his journal, Claris commented: "Very often a little trouble with Yankee sailors. Always an American ship in port." But these minor difficulties were offset by the magnificent shooting in the lower reaches of the Yangtze. At Chinkiang there was ". . . the best pheasant shooting in the river that I have seen . . . as many as fifteen pheasants at a time strutting about. . . ." The birds were flushed over the dikes by beaters, enlisted men from *Thistle*.

This Royal Navy passion for hunting was frowned upon by some of their "in commerce" countrymen, to judge by a November 1910 editorial in Hankow's *Central-China Post*. "Hankow has been honored by a second visit from the gallant British Admiral." sneered the editorial. "We use the adjective, 'gallant' and 'honored' because by long established custom they are those which are generally applied to men of that rank . . . but judging from what we hear of the real opinion of the British community they must in this instance be understood in the Pickwickian sense."

After considerable verbiage directed at the "gallant" British admiral, apparently for his failure to reinforce Hankow's volunteer defenders to their satisfaction, the editorial got in its wickedest digs: "But we must not be too hard on the Admiral, for they say that the unreasonable and unexpected outbreak of hostilities had hurriedly called him away from a shooting expedition when he was bagging pheasants galore—not peasants as another admiral did a few years ago in the same district. Now interference with his sport is the one thing which no gentleman can be expected to tolerate. . . ." Adding insult to injury, as far as Hankow's British community was concerned, was the declaration that HMS *Thistle* had been withdrawn from the troubled city" . . . in order that she may be used at Tatung as a telegraph post . . . ," referring to the little ship and her newly installed "wireless telegraph" set, posted as a relay station at some benighted halfway port.

The *Central-China Post* doubted that radio was here to stay: "If the new wireless on the British boats cannot carry the whole distance from one port to another on the Yangtze, then the British taxpayer has been badly swindled." Steam up, the writer went on: "This new wireless which has been installed on the gunboats promises to be a doubtful boon to the people for whose sake the boats are really here. If they are to be set apart for the exclusive duty of providing the Admiral with telegraphic information, and if their service in this connection is to be strictly denied to all besides, as has hitherto been the case, then the sooner the wireless is unshipped the better." The latter part of his tirade referred to the fact that British civilians were not allowed to communicate via the British naval radio, while the German naval forces were cheerfully transmitting the private messages of their widely dug-in citizens with the help of the German high power military radio station at Tsingtao, the flourishing new port wrested from the Chinese some twelve years before.

Claris didn't mention whether or not the free-spending Americans had been tossing their gold around Ichang, but apparently life at the headwaters of the middle river was something less than hilariously gay in 1911. There was a small recreation club and customs club ". . . where naval officers are very well received," and a "small naval canteen for the bluejackets." It was cautioned that very little beer was kept in stock, however, so that, "it is best for ships to bring up about 40 dozen quarts from Hankow if the ship is going to stay in port for anything over five days." Clearly, if the British tars' kidneys were not kept properly flushed out, it wouldn't be because Lieutenant Claris hadn't warned them.

If by any chance a critical mass had been established by the meeting of Yankee sailors and forty dozen quarts of beer, the results could have been coped with: "A good hospital for men and officers, owned by the French fathers, is situated just below Ichang. You cannot help seeing it as you go up the river as it is a huge block of white buildings."

At that time, Ichang was the head of merchant steam navigation, where cargo brought upriver by foreign and Chinese ships was transshipped to junks for the tortuous and frequently disastrous thirty- to forty-day struggle up the rapids and gorges to Wanhsien, Chungking, and beyond. "People who have business at Chungking," Claris explained, "go from here in junks or better still in 'red boats' which are employed by the Chinese government as lifeboats and are under the charge of the Customs." Claris of course knew that the Chinese Maritime Customs was under a British director general, and was one of the few organizations in moribund Imperial China that still reflected efficiency, honesty, and reliability. Thus, even by 1911, travel up the gorges to Chungking had changed not at all from those days in 1861 when Captain Blakiston, of the Royal Engineers, had blazed the way.

Lieutenant Claris did not spend all his time in Hankow fending off hard-drinking types at the club or knocking golf balls around. Like all gunboaters since the advent of foreign warships on the Yangtze, he was sitting on a powder keg. Sometimes one managed to stamp out the sputtering fuse before the fire reached the powder. Other times one scurried off to a safe distance and watched in fascination as things exploded. And at still other times one perforce remained and took the blast.

A typical case of fuse-stamping took place at Hankow on 21 January 1911. No American gunboat was present; only HMS *Thistle* and SMS *Jaguar,* of the German Kaiser's Yangtze River patrol. Had there been a U.S. ship on hand, assuredly its landing force, the last bloody pub scrap forgotten, would have been shoulder to shoulder with their opposite numbers. On the Yangtze, that was the way it was done.

Chinese mobs are much alike, whatever their target: the foreign "factories" in Canton in 1840, the foreign concessions in Yangtze ports in the early nineteen hundreds, or the western embassies in Red Peking of current times. Sometimes spontaneous but generally well directed by expert agitators in strategic spots, the typical Chinese mob is indeed "a great beast," with its own characteristic shrill compounded scream and its coordinated movements and actions which build on themselves like the ever-increasing waves of a typhoon-blown sea. There is no American version of the 1911 Hankow incident, but Lieutenant Claris recorded it as follows in his journal:

It was near the Chinese new year, and they are always disturbed and excited at this time. Thousands of farmers and beggars from the surrounding villages flock to the cities either to spend or make a little money during the holiday.

On Saturday afternoon the chief of the British police [of the British concession, Hankow] saw a ricksha coolie sitting in his ricksha in the way of traffic. When he saw that he was ill, he put him into another ricksha and took him to the municipal buildings, by which time the coolie had died—according to the doctor, of heart failure. The Chinese authorities were informed and the body placed outside the native city, as is the usual custom. The parents of this unfortunate coolie, hoping to make a little money, carried the body back to the municipal buildings and claimed money for it, saying that the British police official had kicked the coolie to death. A crowd of about 300 ricksha coolies collected and declared they had seen it done, but were soon dispersed by some 30 policemen and it was thought that was the end of it.

At nine the next morning an urgent message came on board for us to land at once, as a crowd of between three and four thousand was attacking the police station. We put ashore 22 seamen and 9 marines armed with rifles and marched to the municipal buildings where we found a tremendous crowd. When they saw us coming with fixed bayonets they started to retire and the police, taking advantage of this, broke up the mob, using their sticks and batons.

We now halted and awaited orders from the Captain [Commander Bailie-Hamilton], who was now in charge of the local volunteers (60), the police (40), some German seamen, who arrived doing the goose step (50), and our own party (40), now increased by the arrival of nine stokers armed with cutlasses.

A message now came that the natives were pulling up the fences and arming themselves with bits of it and stones and bricks and assembling at the back of the British concession. On our arrival there we were met by a crowd of some five or six thousand, who immediately stoned us. One seaman was badly hit and knocked out by a stone, and most of us were hit if not hurt before long. We then charged the mob with fixed bayonets, which dispersed them. But they were in such vast numbers that we soon had an enormous crowd behind us and were being stoned from all directions.

A message then came that an enormous mob was attacking the custom house and that the volunteers (a squad of 11 men) was being very hard pressed. Leaving the marines and the cutlass party with the Gunner, I double-timed my party down to the custom house. On arrival there I found a crowd which afterward was estimated as between 20 and 30 thousand natives. Eleven volunteers were trying to keep them out of the concession. The mob was divided into three parts; some in the Tai Ping Road, some in the street leading into the Chinese city, and the larger part on the river bank at the end of the bund.

I was forced to divide my party of 22 men and station 11 to face the two streets, with the other 11 at the end of the Bund so as to keep all three parts of the mob from joining. Just then, the mob on the river bank rushed the 4 volunteers who were there and Mr. Ramsay (in charge) gave the order to fire over the heads of the mob. This was done, but a native was shot. The mob retired, but at once became much more violent and the stone throwing was soon much more furious than before.

By this time, my men had been in position for an hour and a half and by a little prodding with bayonets and the gentle kicks of the seamen we managed to hold our ground.

The yelling of a Chinese mob is the most ghastly thing one can imagine. It causes one to have a most extraordinary sensation; shivering down the spine and a dull kind of stomachache. Whether it is like school boys cheering, ending up with a dreadful wail, I do not know, but most people agree with me that it is most distressing.

At 1:30 P.M. the Taotai (chief official of the native city and allowed by the Emperor to punish anyone even with death without any questions being asked) arrived and told the consul that if we could get our seamen withdrawn out of sight of the mob, he would speak and guarantee to disperse it.

The captain gave us the order to retire, which no sooner had we started to do than the whole mob from the three different positions rushed on to the Bund and gained the 200 yards we had been holding the whole day.

The Taotai attempted to speak, but received a brick in the eye and was rescued by some of his own men. The natives were delighted with the position they had gained by our retiring and started again to storm the custom house, wherein were two American ladies who for a few moments seemed to have rather a rotten chance of getting out alive.

At 3:15 P.M. after a extra severe stoning, the mob gave a most frantic yell and rushed us. Two of the cutlass party on the river bank were knocked down and the whole of that party was in a very dangerous position. The captain then gave the order to fire. The mob turned and ran, leaving between 30 and 40 dead and 100 or more wounded. This certainly had the effect of quietening things down a bit, and we now simply faced a mob of yelling devils who thought twice before they threw any more stones.

The German captain of the Jaguar was standing just behind me when our captain gave the order to open fire and I heard him say, "Ach! Dot vass vell done!" I think the captain had timed it just about right as in another second I am certain that we should all have been down.

At 3:55 P.M. the Chinese troops arrived (after everything was over) and started to drive the mob back to the native city. This they easily managed to do, as there were 3,000 of them. "So ended the most serious riot in China," to quote the consul general's words, "that has taken place for ten years. Had it succeeded, the whole of the Yangtze Valley would have turned."

Like an oft recurring nightmare, the crowds would come back again and again, greater or smaller, until in the first week of 1927, communist inspired mobs overwhelmed Hankow's British concession once and for all. Turned back to China, it would remain headquarters for British business in the Yangtze Valley. But never again would the Royal Navy's sailors and voluteer civilians in weekend khaki stand guard against howling Chinese coolies. By 1927 the leak in the dike of foreign imperialism in China had widened to a torrent which in another twenty years would wash the despised *fanquei* completely from China's soil.

By 1911, China was overripe for major change—the slightest jiggle could bring about an instant crystallization. Since the Boxer Rebellion of 1900, the Imperial court no longer effectively represented any section of the Chinese population. Chinese merchants and capitalists had begun to assert themselves. Even more indicative of the potentially explosive new situation, the common people were for the first time in history actively displaying in strikes and boycotts, mass reaction to both Imperial do-nothingness and foreign influence.

Plans were afoot to build a railway connecting Canton, Hankow, Changsha, Ichang, and Chengtu. The Americans had lost the concession to build the Canton-Hankow link, as a result of anti-American feeling in 1905 over U.S. immigration policies. Chinese capitalists had mobilized their resources to build these railways, but venal officials in Peking, growing fat on profits from granting concessions to foreigners, had found ways to use foreign money to insinuate foreign interest in the new projects. The incipient Chinese railway magnates were naturally enough inflamed over the issue, exacerbated by the secret societies who linked the whole thing with the hated Manchus.

As for the Ichang-Chengtu link, the traditionally independent Szechwanese saw in the railroad a threat to their splendid isolation. It would bypass the Yangtze gorges which they controlled and from which they derived much profit. In 1911 a few miles of the railroad had already been built, a station constructed at Ichang and tickets sold for the benefit of Imperial inspectors. These gentlemen, after first being made thoroughly goggle-eyed on *samshui,* would be taken for a long ride to impress them with the progress made. They were taken for a ride in more than one sense of the word, because the track ran in a circle.

In early September the USS *New Orleans,* bristling with ten 5-inch guns, lay at Shanghai, her crew excited over the projected trip up to Hankow. Rear Admiral J. B. Murdock, CinC, Asiatic, had decided the explosive situation bore looking into, and felt the twenty-foot draft of the *New Orleans* made her a better choice than his flagship, the armored cruiser *Saratoga,* which drew twenty-eight feet, even though the river was higher than at any time since 1838. On 10 September *Saratoga, New*

Orleans and *Helena* swept up the river to Nanking, where Murdock shifted to *New Orleans* and sent *Saratoga* back to Shanghai. Aboard his new flagship were at least ten unhappy men who would miss the charms of Hankow. They had been restricted for a month just before departure for "being ashore in civilian clothes on foreign station."

Piloted safely to Hankow on 15 September by an old-timer on the river, an American named Langley, they found the roadstead crowded with men-of-war, including *Villalobos, Samar,* and *Elcano*.

Apparently the potables ashore left something to be desired from the medical officer's standpoint; eight of *New Orleans'* crewmen soon were hauled up to mast for having lost their canteens while exploring the city. Punishment: restriction in any port where canteens had to be taken ashore.

With the magnificent Hankow Race and Recreation Club near its zenith, it is doubtful if Admiral Murdock found it necessary to carry a canteen. But he did distinguish himself in one respect in that he was the last U.S. flag officer to exchange amenities with an Imperial excellency. On 19 September, nineteen guns announced the arrival on board of the Viceroy. In another four months he would find himself among the ranks of the new Republic's unemployed.

On 23 September, Murdock departed for Hongkong and Manila. Even though the revolution was felt to be inevitable, he scarcely could have been expected to divine its imminence, no more than could the plotters themselves. When the shooting started a fortnight later, Murdock was far away.

At least the *New Orleans* saw some action. Hurried back from Manila, she spent a month on the Yangtze, "in obedience to orders of the Senior Officer Present, Yangtze Valley, to proceed to Nanking and maintain wireless communication with Shanghai." She arrived in time to join the German armored cruiser *Gneisenau* and other warships in probably the last national salute fired to Imperial China, on 20 October. Soon, the Germans would be shooting quite a different brand of salute, when SMS *Emden,* which later achieved fame as a commerce raider, whanged some seventy projectiles into a Chinese fort which had given affront, thus accounting for the only substantial foreign intervention in the Revolution.

On 9 October 1911, antimonarchist conspirators were clandestinely making bombs in the Russian concession at Hankow when somebody miscued and there was an explosion. Police immediately descended on the place, from which the plotters had precipitously fled, but left behind in their haste a list of names implicating officers of the local garrison. To save the skins of those so betrayed, the planned uprising had to be set in motion at once.

By late morning of 12 October, *Helena* had returned to Hankow from upriver to find the roadstead looking like the rendezvous for an international naval review. Present were USS *Villalobos* and *Elcano,* six British warships, four German, three Japanese, one Austro-Hungarian, and one French. Two Russians arrived soon after. The "big guns" were not all artillery: also present were three vice admirals—British,

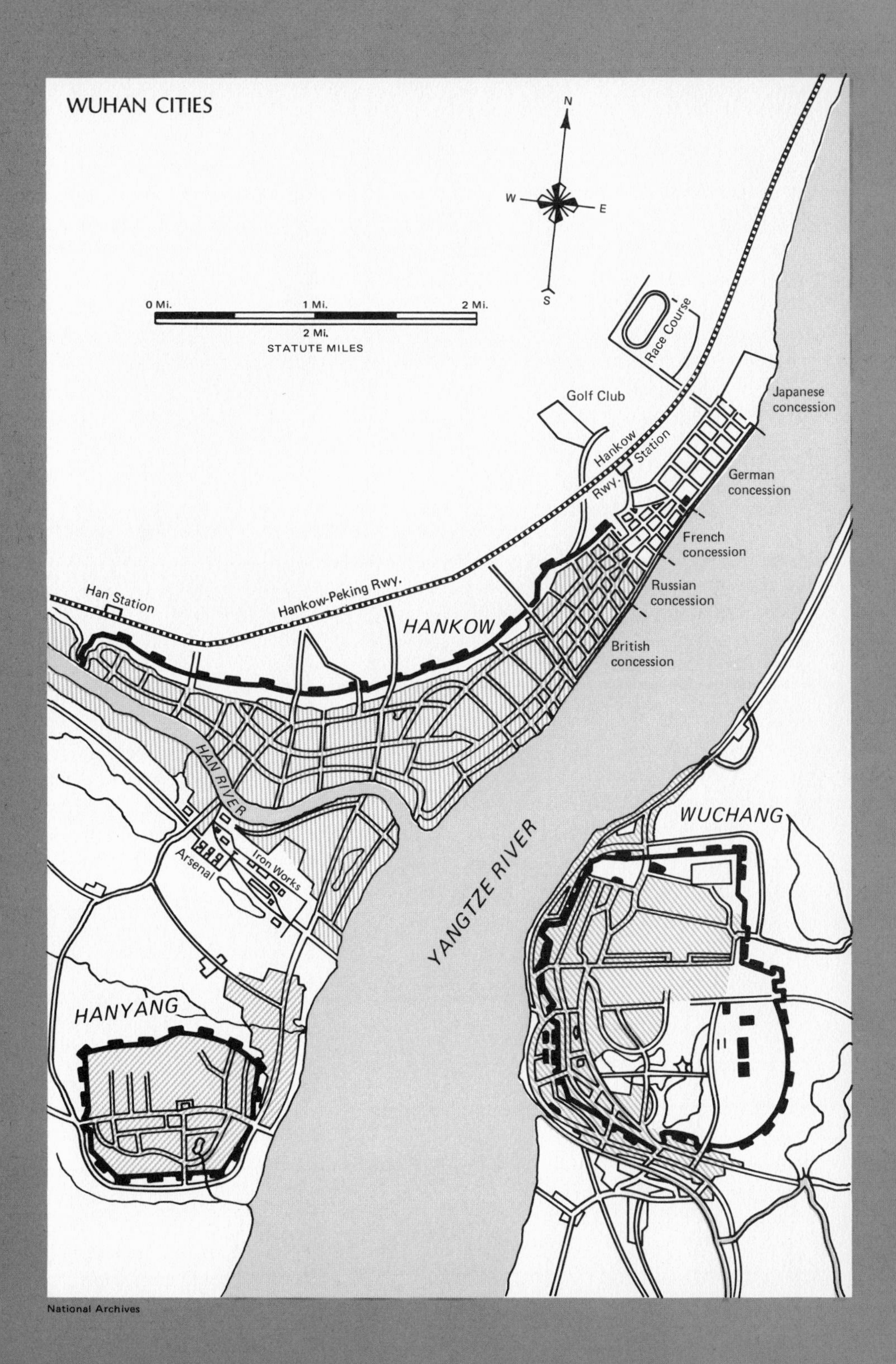

National Archives

German, and Japanese. Nine assorted Chinese cruisers and gunboats were there, but none could safely say on whose side they would be five minutes later. The senior Chinese, Admiral Sah, had been a captain at Chefoo in 1900 when the intervention of the USS *Oregon* probably saved his ship from seizure by a Russian cruiser.

The *Helena*'s first move was to ferry U.S. missionaries from Wuchang across to the comparative safety of the Hankow concessions. That evening, on the orders of the senior officer present, Japanese Vice Admiral Kawashima, she put ashore a landing force made up of Marine Lieutenant Arnold and Second Lieutenant A. B. Miller with twenty-seven Marines, and Ensign A. D. Denny and twenty-three bluejackets with two Colt machine guns. In less than an hour, they were in action, supporting the German volunteer detachment in suppressing looters and putting out fires in the German concession.[22]

The following three nights, *Helena*'s men patrolled the French concession until relieved by sailors from the French gunboat *Decidée*. On 16 October, Miller took twenty Marines to guard the American consulate general and Denny's sailors stood by the Russian municipal building until they were relieved by Italian marines. Two days later the *Villalobos* and *Elcano* sent landing forces, buttressed by six Colt machine guns, to join *Helena*'s men in guarding the Japanese municipal building. It is doubtful whether such a game of international military musical chairs has ever been played before or since.

The fighting by that time had become general, with the whole surrounding region a battlefield. All the native city except for a quarter-mile perimeter bordering the British concession had been destroyed by fire. An element of comedy, frequently an ingredient of Chinese warfare, involved a large quantity of cartridges with wooden bullets, intended for a review at Peking but shipped by mistake to Hankow where the rebels seized them. To the enthusiastic amateurs in the rebel army, one cartridge looked like another. The wood-loaded ammunition made a highly satisfactory bang and the Imperials ran away in as great a fright as they might have from the real thing.

By 23 October, Changsha, Yochow, Ichang, and Kiukiang had all quietly joined the rebel side. Shanghai fell on 4 November. Most of the Chinese warship crews had gone over to the rebels. Their belief in their cause more than made up for the superior equipment and professional training of the regular Imperial troops, whose hearts were not in the fight.

Hankow was the focus of action. Elsewhere in China, particularly on the Yangtze, the rebels satisfied themselves with liquidating the Manchu customs collectors and those anachronisms, the Tartar generals and their bannermen. At Ichang on 14 November, the U.S. station ship enjoyed some small excitement when the SS *Kiangwo,* about to leave town, was held up by the rebels, who were looking for Manchus. But "Ensign H. H. Forgus, of the USS *Samar,* with a few bluejackets turned them away."

With the usual hysteresis lag by which bureaucratic action follows stimulus, Washington concluded in early November that something must be going on in China and that the U.S. forces there needed beefing up. Consequently, the Pacific Fleet commander was directed to prepare five armored cruisers, the pride of the American fleet, for an extended cruise. The old Civil War monitors *Monadnock* and *Monterey*, out of commission at Manila, were designated for reactivation, but the revolution ended before they were ready to take part.

Of almost the same vintage as the monitors, fat old General Yuan Shi-k'ai was dragged out of retirement as a last hope of rallying the fast-crumbling Imperial forces. In classic Chinese sequence, Yuan first defeated the rebels at Hankow and Nanking, then treated with them, and then, in the traditional double double-cross, he joined them. In the face of this coalition, the thirty-century era of the Sons of Heaven gently faded into history. On 12 January 1912, just three short months after the accidental bomb explosion in Hankow, the child emperor Hsuan T'ung, better known to Americans as Pu-yi, abdicated.

At Hankow, with the Race and Recreation Club grounds no longer a battlefield, the gunboaters could get back to the normal routine of golf, billiards, tennis, and light conversation. The businessmen commenced sorting out the jumbled mess of Yangtze trade. Security of the concessions had been enhanced in December by the arrival of 162 officers and men of the British Yorkshire Light Infantry, plus 272 strapping Cossacks from Vladivostok, whose officers' dashing uniforms and penchant for champagne added an exotic touch to the social scene.

As for Shanghai, the Revolution created very little upheaval or change in *fanquei* mode of life. In 1863, the first cross-country race had been won by Mud, with Bog-trotter second, suggesting the character of the course. Since that time the *taipans*, in the English tradition, had carried on, improvising here, substituting there. The year of 1911, Revolution or no, maintained the pace. The Shanghai Otterhounds, a group of some thirty-five Englishmen addicted to drag hunting, were out thirty-six times during the year. Their Honorable Secretary closed the annual report on a note of mild disappointment tempered by a sense of accomplishment: "Although no otter was killed, one was marked to ground in an inaccessible place after one and a half hours of the very best, during which he was frequently viewed and would no doubt have been killed but for the crowd of excited natives which baulked the huntsmen at every turn." One marvels at the variety of game these dedicated sportsmen managed to scare up in a country so densely populated: "5½ brace of gray fox, 4½ brace of badger, 1½ brace of ringtailed civet and 4½ brace of hare."

Few visitors to Shanghai have failed to attend the "race meetings," where ponies thundered clockwise around the track, ridden by gentlemen jockeys, usually the owners. These pint-size horses were caught semiwild in Manchuria, brought to Shanghai by the shipload, and sold at auction in a sort of Russian roulette. For the equivalent of US$15.00, one might acquire a winner on the track or steeplechase, or rent one by the month, all found, including groom, for the equivalent of US$6.00. But sometimes the "horse in a poke" turned out to be a wholly savage little beast that would kick and bite its way out of any human association, or at the very least develop a questionable sense of humor satisfied only by sending his owner heels over appetite into a mud puddle.

In a more serious vein, and at the risk of anticipating events, it may be useful to outline briefly what followed on the Chinese side after the abdication, while the *taipans'* hounds were yapping after otters. Predictably, Yuan Shi-kai, the man with the guns, seized power from the first president of the new Republic, Sun Yat-sen, the man with the ideas. In 1915, with the world preoccupied by the Great War in Europe, Japan presented her infamous "Twenty-one Demands," which would have reduced China to the status of a Japanese colony had they been imposed. Yuan, having illusions of imperial grandeur, resisted. He died in 1916, possibly of a broken heart, after an unsuccessful attempt to establish a new dynasty with himself as first emperor.

In July 1917, Pu-yi, the last of the Manchus, was reinstalled as emperor for a few short weeks, puppet of a local warlord. A month later, China, again a republic, joined the war against the Central Powers. Weak, unassuming but durable Pu-yi was dusted off once more two decades later, by the Japanese, briefly to become the Emperor K'ang Te, of the puppet state of Manchukuo. The era of the warlords had become final.

In June 1914, the Asiatic Fleet flagship *Saratoga,* an old armored cruiser drawing twenty-eight feet of water, was anchored at Hankow, six hundred miles from the sea. She was there, not by invitation of a supposedly sovereign land, or even by mutual arrangement, but simply because Rear Admiral Walter C. Cowles, the new Asiatic Fleet commander, wanted to see the middle Yangtze. To this supine condition the Heavenly Kingdom had fallen.

With her four 8-inch guns, *Saratoga* could have wreaked havoc on a first rank but jerry-built city, while her 6-inch armor could have withstood anything but concentrated heavy artillery fire. These points undoubtedly were not lost on the pragmatic Chinese.

On 6 July 1914, Lieutenant R. A. Dawes left the comfort and cameraderie of the USS *Helena* and took command of the USS *Elcano,* then lying at Hankow. The ship was slightly pot-bellied, had an early Victorian ram bow, and looked as though somewhere in her ancestry there might have been a covered wagon. But her first "cruise" wholly dissipated any misgivings Dawes may have had. Steaming over to the *Ajax* for coal and provisions, he found the ship handled beautifully.

Five days later, the coal dust had been scrubbed off and *Elcano* was once more spotless white and buff, ready for the inevitable exchange of formal calls which every new skipper must at once carry out. In this case, calls were made on the *Helena, Villalobos, Palos,* and the German gunboat SMS *Jaguar,* whose Korvettenkapitän Lüring was, as Dawes described him, ". . . a very pleasant sort of chap who speaks fair English." That a World War was only a fortnight away never entered their conversation, nor their thoughts.

In the exchange of calls, the skipper of the *Helena,* Commander W. C. Cole, left Dawes a card bearing the subscript "Senior Officer Yangtze Valley, Commanding USS *Helena.*" This was a mid-evolutionary form of "ComYangPat," which, with its two-star flag, was still seven years over the horizon.

By 14 July, Dawes had the *Elcano* "in all respects ready for sea." With the informality apparently then the rule, they set off downriver to Huangshikong, port of Tayeh, the latter the site of extensive German-managed iron mines that bore looking in on.

Tayeh, a few miles inland by a pleasant lake, was linked to Huangshikong by a railway that Gilbert and Sullivan could well have worked into one of their lighter pieces. A news clipping in Dawes' journal outlined the recommended procedure for travelers. The first admonition, to those arriving by the Hankow steamer due at 2:30 A.M., was to avoid the Chinese boat house at the pier while awaiting the morning train "unless the visitor be an entomologist."

The morning train, like so many Chinese institutions, was run with only the vaguest appreciation of time. "There are no recognized hours at which trains come or go," warned the clipping, "but as a general rule, the first train arrives about daybreak, and leaves an hour or so later." There were ample trains, even if not punctual: ". . . according to circumstances, sometimes numbering only four each way a day, but oftener eight." The traveler could expect to share accommodations with ". . . a crowd who eat pea-nuts, water chestnuts, sweet-meats and buns." The debris therefrom either was spat on the floor or out the window, more often than not to blow right back into a window farther aft.

Skipper Dawes and thirty sailors found the mines to be an open-pit operation of the most primitive type. With the exception of the use of dynamite, the mines were worked as they had been in the days of the Sung Dynasty, a thousand years before.

Fifty miles downriver from Hankow, Dawes expressed a new skipper's reaction to freedom and at the same time suggested the primitive state of 1914 naval radio: "We are out of communication with the *Helena* and having the time of our young lives!"

Downriver again, Kiukiang was found to be living up to its reputation as the super hotbox of the lower Yangtze. Foreigners languidly played tennis, half-drowned in their own sweat. USS *Samar* on her way upriver tarried only momentarily. Swarms of flies followed the sampans out to HMS *Woodcock,* swinging disconsolately around the hook as station ship.

The Hankow papers, only source of news in those parts, reported that the *Helena* was stuck at Sunday Island by low water. Several days later the same source had her under way again, destination Ichang. Perhaps Dawes himself failed to appreciate the note of wishful thinking in his journal when he wondered how long the flagship would stay in Ichang, where they "may be caught by falling river and find themselves unable to come down."

Dawes took advantage of his new freedom to visit Poyang Lake, where U.S. gunboats rarely put in an appearance. A four-hour run dodging sampans and junks brought *Elcano* to Nankang, at the foot of the back trail to the 3,500-foot-high summer resort of Kuling.

Nankang was the most dilapidated-looking place Dawes had seen in some time. But at a distance, all interior Chinese cities look burned out and abandoned. Not a single brushful of paint profanes the weathered grey of the wooden buildings.

Dawes had first thought of going to Nanchang, across the lake and up the Kan River. But a study of the French charts showed a very narrow and winding channel. There was one, perhaps two, very sharp bends which might not accommodate the 165-foot *Elcano,* and perhaps no space to turn around in at the city. It would have been decidedly embarrassing to wind up aground as something less than an uninvited guest in an area inhospitable since the days of Admirals Evans and Rodman.

The *taotai* at Kiukiang had been notified of the intended visit, but the only official to appear at Nankang was a woebegone 120-pound police sergeant draped in a uniform that would admirably have fitted a man weighing two hundred and fifty. He was mildly interested in where *Elcano* had come from, how long she would stay, and where she was going next. The whole thing obviously bored him to the core.

Elcano's people found little to do ashore. The doctor went fishing. Sails were broken out for the ship's pulling boats so the crew could enjoy some poor man's yachting. The captain took a long walk in town and found the natives friendly. The few missionaries normally about had all trooped off to the cool mountain-top of Kuling for the summer, leaving the local Chinese souls to wander unshepherded until autumn. There was, of course, no club. This was the "guts" of China, unembellished by Western "improvements" of any sort.

Back in Kiukiang on 29 July, Dawes found plenty of mail but "no excitement." Two days later, things were different. "Rumors of war! HMS *Nightingale* was interned this morning with only a Chinese caretaker." There was no time for tennis at the club. Everybody clustered around the bar, excitedly discussing the Triple Entente, the Triple Alliance, what Italy might do, and the intentions of the Japanese. "Those Japs are a bunch of opportunists!" the British merchants all agreed. Yes, there was an alliance with them, but what about their competition up the Yangtze? They were worse than the Germans, they said. And now with the Germans out and the British at war, it was clear who would snap up the loose ends.

Tension in Kiukiang and Hankow ran high. The British naval ratings had left for Hongkong. Germans were streaming out of the interior, headed for Hankow, where British shipping lines refused them passage to Shanghai.

On 3 August, *Elcano* received urgent orders by Chinese telegraph to return to Hankow, where all the banks had closed and business was at a standstill. Unpaid, unemployed coolies, twenty or thirty thousand of them, so Kiukiang gossip had it, were milling around Hankow, ready tinder for incendiaries.

Actually, on *Elcano*'s arrival on 5 August, things were relatively calm, although England had declared war on Germany that day. The troublesome coolies numbered only about fifteen hundred. But banks were closed, and taipans and gunboat skippers gathered around the club bar at noon for gimlets and exchange of news. Rumors were printed one day and denied the next . . . there were supposed to have been great battles in the North Sea . . . the French had invaded Alsace . . . the Russians had plunged into East Prussia in their parade uniforms . . . the Belgians were holding heroically at Liege . . . the Japanese had sent the Germans an ultimatum (contents unrevealed). The German light cruiser, *Emden,* and the Russian cruiser *Askold* had dueled off Tsingtao and sunk each other, it was said, then quickly denied. So the rumors and counterrumors went.

The magnificent German heavy cruisers *Scharnhorst* and *Gneisenau* and the light cruisers *Nurnberg* and *Leipsig* had slipped out of Tsingtao into the unknown, to the terror of Allied shipping in the Pacific. Soon, they would appear off Chile and destroy the British cruiser squadron under Admiral Cradock.

Dawes, as a lieutenant and senior U.S. officer present, had once called aboard the German flagship and had a chat with Admiral Count von Spee, who later met disaster off the Falklands at the hands of British Admiral Sturdee's battle cruisers. "He was a most pleasant man to talk with," said Dawes. "And believe it or not, he started discussing coral formations in the Gulf of Pehchili [Po Hai], where I didn't even know there *were* coral formations."

Back in Hankow, Dawes found the Chinese much exercised over the turn of events. On the theory that when honest men fall out, thieves prosper, General Ting, the commissioner of foreign affairs at Hankow, confided to Dawes that he was deeply worried about the Japanese. They had made an offer to "protect" the foreign concessions in Hankow while the owners were busy shooting at each other elsewhere. "They will never give them up at the end of the war!" said Ting.

The *Quiros,* meanwhile, was struggling upriver to swap stations with *Elcano.* The two met in Kiukiang. Hurrying on downriver, *Elcano* proved the hard way the drawbacks of deep-draft river gunboats; she ran aground in eleven feet of water. Like all the gunboaters, Dawes was doing his own piloting, his store of lower river lore perforce restricted to what he could learn on one trip from Navy contract pilot "Joe" Langley, an enormously fat Maryland eastern shore waterman who could scarcely negotiate a ship's ladder. The SS *Kutwo,* passing by, refused to help.

The following day, SS *Kiang Foo,* in a more traditional application of the brotherhood of the sea, gave a pull. But the hawser broke, and with it, tradition; *Kiang Foo* kept straight on going. Pulled off finally by the SS *Tuckwo, Elcano* reached Nanking, where Dawes found the USS *Galveston,* exchanged gossip, drew stores, then sped on down to Shanghai, arriving two days before the expiration of the Japanese ultimatum to Germany, which was 23 August.

Shanghai was in turmoil. No one knew exactly where the German warships were. Groups of British volunteers were departing for home, their fellow countrymen turning out en masse to see them off.

In very short order, U.S. warships were the only ones, other than Chinese, still patrolling the Yangtze. The French *Doudart De Lagree* had been all the way up to Kiating in late July and had made the 1,700-mile trip to Shanghai in an astounding seven days. Germany's SMS *Otter* had beaten her by one day and interned. Japan's *Toba* had followed suit on Japan's declaration of war on Germany. All seagoing warships of the belligerent nations had of course long since left the China ports.

It soon became clear, after the Japanese capture of Tsingtao and its Shantung hinterland, that this was only a mild forerunner of what the Japanese had in mind for China. This served to draw the many Chinese factions together in mutual protest and protection. The secondary effect was relative peace on the Yangtze, a situation not unappreciated by the Yangtze gunboats, thinly spread and with at least a moral responsibility for protection of *all* foreigners in the interior.

It was more luck than good planning that those remarkable twins, *Palos* and *Monocacy* (sometimes referred to as *Pathos* and *Monotony*), had slid down the ways into the Whangpoo just in time. Until then, the U.S. Navy had nothing whatever capable of negotiating the rapids and gorges of the upper river.

It all started in 1898, when on 4 May, Congress appropriated $260,000 to construct a gunboat for the Great Lakes, to replace the USS *Michigan*, an old crock kept afloat only by many layers of paint over her lacework hull. But in view of the generally peaceable nature of the Canadians and lack of truculence on the part of the redskins in the post-Custer era, the project had been allowed to lapse. Fourteen years later, in August 1912, some diligent fellow in the Navy Department dredged this long-forgotten project from the bottom of the "in" basket and conceived the idea of using the funds to build something for the Yangtze, where every Asiatic CinC since Rear Admiral Bell in 1866 had been clamoring for a suitable river gunboat. The 1898 Great Lakes gunboat plans were felt to be somewhat obsolete, as well as unsuitable. But the Mare Island Naval Shipyard, using plans based on HMS *Widgeon*, discovered that for the $260,000 appropriated, they could build *two* gunboats. Mare Island also saved money by using some odds and ends of machinery out of old torpedo boats, an economy which many a *Palos* and *Monocacy* skipper would regret when the feed pumps failed in a rapid.

The *Monocacy*'s keel was laid 28 April 1913. She was set up, torn down, and shipped in parts to Shanghai in the SS *Mongolia*, where her keel was relaid 27 February 1914, and she was launched 27 April. On 29 May 1914 she was under way, headed for Kiating, 1,700 miles west. The timetable for the *Palos* varied by only a few days.

There were some sticky moments as American labor organizations howled about downtrodden Chinese workmen—the Shanghai dockyard doing the reassembling had only dimly heard of the fourteen-hour day, let alone the eight. Then it was discovered that the Shanghai contract's fine print said nothing about insurance while on the ways, so *that* was an unexpected extra. But when all the bills were marked "paid," the two little ships had cost a modest $275,000.

By early 1915, the Great War had settled down to routine, both in Europe and the Far East. The original *élan* and excitement no longer marked the gatherings in the clubs, where the younger men were sometimes apologetically at a loss to justify why they were there instead of in France. Things were so quiet that wanderlust got the better of Skipper Dawes and he took what appears to be the only U.S. Navy trip on record up the Grand Canal. With his wife and children, in a borrowed Standard Oil launch, he followed the twists and turns of the canal to Yangchow, a walled city some fifteen miles from the Yangtze. Yangchow had been on the seacoast thirty-five hundred years earlier, but in 1915 it was one hundred miles inland—a demonstration of the silt-carrying powers of the Yangtze.

In 1915, Shanghai newspapers were expressing candid views. George Bernard Shaw, labeled a traitor by the British, claimed that if France and Britain had invested more of their capital in alleviating distress at home instead of adventuristic and colonialist expansion in Asia, things would have been better. The worst mistake had been to encourage the paranoia of Potsdam by making an alliance with Russia in concert with France, according to Shaw. As for the Japanese, Tokyo was emitting clear sounds of displeasure with Britain, who she felt was interpreting treaties to suit herself and to her own advantage. The remaining British *taipans* in Hankow and Shanghai had no doubts that the Japanese demands on China threatened British commercial control of the Yangtze Valley. The Americans, with no army, not much navy, and a president intent on keeping out of it, carried little weight with anyone. But nevertheless, Washington did its best; on 12 March 1915, Rear Admiral Cowles, CinC Asiatic Fleet, was promoted to full admiral, the first incumbent to fly four stars.

The massive and desperate buildup of the U.S. Fleet in the Atlantic had drawn heavily on Asiatic Fleet personnel. The *Quiros,* at Hankow, was down to one line officer, Lieutenant (junior grade) B. A. Strait. The war had depleted the French, German, and Russian volunteer detachments in Hankow, Strait reported to the Navy Department. Included among the 140 British volunteers were many Americans who would have preferred to act under American officers. There were two hundred Americans in the tri-city area. The Russian consul general promised barrack space. How about Cavite furnishing forty Springfields, ten Colt automatics, and twenty-five thousand rounds of ammunition? Admiral Strauss, Chief of the Bureau of Ordnance, who must have been hard up for weapons, vetoed it, reminding that the various volunteer corps were formed for protection of the concessions. He didn't seem to recall any American concessions, he said. To everybody's surprise, Strauss's veto was vetoed; acting secretary of the Navy Franklin D. Roosevelt thought it was a good idea. By Christmas, 1916, six months after Strait's appeal, thirty-two American volunteers, in U.S. doughboy khaki, were drilling in the British Asiatic Petroleum Company compound. Their ammunition was stowed in the French armory, and they had quarters in the Russian barracks.

Concurrently, Strait was concerned about the state of the club established at Hankow for the U.S. enlisted men. It "... lodged men overnight, served wholesome food at reasonable prices, rented bikes and had tennis courts." The alternative, according to Strait, were "low and dirty drink shops dispensing cheap liquor at exhorbitant prices. Attached to these dives and living in insanitary rooms are diseased, immoral European and Chinese women." The CinC complained it was "unfair to take money from the funds of ships which never visit the Yangtze to benefit the gunboats of the 2nd Division. $1,500 has been allotted already. Now $500 more. That is *all.*" There is no record of what became of this club, but with the arrival of the White Russians and proliferation of cabarets, one suspects the competition was lethal.

An officer aboard USS *Brooklyn,* visiting Shanghai in 1916, described the Carleton as the only decent place in town for a meal or dancing. The elite of Shanghai gathered there—*taipans,* swains, wives, daughters, and officers from the U.S. warships. Around 10 P.M., a sort of silent curfew sounded. The senior lady took her departure, her party closely followed by those of the others who were escorting. The bachelors and the unattached had meanwhile drifted in to fill the vacated tables. By eleven, the girls—Charlene and Yvette and Iris and Monique and Betty and others—would have arrived in sweeping gowns, casting smiles at old friends and accepting invitations for supper, champagne, and dancing. These girls of the evening had a strangely ambivalent nature. Endowed with wit and animal cunning, they were also charming and kind. The young man with down on his upper lip and a bit too much champagne aboard would be shepherded to a reliable carriage or ricksha and aimed in the direction of home. And if Ellen, or Yvette, or Iris happened to be taking the air on Bubbling Well Road in an open carriage during a summer afternoon and passed a gentleman acquaintance who was escorting a lady, there would be no flicker of recognition. That was the code of the Carleton.

There were, of course, the Astor and Palace hotels, for afternoon tea with music, or stronger things. The Cercle Sportif Français, the French Club, that cosmopolitan meeting place for gay Shanghai of the thirties, was yet to come. Both it and the American Club were built after World War I. The opulent German club, Harmonie, on the bund, went out of business when the war began.

Thomas Woodrooffe wrote a charming tale of British Yangtze gunboats in which he appeared as "Toby," the autobiographical central character. He described the proprietress of "Gracie's," in Shanghai, "who had an attractive southern accent," and one of her girls, "A bulbous blonde who looked as if she had been poured into her dress of electric blue . . . sitting all by herself on one of the glaring green chairs . . . a hideous clash of color." Toby had trouble getting a glass of water in an establishment dedicated exclusively to "bubbly." The water had cost him a "quid," he discovered, on a departure sufficiently ahead of schedule to avoid the *pièce de non-resistance.*[23]

Toby's bill was picked up by his skipper, who had been showing him the town. As he signed the chit, he explained to the wide-eyed young man that absolutely *nothing* other than ricksha or sampan hire was ever paid for in cash, and that included the Sunday collection in church.

Pilot "Joe" Langley occasionally invited some of the Americans to join him for Chinese "chow," the innumerable courses prepared by his Cantonese wife. Mrs. Langley ran a small dairy, from whence came the only pure milk on the China coast, an odd but welcome switch from the whiskey and splash for those who remembered with nostalgia the Naval Academy dairy farm back in sleepy, sanitary Annapolis.

The arrival of thousands of White Russians after the Revolution changed the whole pattern of Shanghai life after dark, especially for the enlisted men. In the modern nuclear-powered Navy, where sailors enjoy movies, TV, and other entertainment undreamed of by the relatively under-privileged bluejackets of 1916, plus a tremendous improvement in pay, it is difficult to realize what a paucity of decent recreation there was for the bluejacket in World War I Shanghai. In April of 1916, CinC Admiral Winterhalter wrote to the Secretary of the Navy about "the low character of resorts on North Szechwan Road extension, in Chinese territory." These dives were located en route to the rifle range and Navy recreation grounds. The liquor was so notoriously bad the admiral had requested the consul to have some of it analyzed. Morphine, made in Europe and shipped via Siberia and Japan, sold for three or four cents a shot. An old *Wilmington* sailor summed it up: "The better cabarets in Shanghai were off limits for anyone in uniform, so most of us went to a section out North Szechwan Road called 'The Trenches.' That sure fit. The entertainment was bad, the singing worse, and if you got to feeling good, they would serve you the dregs from left-over drinks." In the early nineteenth century on Canton's Hog Lane, seamen were sold a potent concoction compounded of alcohol, water, tobacco juice, sugar and a dash of strychnine. Things had come a long way since then, but in "The Trenches," not half far enough.

Visits to U.S. and foreign ships came up every day or so. All calls were official, but some were more fun than others, as Lieutenant Dawes of the *Elcano* found out. Skipper Dawes was about to depart Shanghai in November 1914, when he was invited aboard the armored cruiser flagship *Saratoga*. The CinC, Admiral W. C. Cowles, had three large photographs of himself to be delivered to Chinese generals at Chinkiang, Kiukiang, and Wuchang.

At Chinkiang, the procedure was simple; Dawes found another U.S. gunboat on hand with a junior skipper who would see that the precious mug shot was safely delivered to the general.

At Kiukiang, arrangements were made through the consul for an official call on the general. After all, an admiral's photograph is not handed over to a general like a casual basket of Szechwan oranges. The appointed day was dark, dreary, cold and dismal, as so many Yangtze winter days are wont to be, but Dawes, decked out in fore-and-aft hat, railroad pants, sword and epaulettes, with boat cloak flapping in the breeze, took off in the ship's motor dory. The "motor" was a highly temperamental little two-cycle engine. The odds against its performance this time were less than favorable, as the engine in use was on loan from a Chinese contractor pending delivery of a new one from the States.

The call went off with more precision than might normally have been expected. The engine ran, the general remembered to be on hand and received the picture with appropriate cluckings of gratitude and appreciation. The teacups were ceremoniously emptied and Dawes started home.

Halfway back to *Elcano,* the engine coughed and died. A Chinese gunboat was anchored downriver from *Elcano* and as the swift current carried Dawes by, all activity aboard the gunboat ground to a halt. "Attention" was sounded, the sentry presented arms, Dawes rose, hanging on to the rolling dory with one hand and, with the other, flung back his boat cloak and saluted.

The sweating coxswain got the engine going again and back they chugged, past the gunboat, with all participants again at rigid attention. Then the engine conked out once more, the dory drifted down past the gunboat again, and eventually chugged back up, and during both passes mutual honors were once more rendered. Again the engine broke down and for the fifth time the amenities were observed as Dawes drifted downstream with the swift current.

Finally the *Elcano's* pulling boat was sent to the rescue, it having been noted that the dory was following a somewhat erratic pattern. "I went by the Chinese ship for the sixth time," said Dawes, "this time under tow of the pulling boat—and for the sixth time was rendered the usual honors."

The Yangtze, in the spring of 1916, was "in a generally disturbed condition," with heavy troop movements above Ichang. The historic split between North and South was being exacerbated by northerner Yuan Shi-kai's maneuvering to place himself on the Dragon Throne, to which Sun Yat-sen and his ever pugnacious Cantonese were bitterly opposed.

With no other foreign warships to keep peace on the river, U.S. gunboats were spread far and thin. In April 1916, *Monocacy* was at Chungking, *Palos* in the Tungting Lake area, *Villalobos* at Hankow, *Samar* at Kiukiang, *Quiros* at Nanking, *Helena* and the small cruisers *Cincinnati* and *Galveston* at Shanghai, and *Wilmington* roving.

For lack of men on the station, the *Pompey* was laid up at Hongkong. There had been no drafts of replacements from home for months because of the buildup of the stateside fleet. The *Pompey,* a 3,100-ton, 10-knot craft, was one of the first three gunboats regularly assigned to the Yangtze, and her big complement of eight officers and 106 men together with her poor condition made her the first of the Spanish War acquisitions to quit the as yet nameless "Yangtze Patrol."

By May, traffic on the river between Ichang and Chungking had been suspended, because of lawless bands along the banks shooting at anything that moved. In Hankow, tempers were by no means eased by the presence of nearly a thousand Japanese troops sitting cockily in their concession. Technically, with Japan being a belligerent, at war with Germany, they had no right to be in neutral China at all.

The death of Yuan Shi-kai on 6 June 1916 removed the basic cause for countrywide unrest. The *Wilmington,* at Chenglin, had nothing more exciting to report the following month than the fact the ship was "filled with myriads of mosquitoes and gnats." By August, *Monocacy* felt safe in temporarily deserting Chungking to ascend the Min, her second trip and one of four that U.S. gunboats would ever make up that exceedingly tricky and dangerous tributary.

By February 1917, the effect of the war was reaching east. Crews of German merchant ships that had taken refuge at Manila in 1914 suddenly disabled their machinery. The German gunboat *Tsingtau* interned itself at Whampoa, South China. There was less American annoyance at British stopping of U.S. vessels just outside Philippine waters to search for German nationals and contraband. It was felt they soon would be allies.

On 6 April 1917, the United States declared war on Germany. With China still technically at peace, there was only one action for the last foreign warships on the Yangtze: intern. On 6 May, Lieutenant H. Delano, commanding *Monocacy,* took command of all the U.S. gunboats assembled at Shanghai—*Monocacy, Palos, Samar, Quiros,* and *Villalobos*—and turned over their breech blocks to the American consul general.

The enlisted complement of the little ships was maintained at about seventy-five percent, with one medical officer and a paymaster for the group. Beyond chipping the rust, there was little to do.

The hibernation was short. China, feeling her only hope of coming out whole after the war depended on her having a seat at the peace table, declared herself a belligerent on 14 August 1917. On 16 August, the gunboats were released from internment and five days later the consul general handed them back their breech blocks.

The *Palos,* like an eager beagle released from the leash, was all the way up—to Suifu—by the end of October 1917. In a year the Great War in Europe would be over. But for *Palos* and the rest, another war had already begun: Kweichow and Szechwanese troops were battling.

On the morning of 4 December 1917, *Palos* put ashore Lieutenant (junior grade) E. T. Short and twenty-three men to protect the American consulate in Chungking. Later in the morning, bullets started flying over the ship, and eleven hit her. The battle was over before teatime, the Kweichow invaders having occupied Chungking by 3:00 P.M. On 6 December, Short and his riflemen were back aboard.

In this case, the Americans were strictly innocent bystanders, hit purely by bad aim. Such would not be true much longer. The aim still would be bad, but there would be no doubt that the target was meant to be the *yang kweidzah bingchuan*—ocean devil gunboat.

Concurrently with the opening phases of the new war, the United States was fighting a mild diplomatic skirmish with an ally. What was wanted was a decent man-of-war buoy in Shanghai, now that U.S. warships could once more come into that port without surrendering their breech blocks.

Early in the century, when buoys were allocated, the U.S. Navy ranked after Britain, France, Russia, Germany, Japan, and Italy in the order named, and Americans moored well down the Whangpoo, wherever they could find commercial space. By 1918, the U.S. was no longer a seventh rate power. The Germans, who had held a choice prewar spot off the bund, were now *non persona.* How about *their* place?

A September 1917 effort by American Minister Paul Reinsch brought an answer reflecting the sentiments of a British harbor master working for a British-controlled Chinese Maritime Customs. The Inspector General's letter appeared to Reinsch, "to indicate between the lines that the IG may fall in with the obstructive tactics hitherto pursued by British interests whenever there was any questions of Americans obtaining any facilities whatever in Shanghai."

Nevertheless, direct pressure on the Chinese government prevailed. By April of 1918 the buoys so convenient to the Shanghai Club, which many China sailors have since known, were assigned to the Americans.

1918-1924

The Yangtze "War," 1918–20
The White Russians Arrive
Bandits Up the Yangtze
Wanhsien Interlude
Palos *Visits Shanghai*
Middle Yangtze Tales
Wood Oil and Opium

On the morning of 17 January 1918, the *Monocacy* was fifty miles above Chenglin and working her way up the tortuous channel which narrowed to only fifty or seventy-five yards at times because of winter low water. The ship was sometimes only a few yards from the bank, which towered thirty or forty feet above her deck. Merchant ships had reported being fired on in the area—usually sniping by a squad or so of soldiers—so bags of coal had been distributed about the main deck and the crew had been instructed to stay under cover. To make her identity plain, the ship flew the biggest U.S. ensign in her flagbag.

At 9:00 A.M., a shot from ashore hit the jackstaff. Seconds later there were two more, and then a tremendous volley of perhaps two hundred rifles crashed out. Two bullets hit the bridge. Steering control was shifted to the conning tower and at 9:01 *Monocacy* opened up with small arms. This silenced the volley firing, but at 9:05 a sporadic shot hit Chief Yeoman Harold LeRoy O'Brien, who was standing alongside the captain. By that time the native pilot, in a blue funk, was crouched in a corner performing chin-chins and of no use to anybody.

Chinese troops, spread for two miles along the bank, continued potting away. At 9:25 the Japanese steamer *Tayuen* was seen standing downriver, bullets splashing around her in profusion. Clearly, sterner measures were indicated, so *Monocacy* opened up with her main battery of 6-pounders. The Chinese were such wretched marksmen themselves, sometimes firing from the hip without any pretense of aiming, that they often underestimated the ability of foreigners in this respect and stood around in the open. But a cannon was different. The 6-pounder barked a wicked crack. Its high explosive shell erupted in a big ball of flame and smoke, kicking up a cartload of dirt. The effect of the cannon fire was electric. The Chinese fire immediately slackened and in another five minutes had died out completely.

The river had widened enough so that the *Monocacy* could turn and follow the *Tayuen* at about three hundred yards. The Chinese had had enough. There was no further firing. *Monocacy* had lobbed half a dozen 6-pounder high explosive shells into the banks, plus three thousand rounds of rifle and machine gun ammunition. It had been a busy half hour.

Two men had been wounded, one seriously. Chief Yeoman O'Brien died. At this stage, it was considered unsporting to shoot at a foreign gunboat, except perhaps in Szechwan, where different house rules prevailed. A vigorous protest was delivered to Peking. So sorry, was the reply. *Monocacy* was not painted buff and white, like a proper foreign gunboat; she was war color, a hangover from the recent European unpleasantness. The soldiers took her to be Chinese, of the opposing faction and therefore fair game. They had paid for their lack of finesse with an estimated forty casualties, as counted from the gunboat. But eighteen months after the event, the Chinese government settled with an indemnity of US$525.54 to the seriously wounded man and US$25,000 to O'Brien's widow, a custom that would soon disappear as river firefights became more common.

With the end of World War I, a new era opened in China. The trading nations, less Germany, came back stronger than ever, determined to be their own policemen, as it was evident that the former amorphous Imperial authority had been replaced by near anarchy. The Yangtze Valley was a cockpit of interprovincial warlord strife, overlaid by the primary schism between rice-eating south and grain-eating north, in which the river was a natural dividing line.

Events elsewhere on the river soon foretold the shape of things to come. In November of 1918, Lieutenant Commander Bell-Sayre, commanding HMS *Widgeon,* discovered British wool yarn junks being held by a Wanhsien general who wanted seven taels (about US$4) a bale tax. Bell-Sayre demanded their release. No action. Then he delivered an ultimatum: release the junks or Wanhsien would be bombarded. No action. That afternoon *Widgeon* let go with four blanks and two high explosive shells aimed at an isolated rock, at the same time sending an armed party to cut out the junks. The hint took. Very shortly thereafter the junks were headed upriver, tax unpaid. But the general was miffed. He protested to the British consul at Chungking, who reported to his minister in Peking that the whole proceedings took place without his knowledge or consent and suggested that Bell-Sayre be severely censured for his audacity.

The British Admiralty couldn't have cared less about which Chinese general a Royal Navy gunboat shot up, but Their Lordships were horrified by Bell-Sayre's tactics; he had gone into action with both anchors down and less than full boiler power on the line!

By 1919, disorders were endemic throughout the Yangtze Valley as a result of the north-south hostilities. Neither side was winning any popularity contests with the peasants, who in time-honored custom, were robbed, raped, pressed into service, and otherwise put upon by both sides. The skipper of the *Palos,* bottled up at Changsha in February, reported with some understatement that "the soldiers are becoming more troublesome. A Northerner caught alone by the peasants was boiled in oil. The number of executions has picked up. On one occasion, the beheading occurred on the bund opposite this vessel. The bodies were allowed to lie for several days where they fell as a horrible example."

On Memorial Day, 1919, HMS *Widgeon,* bearing Vice Admiral Sir Frederick Charles Tudor, KCMG, CB, the CinC, British China Station, entered the roadstead at Chungking. Having finally gotten it straight why *Monocacy*'s colors were half-masted and directing HMS *Widgeon* to "dip and remain so until the American colours are two-blocked," Sir Frederick turned the American boarding officer over to his aide, Lieutenant Handley.

"Your admiral must have been in every war during the past fifty years," said the boarding officer, having admired the rows of ribbons which, after the British fashion, perched high on the admiral's shoulder.

"No, never," said Handley, "until we went through the Wushan gorge. . . . Some Chinaman tried to drill the old man. The bullet struck the very stanchion he was leaning against. Right now the shipfitter is removing the stanchion so he can take it back with him."

With rare instances, Asiatic Station officers had lived up to the standards attributed to them by historian Tyler Dennett. But by 1919, there are indications that some superiors were commencing to have second thoughts, at least as far as *junior* officers was concerned. In a delicately turned piece of understatement which named no names, the Asiatic Fleet medical officer sounded the alarm:

"It is well known," he warned, in the Secretary of the Navy's 1919 *Annual Report,* "that rules of conduct do not obtain the same strictness in the Far East; that temptation is more insidious, and that slips in moral conduct are more easily hidden and if discovered more readily condoned than in the United States." The good doctor's advice was to send no more officers to that wicked station who had not been well dried behind the ears by having been at least four years out of the Naval Academy. He would undoubtedly have closed his report with a hearty, Amen! had he witnessed some of the parties in the early nineteen thirties, aboard HMS *Medway* and in the "Grips" Hotel at Hongkong, in which young British and American officers of the Anglo-American submarine squadrons participated—or more accurately—managed to survive.

As 1919 drew to an end, U.S. naval power on the Asiatic Station was more impressive than at any time since the immediate post-Spanish War period. There were *seven stars:* a four star CinC, Admiral Albert Gleaves, in the armored cruiser flagship *South Dakota;* and Vice Admiral W. L. Rodgers, just relieved as CinC, commanding Cruiser Division One—*Brooklyn, Albany,* and *New Orleans.*

The Yangtze gunboats still were under the senior skipper, Commander L. S. Shapley, in *Wilmington.* Strung out along the river were *Elcano, Monocacy, Palos, Quiros, Samar,* and *Villalobos*—Division Two. *Helena* and *Pampanga*—Division Three—had gone south to Hongkong and Canton.

In February 1920, Admiral Albert Gleaves, CinC Asiatic, who took more interest in the Yangtze than had most of his predecessors, urged the Navy Department to order a flag officer to command YangPat. The British had just done so, and the United States should do likewise, to maintain equal prestige, said Gleaves. The Department was sorry; they had just been forced to fill four new flag billets at home and couldn't spare another admiral. Anyway, the Department felt YangPat was not prestigious enough to warrant a flag officer.

But the gunboats got around just the same. For the fiscal year ending June 1920, *Villalobos* steamed 2,608 miles, *Wilmington,* apparently with a more accurate navigator, 3,894.8, and *Quiros,* 3,956. But there weren't enough miles steamed where the shooting was prevalent—on the upper river—where only underpowered *Palos* and *Monocacy* could go. Robbers proliferated. Merchants and missionaries complained bitterly, one of the latter caustically writing the U.S. consul that, "It is a thousand pities that the Japanese are not engaged in upper Yangtze steamer trade or we would soon see some action taken!"

Aside from a flag officer on the Yangtze, Gleaves wanted *more* gunboats. The British had fifteen modern craft, but the U.S. had only two, he wrote the Navy Department. Destroyers were not suitable. Their ground tackle was too light, they were too long to negotiate the bends in the river, fuel was not guaranteed, and their battery was exposed to rifle fire from ashore. As for *Palos* and *Monocacy,* he wanted to "sell them down the river"—literally and figuratively—send them to South China, where the current and the weather both were mild, and those of the crew for whom there was no room below decks would not have to freeze sleeping behind makeshift canvas screens topside in winter.

Life might have been rough on the river but it was steadily improving for the cabaret aficionados and the married junior officers struggling to make ends meet ashore. In February 1920, the "Mex" had cost US$1.11; by May, it had dropped to 83 cents.

In a 30 June resumé, Admiral Gleaves once more cried out for new gunboats, but warned that until they were available, the old ones must be retained. The *Samar,* meanwhile, had subtracted herself from the total by smashing bow on into an oil tanker. As her remains were not worth repairing, she became the first of the Yangtze gunboats to be lost by violent means in a violent habitat.

The *Annual Report* of the Secretary of the Navy for 1920 included the first mention of a title for the Navy on the Yangtze as it was generally known: "The Yangtze Patrol was reorganized last December, Captain T. A. Kearney being placed in command." It also said that "the *Palos* and the *Monocacy* were active last spring in defending vessels from river pirates and lawless elements who were holding up and looting steamers and junks and firing on passing craft." Captain Kearney assumed the command on Christmas day, 1919. At long last, "YangPat" was officially the Yangtze Patrol, succeeding "Senior Officer Yangtze," "Second Division," and various other titles, some informal, some unprintable, under which U.S. warships had operated on the Yangtze for more than half the life of the Navy.

In June of 1920, the new ComYangPat made a tour of his upriver domain. At Ichang, he shifted from flagship *Wilmington* to *Monocacy,* along with a suite of two sailors, two U.S. Marine bodyguards, one mess attendant and an aide, Lieutenant B. O. Wells. At Wanhsien, with five more bodyguard Marines supplied from the *Monocacy,* the Commodore boarded *Palos,* Chungking bound. Six days later, after a round of calls on Chungking dignitaries, he was back at Wanhsien, aboard HMS *Widgeon.* As confirmed by the logs of both *Palos* and *Monocacy,* Captain Kearney had flown his broad command pennant in HMS *Widgeon,* thus not only setting what is believed to be a precedent for an American officer, but also demonstrating the prevailing solid state of Anglo-American upper river relations.

In a mere six years after launching, the hard river life had begun to tell on *Monocacy*—she was feeling old-age pains. Or was it that second-hand machinery Mare Island had installed to save money? In December 1920, an inspection revealed a few deficiences. The boilers couldn't keep a full head of steam for more than three minutes with the engines running full speed. Clearance between pistons and cylinder walls was as much as 3/32 of an inch. The radio set wouldn't operate, which didn't really matter much, as there was no operator on board and the antenna had been housed. The ice machine was out of whack. Main engines and throttles had leaky valves. One air pump had a broken piston and the other an unserviceable valve rod. The port circulating pump had a broken piston ring. The forward main feed pump valve rod was broken. The steering engine had a broken piston ring. In the theme of the old adage, she was a damn' fine ship for the shape she was in. But obviously, no genuine Old River Rat would be downhearted over such trivia; the inspection report wound up with the opinion that the morale of the crew was excellent.

During the spring rise of 1920, *Palos* at last managed to wrench loose from Chungking and head downriver for Shanghai. It was planned, first, to exercise her 6-pounder main battery off Woosung; then there would be exercise of a slightly different character, involving the various delights of the Paris of the Far East. After a long, dark, rainy winter and with everything ashore knee-deep in mud, every man had squirreled up the equivalent of a half bucket or more of silver dollars, burning to be invested in one way or another. Concerning such investment, a River Rat explained his philosophy: "The most of it goes for likker and wimmen. The rest I spend foolishly."

But not in Shanghai after all. At Nanking, *Palos* was sent back upriver; the natives were restless again. Her return trip logged its usual quota of routine thrills. Anchored at night between Hankow and Ichang, an enormous log raft had been fended off with disaster a near miss. The ascent of the rapid above Ichang cost four hours of desperate churning that left the sweating firemen scraping the last few scoops of coal dust off the floor-plates.

But in the half-mile-deep fissures through which the Great River had cut its way, there were good things, too. A skipper and his ship were on their glorious, independent own, one of the last places in the U.S. Navy where it was possible. Once the

Palos and *Monocacy* sailed beyond Ichang, the radio antennae served chiefly as a convenience for hitch-hiking sparrows. On infrequent occasions when a radio operator was on board, and in the unlikely case the radio functioned, its feeble output would not jump the intervening mountains. Mail went out informally via some cooperative passenger or skipper aboard a river steamer. In the happy event the bags were not rifled en route by some acquisitive Chinese, a reply might be back in a month and a half. "Rapid" communication was by Chinese telegraph. One framed a message beginning with "Unless otherwise directed, I intend to . . . etc. . . ." Then straightaway, he *did* it. At the next opportunity the telegram would be sent. The answer might be back in forty-eight hours, by which time the action would have become history.

The *Palos* no sooner had struggled up to Wanhsien when a cry for help came from Sao Ta Chia, twenty miles downriver, where bandits were holding an American wood oil junk. At Sao Ta Chia, *Palos* put a couple of 6-pounder rounds into the hillside, and the bandits made a lightning decision to get out of the wood oil business. The guns were short in length and hellishly loud for their bore size.

Back in Wanhsien, *Palos* met the American flag river steamer *Robert Dollar,* which announced that against their wishes, they were entertaining on board about a hundred Chinese soldiers, who were repaying this hospitality by demanding Y$500 ransom for the captain.

Palos straightaway steamed up close abeam, but not alongside, as her skipper, Lieutenant George S. Gillespie, had only thirty-nine men in the crew and no desire to be boarded.

"I didn't know just what I would do," he wrote later.[24] "I had a pistol in my hand but kept it below the screen. The *Dollar*'s captain pointed out the Chinese in charge, probably a general. I considered shooting him. He was only about twenty feet away, but very close to *Dollar*'s captain. I was about to suggest to the captain he step away, so that I could get an open field of fire. Then I decided to try diplomacy first, asking the Chinese to leave the ship. And this they did."

If Gillespie had not been a dog lover before, he most certainly must have been converted shortly after the *Robert Dollar* episode. With his ship anchored for the night twenty miles above Wanhsien, Gillespie was snoozing peacefully away after a hard day fighting Old Man River. The ship's dog, "Sooner," an English bull terrier, shared his cabin. Sooner was not, as a general thing, the barking type, but he had a king-sized hate for the Chinese. For their part, Chinese dogs get frantic at the sight or scent of foreigners because, so the theory goes, they eat so much meat they smell like cats. And the hereditary enemy of all Chinese dogs is the cat—the primeval lion, leopard, and tiger of prehistoric Chinese forests.

During the night Sooner's throaty growl awakened the skipper who automatically rang a loud gong for the deck watch and ordered a good look-see all around.

The deck watch was soon back, a little out of breath. "There were some boatloads of soldiers drifting down on us," he said. "But they are going away now."

Not for a long time had an American ship been taken by boarding, and it is just possible that an English bulldog by the name of Sooner kept the record clean.

On 8 July 1920, as *Monocacy* lay in the roadstead of Wanhsien, the postmaster of the city, a foreigner, appealed for help. His safe was full of cash and the soldiers thereabouts were equally full of zeal to pinch it. Could the gunboat accept it for safekeeping? Here was fantasy: the postmaster of a Chinese city asking a foreign warship to defend the Chinese government's cash against pillaging by its own soldiers.

At noon, *Monocacy*'s armed party picked up the postmaster's C$5,600 for safekeeping. By 6:00 P.M., firing between various Chinese elements ashore near the anchorage sent the crew to battle stations. One could never be sure, in view of the quality of Chinese markmanship, where the bullets would land. But in fifteen minutes either Chinese ammunition or animosity had run out—the firing ceased. An hour later, a Chinese officer sampanned out to assure Skipper C. D. Gilroy that foreigners and their interests would be safe. Next morning, bright and early, 6:00 A.M., the Chinese was back for a fifteen-minute chat, apparently with a resumé of the day's planned activities, as the *Monocacy* shortly thereafter hoisted anchor and moved two hundred yards upstream, according to her deck log, "to get out of the line of fire of the Chinese at Chung-ja-pa and Wanhsien."

The schedule of events apparently was faithfully carried out, as shooting commenced on both sides at 7:30 A.M. and continued intermittently all day.

Six A.M. seemed to be standard starting time for military events. At that hour one morning, Ensign Stanley M. Haight, the executive officer, took a small party ashore to retrieve the ship's sampan, which had been seized by the Chinese military at Wanhsien. Either the argument Haight put up was sound or the show of force convincing: they were back with the sampan at seven-fifteen.

By 11 July, things ashore had cooled sufficiently to chance handing back to the postmaster his working capital. Five days later the local revolution was on again. Artillery shells started arching across from the south bank into Wanhsien. Haight buckled on his .45, filled a canteen with coffee, and prepared to hit the beach.

By that time the stuff was flying across the river both ways and Captain Gilroy's mission for Haight was simplicity itself: go ashore and tell those Chinese bastards to watch where they're shooting; they might hit the Standard Oil tanks!

Haight sampanned ashore and started his climb into the hills to locate the offending batteries. The inevitable and indispensable Chinese mess boy traipsed along behind to interpret.

In four hours, the two were back aboard, having dodged considerable flying metal meanwhile. They had located the Chinese artillery commander, impressed him with the wisdom of sparing Standard's precious tanks, and had in fact remained with the guns for some time to insure that the gunners got the message.

The Chinese artillery commander was about thirty or thirty-five years old, over six feet tall, and much to Haight's surprise, spoke perfect English. Haight came straight to the business at hand. Pointing out that he was from the U.S. gunboat *Monocacy*, down there in the river protecting American and British interests, he said they didn't want shells to endanger the oil tanks. "I am here to warn you that if you destroy the oil tanks, *Monocacy* will have to retaliate," announced Haight, with all the dignity of his one stripe and twenty-three years.

"We are interested in stopping those bandit forces across the river who are out to raid and sack Wanhsien," said the Chinese.

Haight peered through his binoculars and announced that the Chinese shells were dropping less than a quarter of a mile from the tanks.

"I don't have any glasses. May I borrow yours?" said the Chinese. He adjusted them to his eyes and had a look. "I think it would be safer if you marked a limiting arc for our fire," he said. Haight walked fifty or so paces out beyond the guns and laid out a cairn of sticks and stones. "Don't lay your guns to the left of the marker," he said.

The two were soon on informal terms. Li Tan-hsai amiably ordered tea for his guest and sampled the contents of Haight's canteen. Between booms of the cannon, the chat drifted to business. "I would like to buy one of your 6-pounders," said Li, "and one thousand rounds of ammunition." A discussion of this intriguing proposal was unfortunately cut short by the arrival of a messenger with the news that enemy troops had crossed the river upstream and headed their way.

Haight and his Chinese messboy (several shades lighter since the latest news) scrambled down to the river bank while Li prepared to limber up and gallop off with his artillery to less unfriendly surroundings. It had been just another day in the life of two men, briefly met, briefly friends, with vastly disparate problems. But there had been that spark of human understanding in spite of the different worlds in which they moved. It was in microcosm, one might say, this senseless world at war or threatening war, over what, no one really knew.

In addition to her pair of 6-pounders, which artilleryman Li Tan-hsai admired, *Monocacy* carried half a dozen assorted machine guns and, from time to time, a 3-inch field piece on the fantail, ready to be rolled ashore and towed into action by manpower. The *Palos* was likewise armed.

In later years on the river, "repel boarders" drill was taken seriously and held often. Unlike the *Ashuelot,* which carried cutlasses, boarding pikes and battle axes, *Guam, Tutuila,* and the bigger half-sisters of the "new six" carried sawed-off shotguns, Thompson sub-machine guns, Springfield rifles, and Colt automatic pistols—a weapon of some sort for every man in the ship's company.

The various moves of Chinese generals around Chungking in 1920 would have delighted the heart of a good chess player. Pawns were traded for silver bullets, checkmates arrived at when the price wasn't right, the board occasionally swept clean in righteous anger when the rules were not observed, then the whole game set up again, generally with the same pieces.

With the indispensable arsenal at Chengtu, plus opulent merchants at Chungking, overlaid by opium everywhere, efforts to levy tribute at the gate—either Wanhsien or farther upriver—were always in order. In July of 1920 such a situation existed at Tang Chia To, eight miles below Chungking. There, an enterprising small-bore general had set up shop as a toll collector.

In olden-day Marine Corps posts, the "word" was, if it moves, salute it—if not, whitewash it. A corollary on the Yangtze was: if it moves, shoot at it—if it stops, tax it. The SS *Alice Dollar,* U.S. owned, as she headed for Chungking, was heavily fired on at Tang Chia To, on 20 July. To prevent any such foolishness on her next run through, *Monocacy* fell in astern as she headed downriver the following day.

Only four miles below Chungking, bullets started arriving from the north bank. Half a dozen rounds from *Monocacy's* 6-pounder put a temporary halt to that. But a quarter of a mile on, the next detachment opened up. From then on to Tang Chia To, the firing from ashore was continuous, replied to in best gunboat protocol by 6-pounder and .30 caliber fire from the *Monocacy*. From there on, the path was clear. *Alice* proceeded on her way, while the gunboat anchored long enough at the edge of No Man's Land to make clear who was senior rooster on the local dunghill. Under way at noon, she chugged on back unmolested to Lungmenhao Lagoon, counting her bullet holes. The operation had covered half a day. She had been hit five times out of an estimated six hundred shots, a fair enough score for Chinese marksmen. Two men had been lightly wounded. The American consul, Mr. C. J. Spiker, wrote the customary blistering note to the Honorable Chiang P'an, commissioner of foreign affairs at Chungking, who eventually sent the customary reply of "so sorry."

Monocacy sailors not engaged aboard swabbing out the guns and sweeping up the debris shaken loose by the firing found their way to the canteen for some cooling beer. The skipper, executive officer, and doctor trudged up to the club for a go at tennis and a bit of gossip about the latest happenings. One could say it had been a very nearly normal, sticky, Szechwan summer day.

With live targets to practice on up the river, it seemed superfluous to go to the trouble of arranging an official practice. However, it did allow a much appreciated visit to Shanghai as a dividend. Such a practice was described by Frank Brandenstein, chief commissary steward in *Wilmington*. A tug had struggled all the way up from Manila with a target raft, which was then anchored in the Yangtze off Woosung, where ships in turn could steam by and blast away. By the time the old *Wilmington* had creaked into position to fire, the raft's convenience as a fishing platform had been taken advantage of by a number of Chinese. A launch was sent to shoo them off, but by the time the launch got back to the ship there were more fishermen aboard the raft than before.

Then Skipper "Dusty" Rhoades tried out a new tactic. Steaming in close, he fired a blank saluting charge in the direction of the raft, at which the squatters took an extremely hurried departure.

"The firing of our starboard battery was really something to see," wrote Brandenstein. "There was only one good breech plug on that side and the gunner's gang was on the run, moving that plug from one gun to the other."

In the early twenties, upriver amity between British and American gunboaters probably was at an all-time high. This was due partly to cameraderie built up during World War I and partly to the pure practicality of cooperative enterprise. The U.S. gunboats *Monocacy* and *Palos,* the only ones which could operate on the upper

Yangtze, were close to being amalgamated into one joint patrol with British *Teal* and *Widgeon,* in an era and area where the pace of events was increasing. U.S. and British armed guard detachments and crew replacements rode back and forth on each other's ships. Whoever was nearest a piracy or piece of banditry rushed to the scene, regardless of whether the victim flew Britain's Red Duster, or the Stars and Stripes. Other foreign flag ships engaged in so much marginally legal activity that the Anglo-Americans left them to their own devices and protection.

A prime example of the unorthodox results of this upriver solidarity took place in the summer of 1920, at Hankow. The roadstead held four British, one Italian, one French, and two U.S. men-of-war. As usual, the British were senior, by virtue of their 10,000-ton cruiser *Hawkins,* with seven hundred and fifty men and seven 7.5-inch guns.

The ex-German and ex-Russian concessions had been released in 1917 and 1920, respectively; the former when China declared war on the Central Powers (to save herself from Japan), and the latter when China withdrew recognition of the by then defunct Imperial Russian government. With China's state of near anarchy in 1920, these choice pieces of real estate might suddenly be up for grabs. With the exception of the Americans, who had no territorial ambitions, the foreign warships were assembled like so many waiting birds of prey

One steaming August night in Hankow, where temperatures soar constantly above one hundred degrees for days and nights on end, the executive officer of USS *Elcano,* Lieutenant Stanley Haight, had abandoned his hot box of a cabin for a cot on the not much cooler deck aft.

His fitful slumber was interrupted by the deck watch. "There's a British commander on the quarterdeck and he wants to see the commanding officer *immediately!"* gasped the sailor.

Haight slipped into his shoes and padded forward. The commander was an old school Britisher at his overbearing best. He announced himself tersely and got straight to the point: he was from HMS *Hawkins* and demanded to see the commanding officer, instantly. Haight explained the skipper's absence—ashore for the night at the Wagon-Lits hotel, luxuriating under a revolving overhead fan—and invited his caller to the cabin for a cup of coffee while the apparently highly explosive matter at hand was gotten into.

The Britisher was all give and very little take. "You Americans have *got* to stop assaulting our sailors!" he shouted. "Last evening a dozen or so of our men on shore leave were badly beaten by your people. One of them is not expected to live. Besides the punishment you may give the perpetrators—which is your affair—there *must* be an understanding and some control over your men!" In his apoplectic excitement, the commander neither drank his coffee nor lighted the cigarette he had taken.

Haight mentally reviewed the whole affair while the Britisher sizzled and complained and mopped his florid, perspiring brow. He knew that the men of *Teal, Widgeon, Palos,* and *Monocacy* were on the best of terms. This good feeling spilled over onto the *Elcano,* which saw a good bit of the upper river warships at Ichang.

The U.S. officers, when invited aboard His Majesty's gunboats, contributed ice to cool what otherwise would have been lukewarm pink gin. There were frequent turnabouts at dinner invitations.

The sailors enjoyed each others' hospitality, too. "When are you having Australian mutton again?" a U.S. gunboat bluejacket would amiably shout across to his Limey opposite number. "Warn me so's I'll lay off dinner with you swabs that day!" Then loud guffaws from all hands. "I say!" the Limey might yell in return, "when you Yanks get a tot of rum a day from His Gracious Majesty, then we'll knock off the mutton!" And so, the banter jelled into a united front of sailors from both navies—the Yangtze Rats versus the common enemy. The common enemy, naturally enough, was the big ship, *any* big ship, *any* nationality, from "outside."

Haight had been aroused from his topside snooze the night before by a commotion on the quarterdeck created by *Teal* sailors and a couple of *Elcano* bluejackets, all considerably the worse for wear. The next day, Haight managed to worm the story out of his men, and this he passed on to his angry guest.

The gunboaters had a special hangout ashore, explained Haight, where they met their "pigs"—the Chinese girls who furnished feminine companionship in the days before the White Russians came. Onto this bit of sacred gunboat soil barged a group of Hawkins men, shouting, "You Yanks and you Blokes too! Clear out!" There had been a few punches swapped before the greatly outnumbered Anglo-American gunboaters were bodily ejected. The *Elcano,* with twice the complement of HMS *Teal,* clearly was the logical place to sign on recruits for a counter-attack. It was these recruiters that had awakened Haight, who, seeing that the *Teal* men were aboard under friendly auspices, headed back to his cot on the fantail.

The recruiters, meanwhile, had done well. About a dozen *Elcano* avengers soon jumped in the sampan alongside, passed by *Teal* en route for more reinforcements, then headed for the establishment from which the earlier group had so humiliatingly been given the bum's rush. There, they found the *Hawkins* men, swilling beer and happily pushing the *Teal*'s and *Elcano*'s "pigs" around the dance floor. The conquest had been almost too easy.

Beer mugs and fists began to fly and the outnumbered *Hawkins* party made a strategic withdrawal to the men's room, thoughtfully bolting the door behind them. Some tactician in the River Rat detachment shouted, "Let's break the door down!" This they proceeded to do, using an upright piano as a battering ram. The piano screeched across the dance floor, took out the door and the whole wall, and crashed in a mass of woodwork and *Hawkins* sailors against the outside wall of the building.

"Those, I suspect, were the *Hawkins* men who were badly hurt," opined Haight. He felt sure that no lethal weapons other than the badly abused piano and a few beer mugs had entered the action.

The commander tersely called for his boat and left, with a formality and dignity that could be mustered only in an old-time Britisher stirred by deep emotions. "Those bloody damned colonials!" one might suspect he mumbled as he lowered himself into his gig.

Palos had missed her spring overhaul, spending another blistering Chungking summer making jury repairs to her ailing machinery. Another winter trapped by low water in Chungking would finish her off.

On 9 September 1920, she dropped her mooring chains and headed downriver, but that time in the role of convoyed instead of the convoyer; the Standard Oil Company's tug *Meitan* sailed with her, an oil barge lashed alongside and a Navy signalman aboard for quick communication.[25]

In spite of an exhaustive overhaul of her feed pumps, using all the poor facilities of Chungking, everybody was prepared for the worst. After two hours they were justified; two feed pumps stopped. *Meitan* steamed on ahead to rendezvous ten miles downriver where maneuvering room was available to take *Palos* in tow. Meanwhile, *all* feed pumps had failed and water disappeared from the boiler gauge glasses. Fires were hauled from under both boilers, and in short order the few remaining buckets of steam were bled off to the steering engine. Cold, dead, and out of control, *Palos* drifted with the current, sometimes within several yards of a jagged river bank. Boats were made ready and the men quietly put on life jackets.

A very little short of disaster, *Meitan*'s Captain Miclo, in a marvelous exhibition of seamanship, whirled the tug around, crashed alongside *Palos* and secured her on his free side. It was so close that *Palos* brushed her stern against the rocks. Lashed three abreast, the two helpless hulks outboard, the clumsy trio made it miraculously to safety at Wanhsien, the first time in recorded Yangtze history that such a deed had been accomplished.

In the mysterious East, destiny could take many forms. Almost a decade later, Captain Miclo took his ship from Hankow at midnight, en route to Ichang. Clear of the pontoon, he turned the deck over to a mate, strolled aft for a smoke, and was never seen again. He was not the first, nor would he be the last, whose disappearance was never explained. The opium syndicate had long arms and a long memory; very few Yangtze ships failed to become involved with it.

After an American sailor was slain during a Shanghai bar fight, the *North China Daily News* reported the full details on 4 January 1921. It began with the peaceful entrance of a half dozen *Wilmington* sailors into the Victoria Bar, where twice as many Italian sailors were already several drinks ahead. The women in the bar soon expressed marked preference for the Americans, a choice no doubt influenced by their knowledge of American and Italian pay scales.

"With this development," wrote the local reporter, "a certain liveliness prevailed." Next a merchant seaman declared that "there are enough Americans here to fight the whole Italian Navy." More Italians rushed in to give the Americans their money's worth. Some U.S. Marines, who disliked Italian sailors even more than American sailors, joined the fray, and chairs, tables, knives, and a few bullets accented the liveliness. The Americans soon dug in behind the bar and the Italians, having swept the local field, departed for further conquests, recruiting reinforcements on the way, and burst into the Tivoli Bar, where they met an equal force of more American sailors. "Knives flashed," wrote the man from the *News,* breathing heavily by now, "and a furious assault was made on the Americans." Evidently the Americans assaulted right back, for "worse scenes ensued than at the Victoria Bar" and had not the police arrived and restored order, "there must have been a terrible tragedy. The outer clothing of the *Wilmington*'s men was slit to ribbons." The article closed with statistics on the dead and wounded and a note that the Americans and Italians had suspended further shore leave as a precautionary measure.

In such informal affairs did international sailors sharpen their teeth on each other in preparation for dealing with bandits, pirates, revolutionaries, et al, when and if the time came.

The average brawl ashore was usually never such a serious affair—a sort of Marquis of Queensbury thing. There was, for example, the 1921 fracas in Hongkong, during which some happy British tars flung a *Wilmington* bluejacket off a dock for an unscheduled bath. The *Wilmington* men, outraged at this affront to American dignity, naturally enough beat up the British, and this warmed up feelings to the point where Admiral Joseph Strauss postponed a scheduled visit of the Asiatic Fleet to Hongkong.

But the *gaffe* was soon atoned for; *Wilmington* was the only ship in port to greet a world-famous American, Madam Schumann-Heink, on a world tour. Sailors manned the rail for her, bunting flapped in the breeze, and the big Sunday colors were broken out. Schumann-Heink came aboard, with tears in her eyes, to thank them for the welcome. Like many American parents in the Civil War, she had sons fighting on both sides in World War I.

By the 1930s, the bloody flaps between the international sailormen had virtually ceased. Perhaps the hardening of a common front against the intransigeance of the Chinese had done it in the twenties and similarly against the Japanese in the thirties. The British bluejackets and the Americans were now divided only by a common language. On the Yangtze they shared each other's sometimes scarce beer. Americans bumbled through cricket games and British fumbled the baseball. There were so few French and Italian sailors who spoke English and their tastes in the basic essentials—wine, women and food—were so different that this tended to accentuate the gap which already existed through the difference in color of the money in their respective pay envelopes. Like dogs strange to each other, they might sniff in passing, perhaps growl a little *pianissimo,* then pass on in peace.

In 1843, when Shanghai was laid out, and for twenty years thereafter, European women were scarce. Nor did dependents follow the fleet. Chinese women, euphemistically termed "housekeepers," sometimes became common-law wives, or simply remained as mistresses of the *fanquei* businessmen.

The U.S. consular archives reveal that between 1879 and 1909, legal liaisons still were seldom registered, even though the foreign population was bounding upward in numbers. In that 30-year period, there were recorded only 221 American marriages, of which 15.4 percent were to Asian women. The Americans included 11 mariners, 2 policemen, 2 sailors, 3 customs clerks, 1 engineer, 1 missionary and 14 who kept their profession, if any, to themselves.

From 1910 to 1918, there were 202 marriages recorded, but the percentage marrying Asian women had dropped to 8.9. During all this period, no American woman had married an Asian. But from 1920 to 1932, the percentage of new Asian wives of Americans had plunged to 2.2. What brought about this tremendous change? It was not skin color nor "foreign-ness," because between 1946 and 1968, 53,864 Japanese women were married by Americans in Japan. The answer is simple: in 1920, something new—more glamorous and exciting—had appeared on the scene: the White Russians!

There had been Russians in China since the sixteenth century. In the days of Peter the Great, Russians sent tea from Hankow, up the Han River, portaged it across to the Yellow River, then boated it upward as far as possible. Thence via camel caravan this vital ingredient of Russian existence went to Nijni-Novgorod for further distribution all over the country. By 1896, the Hankow tea trade was largely a Russian operation. Tea went to Odessa by the shipload and tons of brick tea found its way overland to Siberia and Mongolia. By 1900, the Hankow tea trade was *wholly* Russian. Temporarily abstemious British tea tasters sipped the various brews during the new season, leaving their Muscovite masters free to blunt their taste buds with champagne. This, they could well afford. So great was their middle Yangtze influence that one might expect a bill made out in rubles instead of the formerly universal British pound, or Chinese tael.

Shanghai in 1910 held 324 Russian diplomats or businessmen of substance. By 1920, the number had swelled to 1,476 and by 1930, to 7,366. These were not prosperous businessmen or pompous diplomats, but refugees from the Red Terror, which like a slow but relentless tidal wave, had swept the last of the White armies against the sea at Vladivostok or into the neutral haven of Manchuria.

The Japanese withdrawal from Siberia in the autumn of 1922 left the White Russian remnants without support. The first evacuation ships sailed from Vladivostok on 16 October 1922. Twenty-nine Russian vessels, the largest the 3,000-ton gunboat *Manchuria,* the smallest not much larger than a harbor tug, plus two Japanese ships, brought off 10,300 fighting men and their families.

The *Manchuria,* designed for not more than three hundred passengers, was packed with 1,260 soldiers, sailors, old men and women, and children. They were not going home, but to some strange unknown place, perhaps forever. Everybody took along all that space and strength permitted: bundles, bags, parcels, trunks, guitars, clocks, samovars, ikons, vodka, homemade pickles, and photo albums.

With some ships towing others and passengers rationed to three glasses of water a day, the "fleet" finally reached Possiet Bay for the first nose count. By 31 October, they anchored in Wonsan, Korea. There, the half-starved emigrés were put up in old Japanese barracks, where ten or twelve children died daily. Others went on to Port Lazarev, where more Japanese barracks were available. In the summer of 1923, the second stage of the exodus began to Shanghai, where many Russians already had arrived overland, via Mongolia and Manchuria. By Chinese presidential mandate of

23 September 1920, the moribund Russian Empire's consulates, along with the concessions in Hankow and Tientsin, had been taken over by the Chinese authorities. From 200,000 to 300,000 Russian subjects—in Sinkiang, Mongolia, Manchuria and the port cities—had not only lost their extra-territorial rights; they were stateless.

No one who visited Shanghai in the thirties could forget the sights and sounds of Avenue Joffre, the center of the Russian colony—crowds of white people, signs in Cyrillic and the balalaika and accordion music in the background. This was old Russia in miniature, where restaurants were filled with strong odors of cabbage soup and chicken cutlets stuffed with butter.

Nor can the Old River Rat forget the gayety, lights, and music in cabarets along Hankow's "Dump Street," where hundreds of throaty-voiced White Russian girls put on a brave front and a bright smile to the world, somehow managing to keep patched together the frail fabric of mere existence.

These near-penniless but generally well-educated people, on initial injection into Chinese life, found themselves in potential competition with hordes of coolies happy to work for bare survival. With magnificent organization and self-help, the exiles formed some sixty-nine different societies, affiliated with either the Russian Emigrants' Committee or the Council of the United Russian Public Organizations (SORO). The latter solemnized and registered marriages, issued birth certificates, granted divorces, drew up wills, and registered business agreements, in effect functioning as a consulate *vis-à-vis* the Chinese authorities. Both organizations had to take care not to be overtly anti-Soviet, to avoid repressive Chinese action through Soviet pressure on Nanking. But in the Russian Officers' Club, modest and poor by Shanghai's opulent standards, old Czarist diehards drank gallons of tea and argued over past campaigns and future victories when they would supply the military push that would bring the ramshackle Soviet anathema down. Six hundred of them served in the paid detachment of the Shanghai Volunteer Corps, a full general commanding one platoon with a mere lieutenant general as his assistant.

Russian men worked as chauffeurs, mechanics, bus drivers, tram car inspectors, bodyguards for rich Chinese, and such other lower grade occupations, for which they received the equivalent of US$30 to US$60 per month. The women were nurses, cinema attendants, shop assistants, and—best known to the sailormen—dancing partners, for a few cents a dance and a cut on their "champagne" (apple juice) cocktail.

With the Great Depression, many Russians replaced Europeans in foreign firms. They took no expensive home leave and worked for lower wages, perhaps the equivalent of US$100 a month. But life was never easy. A Russian teacher, once a high school principal in Old Russia, built a fire in his little apartment stove only on lesson days, living the rest of the chilly Shanghai winter in much mended gloves and overcoat, indoors and out.

One of the unusual aspects of this country within a country was that the Russians did not intermarry with Chinese. Contrarily, they provided the brides which reduced the percentage of Asian women marrying Americans to the near vanishing point. In 1920 to 1922, 45.2 percent of the Russian women married to foreigners had American husbands, 28.6 percent had British husbands, and 3.6 percent had Asian husbands. Ten years later, American charm and dollars had won out; 62.1 percent of the husbands were American, only 12.1 percent were British, and 5.6 percent were Asian.

This predilection for American men, flattering as it was, had its potential drawbacks. When the cabarets, normally alive with sailors drinking, dancing, and chatting with Tanya, Nina, Olga, Vera, and so forth, were singularly devoid of customers, it was with good reason, such as the time a jealous young Russian swain, fed up with American competition, had fired a shot through both the transom and his wealthy rival. But Shanghai had a short memory. In a week, things were normal again. New ships had hit port, their crewmen innocent of any knowledge of the unfortunate *affaire d'honneur.*

A standard prewar Shanghai anecdote concerned the Russian who returned to the Soviet Union. He obviously wouldn't be able to write openly how things were, he told his friends, but he would send a photograph. If he were standing up, conditions were fine; sitting down, not so good. The photo eventually arrived, with the subject flat on the floor. But World War II changed attitudes enormously. Once-bitter anti-Soviets had found pride and patriotism in the Motherland's heroic defenses against the Germans. The Soviet consular establishments in China urged the expatriates to return "home," and many did.

One of these was Alexander Vertinsky, the famous singer in the Renaissance restaurant, who waited imperiously for dead silence before beginning one of his sad, haunting melodies. On reaching the Moscow station, Vertinsky put down his suitcases and lifted up his expressive hands in a typical Vertinsky gesture. "I have returned, Mother Russia!" he said with dramatic feeling. Then he turned to pick up his suitcases, but they were gone. "I recognize you, Mother Russia," he added sadly.

And what became of the other thousands who once trod Avenue Joffre, then returned to Mother Russia? Or those who spread across the world, streaming in a second terror out of Shanghai and Harbin and Hankow as the Chinese communists took over? Sasha, Nina, Alexandra, Tina, Olya, Nadia, Tatyana, Fanya—dear, gay girls who filled the evening hours with fun and happiness—where are you now?

After World War I, China had progressively fallen apart, and by 1921 affairs were back to normal; that is to say, worse. On 3 February of that year, American Minister Jacob Gould Schurmann wrote to the Secretary of State that "American standing in the Yangtze Valley at the present juncture is undoubtedly at a low ebb, due to a series of events in which the American authorities were not successful in demanding and obtaining the respect and security due to American persons and property." One needed no course in statesmanship to translate "American authorities" as "U.S. Navy."

Secretary of the Navy Denby lamely apologized that he had only six vessels available, but was considering detailing a flag officer to command them. *That* would throw a proper scare into those pesky bandits.

The American consul in Shanghai, M. F. Perkins, smarted from the bad time the businessmen were giving him. Carry a big stick, was his motto, as he wrote that "There must be created in the Chinese a state of mind which instinctively senses the certainty of retribution . . . and that officials, military and civil, will be held personally responsible . . . ." The consul further had a jaundiced eye for YangPat and its superannuated "force," particularly when compared to the British Yangtze flotilla, doubled in the previous two years. He complained bitterly about the YangPat practice of withdrawing *all* the gunboats to the mouth of the river for annual target practice, leaving the ports uncovered. "Of little avail that ships arrive *after* catastrophes have occurred," he concluded.

Admiral Gleaves' importunities had at last borne some fruit in that by 21 March 1921, the Department was not only considering a flag officer for YangPat, but a "suitable flagship" to carry him. In mid-1921, Captain D. M. Wood had reported as ComYangPat and was flying his broad pennant in the *New Orleans,* a small, 3,400-ton cruiser.

Also in March, *Elcano, Villalobos,* and *Quiros* had their home port changed from Manila to Shanghai, suggesting that after twenty years, Washington had concluded they were on the Yangtze to stay.

The new CinC, Asiatic Fleet, Admiral Joseph Strauss, who had assumed command in February, sailed the Yangtze in July aboard *Wilmington* for twenty-six days, believed to be the longest such tour on record. He covered the river even to perennially disturbed Changsha and in a smaller ship fifty miles upriver beyond Ichang. But he did not reach Chungking, cut off by the then current "war" between Szechwan and Hupeh. The cruise made a marked impression on Strauss, who had already involved himself closely in YangPat operations.

On 5 August, the Chief of Naval Operations ordered a name change for "Yangtze Patrol," as recently christened, to "Yangtze Patrol Force." The new title involved a good bit of work throughout the Navy amending records, blueprints, and blank forms, but was of otherwise questionable merit.

A move which made far more sense—in fact, second best only to the establishment of the YangPat command itself—took place in September 1921. The Patrol's supply activity was detached from Shanghai and moved to Godown Hankow. Godown, of course, was pidgin for warehouse. But this was no ordinary warehouse. Presided over by a chief pay clerk under the direction of the patrol supply officer, Godown Hankow carried every conceivable thing necessary to keep a gunboat and her people alive and happy—spare parts, food, pharmaceuticals, paint, clothing, cleaning gear, toiletries. For those who had found the strain of life on the Great River too telling, there was even a small stock of coffins. When Godown Hankow was set up, the Patrol Medical Office at Shanghai was closed, thus leaving the only excuse to visit the Paris of the Far East the periodic drydocking of the ships and the spiritual revival of gayety-hungry River Rats.

On 12 October, Rear Admiral W. H. G. Bullard, the flag officer, arrived at last.

He soon was established in the flagship *Isabel*. This dainty little 900-ton yacht, named after the daughter of former owner-industrialist Willys, had filled the role of small destroyer in World War I, armed with two torpedo tubes and four 3-inch guns. For the next two decades after her arrival in China, minus the torpedo tubes, this white and buff ship ranged the Yangtze as flagship of the Patrol, then the whole station as the CinC's yacht.

Commodore Wood, before his relief by Rear Admiral Bullard, had immersed himself with enthusiasm in riverine affairs. First he made a recommendation that six Eagle boats augment the Patrol. These coffin-shaped atrocities, left over from World War I, drew over eleven feet, counting the sweep of their low-hung propellers. Then he wrote a very complete, knowledgeable report on what was up the Yangtze. As for the old ex-Spanish gunboats, he felt that *Quiros* and *Villalobos* were hopeless cases. Lacking cold storage, their people had to live hand-to-mouth. The radio installations were poor. The main batteries were unimpressive. Electric plants were marginal. *Elcano* would pass. She could carry a large crew and thus provide a sizeable landing force. Her radio was acceptable and she was built of old-time, slow-corroding wrought iron.

Wood had another idea that got off the ground only far enough to be promptly batted down by watchdog Strauss, but not before it had provided a most unusual incident unique in Patrol history.

In the course of the current Szechwan-Hupeh "war," both banks of the Yangtze from fifty miles above Ichang to Chungking were spotted with irresponsible, trigger-happy groups of bandits who had brought steamer traffic to a halt. As a result, Standard Oil at Chungking was scraping the bottom of its tanks, and appealed to the skipper of *Monocacy,* Lieutenant Commander G. E. Brandt, to try something new—convoys of oil-laden junks flying the American flag by virtue of their U.S.-owned cargo, guarded by a posse of *Monocacy* sailors riding a houseboat. Wood thought highly of both Brandt and of the idea. When one was in intimate daily contact over a beer at the club with the commercial people, friendships were closer and sympathies more deeply felt than in the impersonal atmosphere of great Shanghai—or even of Hankow. Both Wood and Brandt thought Standard Oil, which eagerly offered to stand the cost of upkeep of the sailors and their transportation downriver to the oil junks, deserved American naval protection.

Thus, in May 1921, two houseboats, each carrying three *Monocacy* enlisted men under the command of a chief petty officer, armed with a Lewis machine gun and rifles, started creeping upriver. Each houseboat convoyed about half a dozen oil junks at the end of tracklines towed by the usual eighty or a hundred coolies. At every bend in the river, on the approach to a town or before tying up to the bank for the night, the sailors let go a tremendous fusillade of gunfire to let it be known that the convoy had teeth. And by the volume of fire, it would be apparent they were Yankee teeth and not a bunch of lousy, villanous "guards" from some local bandit outfit out for "squeeze."

In the conduct of their duty, some earthy, pithy, lengthy, sometimes humorous and sometimes anxious reports were written down by the chiefs in their cabins aboard the houseboats, recounting what went on during the tense days and tenser nights. . . . Please excuse the heavy expenditure of ammunition, one asked. And also, could some sugar and coffee be sent by some downbound conveyance? One apologized for the use of pencil; there was no more ink. Finally, on 17 May, the luck of one of the parties ran out; tied up for the night, the houseboat was jumped by from twenty to a hundred armed men. In the dark and the excitement, it was difficult to count accurately, the chief said. Two Chinese were shot when they came aboard. A third made a lightning re-estimate of the situation and jumped over the side. A bullet struck the Lewis and jammed it, but with rifle fire they drove off their attackers and felt they had killed six or eight and wounded twenty more. But unfortunately, seaman Everett Conley was hit in the leg, which on his arrival in Chungking aboard a British gunboat, had to be amputated.

News of this affair left Admiral Strauss furious. "The employment of armed guards will be confined to those vessels legally entitled to fly the American flag," he decreed, pointing out that he strongly objected to depleting gunboat crews for any armed guards, let alone Chinese junks, which even the diplomats agreed could not qualify to fly the Stars and Stripes.

Nevertheless, Strauss was pressed to do something. Consul General E. S. Cunningham at Shanghai, having got nowhere locally, went to Minister Schurmann in Peking. He felt Standard Oil's demands about forcing the upper river with foreign gunboats were valid, he wrote. Also, he wanted to have the *Monocacy, Palos,* and *Elcano* furnish armed guards for the merchant ships, after the gunboats had blasted the "bandits" off the gorge cliffs. HMS *Widgeon* tried a one-ship forcing job in October at Kuan Tu Kou, and had had to fire sixty-two 6-pounder shells and three thousand machine gun rounds to get the Chinese heads and tails down.

Strauss wasted no time straightening out the diplomats as to who was running things up the Yangtze. He replied to Minister Schurmann that he hoped he might be forgiven for disagreeing with Mr. Cunningham ". . . in that the latter is stationed six hundred miles from the scene of action and by education and experience is not so well fitted to suggest measures to meet a military situation as are those whose business it is to do so."

Slightly downstream, Ichang was being tenuously held by Wu Pei-fu against superior numbers of more poorly equipped Szechwanese. The city had been looted on 29 November 1920 by out-of-hand soldiers and on 11 June 1921 was once more systematically looted by troops under command of officers, to the tune of an estimated C$5,000,000. HMS *Gnat,* two big 6-inch guns belying her innocuous name, was the only foreign warship present. But her skipper, possibly remembering the unfortunate experience of Bell-Sayre at Wanhsien, would not take independent action without prior consular approval. This was so late in coming that it was 3:00 A.M. before *Gnat* managed to land twenty men, by which time things had gotten so far out of hand that they could only hole up in the compound where all the foreigners had gathered and defend the walls while Ichang blazed.

In fairness to the British, it must be remembered that their gunboats, unlike the

American gunboats, had far fewer white personnel, sometimes as much as a third of the ratings' billets being filled by Chinese. They thus had far lower potentialities for furnishing landing parties. However that may be, a report from the *Palos,* which arrived too late for the action, took an exceedingly dim view of *Gnat*'s non-characteristic performance: "The fact that *Gnat* made no effort to protect foreigners is expected to convince the soldiers that they can loot with impunity . . . without interference from the gunboats."

As an interesting commentary on what might be expected from even less well-disciplined troops, Wu Pei-fu, the commander of these rambunctious defenders (and looters) of Ichang, was a veteran warlord. Rear Admiral W. W. Phelps, who relieved Rear Admiral Bullard as ComYangPat in July 1922, visited Wu and gained the impression he was unpretentious, sincere, honest, and efficient. He even forgave Wu the mild discourtesy of continuing to shuffle papers during the visit, on the excuse of press of business.

Later, Wu made the cardinal error of going off to fight in Manchuria, leaving his trusted subordinate, "Christian General" Feng Yu-hsiang in charge at Peking. Feng promptly defected, taking over the rear area and the funds. This left Wu with no base and no alternative except retreat to the Yangtze by sea. With exquisite Oriental finesse, Feng had engineered the break by sending heavy drinker Wu a bottle of mineral water for a birthday present, an intolerable loss of face for the recipient.

The sometimes breathtaking landscape of the Yangtze, the frequent intimacy with mortal peril, the utter *differentness* of the whole scene, is such that many a traveler there has failed to express such impressions in words. One who did report vividly was a missionary who traveled with his family on the upper river in 1921:

The encounter was on the Yangtze, two days below Chungking, on a well-found, three-room houseboat with a good skipper and crew. Then a shot rang out, but skipper was inclined to go on and ignore it. But shots followed shots and as our deck was closely packed with twenty-five men presenting a solid target of human flesh, which the most amateur rifleman could scarcely miss, the skipper changed his mind. The two men of our party got on the roof in order to try to persuade the robbers to cease fire, but as the boat neared the shore, large stones were hurled at the crew and us. Some of the fellows then rushed on board with their huge knives, while others stood sentry on shore with pointed rifles. They searched our persons—a proceeding particularly distressing to the ladies—not even omitting the baby a year old. Bedding, boxes, bags, and everything on deck was all ransacked, slashed open with their big knives. Fortunately for us they did not want heavy loot, but only smaller articles, such as watches, handbags, dollars, and copper cash. Some things they took which could be of no value to them but were of much account to us, such as correspondence, notebooks, pens and so forth. Other things were damaged beyond repair. With everything emptied onto the deck—bedding, clothes, medicines, stationery, books, toilet ware—our compartments presented one glorious mess.

Most providentially for us, while the men were in the middle of their game, HMS Widgeon *came down river, and we signalled for help. When the sentries saw the* Widgeon *turn, the fellows on deck quickly cleared off with what they could carry. The* Widgeon *came alongside, and when it was ascertained that we had been actually robbed, three shells were sent after the retreating gang, but with what effect we know not.*

Such an experience indisposed us to face another of the same kind, or perhaps worse. So when we came to Feng Tu Hsien, and were informed by the city magistrate that the next section of the river was held by over 1,000 ex-soldiers, all well armed, and that he had only 200 men protecting his city and could give us no escort, we could only wait, hoping that something would turn up. And something unexpected did turn up not by chance, we think, but by God's good Providence for us—after waiting two days. The USS Monocacy *came upstream, as the captain had some business with the magistrate. So we sought advice from Captain Brandt, and to our great relief, he kindly invited us to go down on his boat to the place where he was standing by the Standard Oil steamer* Meitan, *and also put an armed escort on our boat.*

The next morning we were three times fired upon, but instead of pulling in, a reply was despatched by machine gun and we passed on.[26]

The *Villalobos,* patrolling the middle river in 1921, reported on 22 July that business was at a standstill there and on the upper Yangtze. But the staff of life, opium, still moved. On 15 October, the French SS *Ku Kin* was nabbed by the Customs at Ichang with *six tons* of the stuff aboard. About the same time, the U.S. consul at Changsha discovered to his chagrin that a well-known American appliance company had deviated slightly from its mission of uplifting the housewife's burden in that it was providing a repository for a mere ton of opium. *Villalobos* sailors were nervously standing guard over this hoard of tarry dynamite until the Customs could make arrangements to take custody without being highjacked by the local general or the Syndicate.

Szechwan supplied the greater part of this opium, long a major crop in the Empire Province. The process of making opium was carried on just as Alexander Hosie described it in 1890.[27]

In this valley [near Chungking] which extends for miles, I made first acquaintance with the poppy in full bloom. Fields of white and purple equalled in number the patches of wheat, barley, and rape. Where the flowers had fallen, the peasants, principally women and children, were busy harvesting the juice. Towards evening, the peasants may be seen moving in the poppy fields, each armed with a short wooden handle, from the ends of which protrude three and sometimes four points of brass or copper blades. Seizing a capsule, the operator inserts the points of the blades near the top of the capsule, then draws them downward to the stem. From the incisions a creamy juice exudes. This is scraped off and put in an earthenware jar, to be fired or left in the sun to dry. Thus, the weight is reduced to half, and the opium is ready for boiling.

Aside from strong opium, Szechwan produced some tough Chinese. As late as 1921 they still were carrying on their boycott of Mackenzie and Company (British) and the Robert Dollar Company (American), a hangover from the 1919 boycott of Japanese goods which followed the shooting of some Chinese laborers at a Japanese-owned Shanghai cotton mill. That year the SS *Robert Dollar* was shot so full of holes during the summer and fall that she had to spend the winter in overhaul. Four years later she hit a rock and sank with considerable loss of life.[28]

The year of 1921 on the Yangtze might be summarized by Admiral Strauss's report to the Chief of Naval Operations:

China is torn with strife and dissention, he wrote, with mild understatement, . . . *partly a question of [provincial] self-government and partly to undisciplined troops who get out of control and mutiny because they have not been paid. On the upper Yangtze River . . . Ichang has been looted twice within the year and Wuchang across from Hankow once. Steamers flying foreign flags have been fired on and have had to be escorted by river gunboats. Landing forces have been landed frequently and some detachments are still on shore at Ichang.*

As a tailpiece to 1921, an obituary for the *Samar* is fitting. Launched at Manila in November 1887, she was taken over by the U.S. Army, commissioned in the Navy in May 1889, and wrecked and sold on 21 January 1921. A former skipper noted that a small imbalance in coal stowage or consumption brought on a list out of all proportion to the weight involved. "We frequently sidled up and down the river, much to the amusement of the British gunboat sailors, whose ships were much more modern," he said, adding that all U.S. river gunboats were the subject of jocular comment due to their being antiquated, and for the most part, unsuited for the work.[29]

With two officers and forty-three men, ten of them Chinese, ready to spring to action behind four 3-pounders carried into battle at eleven knots, *Samar* might not have been a world shaker, but she undoubtedly broadened the education of her inhabitants along lines not met in Hampton Roads.

When Wells reported aboard the *Samar,* the only other officer attached to her was in Changsha, mortally ill of appendicitis, leaving Wells to chart his way alone and uninstructed. One night, awakened by a huge din, Wells found draped around the bow one of those enormous floating villages which navigate the Chinese rivers, depending almost wholly for guidance on a beneficient fate—a half-acre of floating timber en route to market populated by a swarm of coolie families, dogs, chickens, and pigs. Swept along like a chip, *Samar,* dragging both anchors, was soon afoul of a Japanese gunboat moored below her. Japanese and Americans, temporary allies, combined to bisect the raft which, like a split amoeba, continued its voyage in a chorus of yapping dogs, squawking chickens, and shrieking humans.

Samar was neat, but not roomy. All the way aft under the main deck were two small staterooms and the wardroom. Topside were several large wicker chairs from which the two officers could survey the passing scene, take their morning coffee, and admire the shiny brass brightwork of the 3-pounder sternchaser. Was it ever loaded with a hatful of BB shot and a pound of black powder to bag a duck dinner? Not to his knowledge, said Wells. But it could have been, as many a gunboater well knew.

A 1919 Christmas menu for the *Samar* indicated that by some sort of magic there appeared on the crew's mess tables the delectables identical with those in any U.S. Navy ship on any station at that season: the standard roast turkey and giblet dressing, mashed potatoes, green peas and the inevitable cranberry sauce. But the Yangtze was not always so provident at Christmas. The wardroom officers aboard *Tutuila,* blockaded at Chungking in 1939, found no turkeys available in that otherwise food-rich province. So a large, white goose was substituted. In conformance with an old timer's advice, a dollop of *samshui* was poured down its throat to relax its muscles, so that when its head was chopped off there would be none of the thrashing and flapping that tenses up the victim and makes it tough. The goose, as might have been expected, was soon hilariously drunk, and such an amiable, entertaining, ingratiating type of drunk that no one could bear to part with him until after the New Year, when he finally fulfilled his destiny as a tender goose dinner.

The *Samar*'s roster included a name almost any Yangtze sailor of between-the-wars vintage would recognize—"Ducey," listed as "mess boy." He was one of two small orphans picked up along the river and christened "Acey" and "Ducey," for the name of the Navy's favorite game. "Ducey," Wells wrote, ". . . was undoubtedly one of the finest young men I have ever known, near perfect as a mess attendant and determined to improve himself. He practically never went ashore and most of his non-working hours were spent in poring over books or filling dozens of notebooks with perfectly formed Chinese characters." In 1934 Ducey, all four feet, eight inches of him, served in the USS *Mindanao,* where he still displayed all the attributes credited to him by Mr. Wells.

Sailors everywhere are known for taking up with lost dogs, and if on the Yangtze they were more occupied with *show hai dzah* (small boy) it was because there were more lost boys than dogs. A reporter for the *Shanghai Evening Post and Mercury* learned this one night in Browning's Cafe, at 68 Broadway, where he overheard three sailors in conversation with the proprietor, "Captain" Browning.

The men, William Horstcamp, R. D. McKercher, and Walter Pegg, were all from the USS *Palos,* and concerned with finding Mr. Fu Chung and getting him back to the ship. Curious as to why anyone named Fu Chung had to be taken aboard a navy vessel, the reporter found out that Fu Chung was merely a 12-year-old orphan boy, picked up the year before at Changsha. Scrubbed, fed, and fitted with a small size sailor's uniform, Fu Chung was adopted by the ship. He learned English, ran errands, and chipped rust and painted in corners too small for any American bluejacket.

The story of Fu Chung was printed on 15 June 1931, and Walter Pegg still carried his copy, faded and frail, thirty-seven years later. One can only wonder, and hope that Mr. Fu lasted as well.

When Lieutenant Scott Umsted arrived at Chungking in the autumn of 1923 to report for duty in USS *Palos,* the city had been open to foreign shipping for just a generation. The little underpowered launch *Leechuan* had fought her way upriver from Ichang in March, 1898, the first steam vessel ever to reach the port. By 1923, Captain Cornell Plant, doyen of upper river experts, could look out the window in Chungking and see at one time or another in Lungmenhao Lagoon a half-dozen China Navigation Company (British) ships, the steamers *Shutung, Wanliu, Wanhsien,* and *Kiating,* and the motorships *Suiting* and *Suishan*. There was the Indo-China Steam Navigation Company's *Kiawo* and a pair each of French and Italian steamers.[30] On the Chungking side there might be ten or more small Italian and Swedish motor craft that continued on upriver 238 miles to Suifu. Three Cox's steamers, *Chinan, Chi-Lai,* and *I'Ping,* used the port. The Asiatic Petroleum Company (British) had three tankers, *Shukwang, Tienkwang,* and *Anlan*. Standard Oil and the Robert Dollar Line were there too. Plant thought it was ". . . a goodly company, when one considers that six years ago the *Shutung* and *Shuhun* and two other small steamers alone held sway on the upper Yangtze."

But on Lieutenant Umsted's arrival, the shipping was having a bad time of it. Chungking was popping like a firecracker factory in a five-alarm blaze. "The war between the Ins and the Outs was in full blast," he wrote to his family.[31] "The fighting was going on right over our heads." Six bullets hit the ship during his first day aboard. One man got nicked.

"Then the Outs were put out a little farther and the Ins adjusted themselves," Umsted wrote. But troops continued to flow back and forth along both banks. The *Palos* angled from one side of the river to the other, standing by a Dollar Line or a Standard Oil ship or ferrying U.S. nationals across the 1,500-yard-wide Yangtze. Sampans were fair targets for fusillades of shots from both banks, and thus had lost some of their charm as means of transportation.

This was the perennial game of Chinese military musical chairs. An occasional small cannon boomed out, to accompany the rattle of machine gun fire as the two sides belabored each other from a safe distance across the intervening Yangtze. General Yang Sen,* Peking's man in Szechwan (they hoped), had seized the north bank city of Chungking, muscling out Szechwan's local man Chow Hsi-cheng. The latter, in righteous indignation at losing the rich opium and other revenues of Chungking, was shaking his fist from the safety of the south bank (the Lungmenhao foreign colony side) and making dire threats as to what he would do to Yang just as soon as his 10,000 Yunnanese allies showed up.

"They will probably drive Yang Sen out of the city," opined Umsted. "Then the Szechwanese [Chow Hsi-cheng] will have to turn on his Yunnanese allies to get *them* out of *Szechwan*." China's eternal triangle, one writer called it: "Two generals fighting, whose plans were upset by the treachery of a third."[32]

Lieutenant Umsted had no more than tried on all his seven hats—his seven jobs, from executive officer to special disbursing officer, would have been handled by seven different officers on a cruiser—when he became involved in one of those in-

* Peripatetic, durable Yang Sen was pushed down to Wanhsien, where he tangled in 1926 in a most spectacular way with the British gunboats. He managed to avoid an active part in Chiang's accession and died peacefully in bed in 1932.

cidents which periodically tested the patience, skill, diplomacy, and endurance of Yangtze skippers. That was why a little 175-foot-long spit kit of a gunboat with a 50-man crew carried a two-and-a-half stripe skipper—Simpson. The Navy hoped that an old man of forty would be mature enough not to precipitate some horrible international incident.

The Chungking shooting having momentarily cooled, *Palos* dropped downriver to Wanhsien, a port always good for a sporting exchange of shots with the locals. The week before, the current Wanhsien general had nabbed a cargo of copper aboard a river steamer flying French colors. The event sparked a comment from Umsted that "the American flag means something, and is not a cloak for clandestine trade or other shady operations." He felt the British could make the same claim. But it was of course no secret that the Japanese openly smuggled and carried arms and soldiers and that the French tri-color flew over ships known to be entirely Chinese owned.

In a day or so, a radio message reported the imminent arrival of the SS *Robert Dollar* with ninety tons of copper. That was contraband to the outraged Chinese and they made clear their intention of seizing it.

The question really revolved around the matter of national prestige, of "face," described by an expert as "a religion, a philosophy, a national cowardice. Either you have face, in which case you are respectable, or you haven't it and are mud. To lose face is to lose hope. You bribe for face. You steal for it, murder for it, pretend, whisper, prevaricate. Face is solvency. Face is God!"[33]

The copper was consigned to an English firm at Chungking and was financed by an American bank. That the red metal was headed for Yang Sen's Chengtu arsenal was about as certain as indigestion after a Chinese banquet. The copper would soon be back at Wanhsien, in the form of bullets aimed at the defenders. But if every tin-pot general along the river felt he could vote on what might pass his doorstep—cargoes not in treaty violation—what was the point in having treaties? Or gunboats?

Aboard the *Palos,* Lieutenant Commander Simpson did some furious cramming on international law. He initially felt that the Chinese had a point and proposed to meet them part way by disembarking the copper at Wanhsien and stowing it in the Standard Oil compound. If everybody closed their eyes long enough, maybe the whole thing would go away. "The hell with *that!*" said the Standard Oil man, who was having enough trouble with the general as it was.

Umsted's arguments for prestige tipped the balance. When the *Robert Dollar* arrived next day and anchored close aboard *Palos,* Chinese troops swarmed down to the foreshore, intent on taking the copper, and *Palos* went to general quarters.

In the ensuing armistice, the general himself came aboard to palaver. By three o'clock that afternoon an amicable solution was arrived at which saved everybody's face: Skipper Simpson would sign a statement promising to see that the copper went into tea pots or opium lamps instead of bullets and the general, for his part, called off the looting party.

"Then all hands parted good friends," Umsted cheerily concluded. "We had a big party on the *Robert* that night and she left the next morning at daybreak. Since then all sorts of congratulatory letters and radios have come through to Simpson and he is saved. The copper is now in Chungking and everybody is happy." By "everybody" Umsted evidently did not mean the baffled Chinese soldiers.

This was Umsted's first encounter with river diplomacy. His conclusions coincided pretty closely with those of almost anyone who ever has sailed the Great River: "It is certainly an interesting place, with some new problem to solve every day." That the solutions sometimes did not encourage feelings of endearment toward the natives is suggested by his closing thought: "This is a thieving country. I am getting to regard the Chinese as so many swine."

Foreign social doings in Wanhsien would never have made the chit-chat column of the *Washington Post,* but the gunboaters enjoyed it just the same. Thanksgiving day was charged with as much gaiety and good will and probably better food than that first time the Pilgrims entertained the Redskins.

"It was some party!" wrote Umsted. "All the foreign community of Wanhsien was on hand." The commissioner of Chinese Maritime Customs, an American, was host. His house had been a former temple. The guests were Lieutenant Commander Simpson, Lieutenant Umsted, and Dr. Smith of the *Palos;* Mr. Swift of Standard Oil; the representative of L. C. Gillespie; the Asiatic Petroleum man; the Skipper and Number One of HMS *Widgeon;* the French postmaster of Wanhsien, and his Japanese wife. As the only foreign woman in town, this latter very charming and gay lady was of course the social arbiter of the *fanquei* outside the missionary colony.

Filled to the marks with food and drink and charitable good feelings toward their fellow men, the diners had listened to Skipper Simpson read President Coolidge's Thanksgiving proclamation to a wholly peaceful world. The men were having their cigars and liqueur when the air was split by the staccato rattle of machine guns.

The startled diners rushed to the window to see the British SS *Anlan* (Asiatic Petroleum Company) being heavily fired on from both banks of the river. She was going full speed to escape the bullets, overturning sampans as she went.

One never knew in China what strange anomaly would arise next: Whether by singular coincidence or magnificent staff work, just as the firing died away, a messenger arrived from the local general, Lu Tung Chow, who would be much honored if the officers of HMS *Widgeon* and USS *Palos* would join him for dinner next day.

The dinner was an unforgettable affair of some fourteen courses lasting three hours, with the only non-Oriental touch being the thoughtful substitution of knives and forks for chopsticks.

Having some time ago discovered the hard way that heating arrangements in a Chinese establishment were sketchy at best, the foreign diners were well fortified with two or three sweaters. And for those with previous experience in Chinese gustatory extravagances, it was comforting to note that right next door to the general's yamen and dining room was the hospital.

The roast piglets, stuffed, then carefully sewn back into almost perfect lifelike reconstruction, were impressive enough to be singled out for special mention. The skin, peeled off in strips and dipped in some unidentifiable sauce, was pronounced delicious. The foreigners were intrigued by something tasting like a cross between a clam and an olive. Umsted's emotions were mixed when he learned that he had just swallowed a portion of snake. He had seen them for sale in the market—so much per foot—and quickly downed a brandy to help forget the whole thing.

Following the customary resounding belches on the part of the Chinese, which signified that the parade of dishes was at last at an end, magicians appeared to boggle the imagination with feats of legerdemain not easily forgotten. One of them took Umsted's class ring, laid it on a table, passed a handkerchief over it and caused it to disappear. Then he handed Umsted a box from a table some distance away. It was a nest of boxes, one inside the other. "I went through about nine layers," Umsted said, "until I got down to one about the size of a silver dollar and an inch thick. There was my ring!"

The *Palos* crowd had a good month in Wanhsien. Umsted's only recorded complaint was the diminutive size of his stateroom. "Don't send me any papers or magazines!" he instructed his mother, who had come out to Shanghai. "There is no space for them in my room. Just send clippings. I can stuff them in my pocket and read them in spare moments." He was invited to tiffin aboard *Widgeon,* played chess with the Standard Oil man, and dined with the postmaster—a properly magnificent French meal, equally magnificent French wine, and the proceedings enlivened by the postmaster's "cute little thing" of a Japanese wife.

The skipper of the French gunboat was at the dinner, too, so they arranged a rifle match with the French and British gunboats. The *Palos* crew had been practicing for some weeks at a range set up on the flats bared by the winter's low water level. Aside from giving the men something useful to keep them busy, it was good *face pidgin;* Chinese soldiers were mightily impressed by the Americans' accuracy as well as by what appeared to them to be a fabulously reckless expenditure of ammunition. The Americans casually let it be known that this was only a patch on what still remained aboard ship.

"When they saw our machine guns getting forty bulls eyes out of forty shots in about twenty-five seconds they were bewildered," Umsted wrote. He was sufficiently set up over the whole thing to declare that ". . . we could stave off an army of 10,000 without the least trouble."

Much to the delight of Simpson, whose wife was in Chungking, plans for *Palos* to stay at Wanhsien over Christmas were canceled. But before they sailed upriver, Dr. Smith and Umsted had a rare opportunity for sightseeing. In Umsted's words:

Yesterday, the shooting being over, the Doctor and I took a walk straight back from the ship. I never in all my life saw anything like it. Right back from us was one mass of terraced farm lands. The Chinese did not primarily intend to make their farms beautiful, but they are. We climbed the mountainside, intent on reaching a certain summit. As we neared the place we saw there was a set of steps going up, with a gate at the top. The gate was open. We decided to try it. But when we got there, the gate was closed and barred and we were told in good Chinese that foreign devils were not wanted.

It was the only place we could find to get on the plateau. We stayed there awhile, offering coppers to get in, but nothing doing. The steps leading up, the gate, the method they had of keeping their place to themselves was exactly as I imagine it was in nineteenth century Europe.

Finally, some of their own people came up the steps, and they were afraid we might try to get in along with them. So they would not let their own people come up while we were near the entrance.

We were determined to get in if at all possible, so just stuck around. I was afraid they might take methods of their own to drive us away, but they made no such move.

After a while, the gate opened and a very nice Chinese on the other side beckoned to us to come in. As we did, he barred the gate behind us.

Once inside, we looked around. It was apparently a feudal state all by itself up on top of the mountain. The land was one immense garden. We walked up to the top of the hill, followed by men detailed to watch us, and sat down by a temple to view the farm land and scenery below. Innumerable little lakes, temples, winding pathways, stone bridges, streams, tombs, closely trimmed trees, were a sight to delight the eye.

It was too dark to take a picture. (The sun never shines here in winter.) We hated to leave the place, but hated worse to go home in the dark. On our way out, we were ushered out with the "here's your hat; what's your hurry," theme.

Inside the gate sat a bronze, smooth bore Chinese cannon. "Ding hao!" (Very good!) they told us. I managed to make them understand that we were from the U.S. gunboat that had done last month's shooting. They seemed to be impressed. On the way down, we stopped at the Standard Oil mess for a slight restorative and there related our tale. The Customs commissioner was on hand, a man who knows the country well. He had never been able to get through that gate, he said.

This is a queer country. About two miles from the big port of Wanhsien is a city of some ten thousand people perched on top of a mountain that rises straight up on all sides. No one, not even an "outsider" Chinaman can set foot in it. A religious community, they make heavy clothing in exchange for food sent up from the farms below.

Palos had one more last crack at defending the White Man's Burden at Wanhsien. In mid-December, proudly flying the Stars and Stripes, the upper river vessel *Che Chuen* chugged up the Yangtze on her maiden cruise. While she lay at Wanhsien, a Chinese official of some sort took it on himself to shake down a foreign missionary's baggage. This was an insufferable affront to U.S. dignity under any conditions, especially with a U.S. gunboat present. But the crowning insult was the presence aboard *Che Chuen* of Rear Admiral Charles B. McVay, Jr., the new ComYangPat, whose handlebar moustaches were probably quivering in barely suppressed annoyance. He was on an inspection tour of his upper river domain, with no time or inclination to dally and exchange unpleasanteries with some small bore bandit general. So, taking Skipper Simpson aboard *Che Chuen* to brief him en route Chungking, McVay departed, leaving *Palos* at Wanhsien, and Umsted charged with the duty of cutting the general down to size.

Although the quickest way to a man's heart is said to be through his stomach, a variation, in the case of minor Chinese officials such as ragtag *likin** collectors, was through the local general, especially if his yamen was within three-inch gun range of the river.

Accordingly, acting commanding officer Umsted hied himself ashore. With pains to make his presence and purpose conspicuous, he reconnoitered the responsible party's headquarters, checked out range markers, then went back aboard and moved *Palos* into an ideal firing position. An apology was soon forthcoming, and that was the end of *that.* The finger was off American ships at Wanhsien for a long, long time.

On 23 December, when *Palos* once more moored at Chungking, the water level was down to five feet. Zero level did not mean a dry river bottom; it was the lowest recorded level of some years back and henceforth used as a reference. Late that summer, the level had been forty-nine feet. But in spring, when the central Asian snows came cascading down as melt water, the Chungking level could rise to a hundred feet.

Low water brought two bonuses. First, the river narrowed to half its spring flood stage width, with current slowed to half. This halved the hazard of sampanning across to holiday parties held in the city proper. The second advantage was the baring of a sand and shingle piece of former river bottom on the Lungmenhao side that could be used as a race course. From 1937 to the end of World War II, such a flat on the city side was used as a landing field for Chungking's only air link with the outside.

In Chungking the parties came thick and fast. On Christmas Eve, the Asiatic Patroleum *taipan* hosted sixty at dinner. Christmas forenoon, champagne corks popped aboard the British gunboats, holding traditional open house. Skipper and Mrs. Simpson entertained at Christmas "tiffin," the guests breaking away in time to cross the river for dinner at postal commissioner Greenfield's and to spend the night at the American consulate.

Chungking's gates groaned shut at sunset; once inside after dark, one stayed there until morning.

At Greenfield's big dinner party, coffee and brandy were followed by an appropriately stuffed, bewhiskered, and red-suited Santa Claus. After all the lights had been extinguished, the drawing room door was thrown open to display a lighted Christmas tree. The Greenfield's seven-year-old daughter's enthusiasm and shrieks of glee all lent a touch of home to the expatriate gunboaters, who, along with the child and all the other guests, received presents. Umsted's present, in the opulent Oriental style, was a solid silver match holder and tray, so fine he accepted it with embarrassment.

Not everybody had so jolly a day of it. While others feasted, Dr. Kelly, an American official of the Salt Gabelle (the Chinese government's monopoly, administered by

* *Likin,* introduced in 1852 as a special tax to suppress Taipings, was extended throughout China in 1863. Originally it was 1/10 of one percent of value of goods, but actually was largely at the whim of the individual "collector."

foreigners, and a prime source of revenue) spent most of Christmas discussing money matters with representatives of a general who had recently captured the city. Their deciding argument consisted of five henchmen holding loaded and cocked pistols at Kelly's head. At highway robbery, Chinese forebore much of the procrastination standard in their other dealings; in excess of $100,000 was signed over and the case closed before the U.S. consul or anybody else outside learned the "negotiations" were in progress.

A few days later, the *Tze Sui,* a Chinese vessel with a British captain and under American charter, pulled into Chungking with the news that while anchored a few miles downriver, bandits who had boarded as passengers at Wanhsien had suddenly revealed themselves. The captain had done his best to defend himself but was soon overpowered, filled full of bullet holes and tossed overboard.

There were rumors about that the Chinese were out to "get" the captain. It would not have been the first time that some foreign merchant skipper had disappeared under mysterious circumstances. Earlier, in Ichang, a British Captain McArthur had vanished from his ship following a row over opium. Even gunboats were not immune. In September 1920, a Chinese, clearly above the coolie type, had come aboard *Palos* at Chungking and without preamble had addressed the skipper, Lieutenant Commander Glenn Howell: "Captain, you no search ship on way to Ichang, I give you $10,000." Unbeknown to Howell, *a million dollars worth of opium* actually was smuggled aboard and rode the ship to Shanghai.[34]

At the race track on the Chungking shale flats, business was booming. The first two meets had attracted so many Chinese that the sweepstakes' first prize had totted up to Y$1,500, over US$1,000. The Chinese, inveterate gamblers, would even bet on which way a head would land after the executioner's broadsword had deftly parted it from its owner's shoulders.*

Ponies were cheap in Chungking. Umsted bought half a share in one for about US$15, including saddle and blanket. The monthly feed bill, with the *mafu* (groom) thrown in, amounted to three or four dollars more. "Johnny Walker" was no winner, but he gave his part owner some exercise and a much toughened posterior and, incidentally, further insight into the nature of Chinese man, or, to be more explicit, woman. One day at the track, Johnny Walker, Umsted up, was coming hell-for-leather down the stretch, with the usual crowds of Chinese idlers hanging on the edges. Suddenly a baby girl darted out and crack! into the pony's forelegs she went. Then horse, rider, and girl were sorted out and put back on their respective feet. "The poor kid got an awful slash in the mouth and it looked at least as though her beauty would be ruined, with probable permanent damage."

* John D. Wilson, a signalman aboard USS *Noa* in 1927, described an execution he had witnessed at Nanking: ". . . there were 17 being executed. They were all in a row, their hands tied behind them, kneeling down on the earth, evidently resigned to their fate. The executioner was . . . tall, fat, bald-headed, with gaudy clothes, and a knife about four feet long. He started at the beginning of the line, just letting the knife fall by its own weight and off would come the head. The body would convulse, blood would spurt out, and the executioner would go to the next one. . . . After it was all over, he stood back like a pouter pigeon, with the populace admiring him and his muscles. He actually pounded his chest, like a gorilla does."

The ship's doctor checked her out and patched her up. In the meantime, crowds of Chinese had gathered, all laughing fit to burst at what appeared in their book to be a hilarious joke.* "They have no such emotion as sympathy," wrote Umsted, in mild understatement.

He gave the baby and mother each a silver dollar and was soon aware of his gross tactical error. "Now this may be hard to believe," he wrote, "but since I gave those dollars, many loving mothers have brought their babies to the race track in hopes that their offspring may earn the same! We are bothered worse than ever and the foreign community blames me for it and I guess they are right."

In February 1924, *Palos* was back at Wanhsien when word came that two Gillespie and Company junks were wrecked downriver below the Hsin Lung T'an. At winter low level, the Hsin Lung T'an was in no state for *Palos* to negotiate without ripping her bottom out, so the investigation was launched from above. Suspecting trickery, the gunboaters and Gillespie's man Jenkins poked around in the several small villages en route, looking for telltale wood oil casks bearing the Gillespie chop. No wood oil was turned up, but the party profited to the extent of some first-hand education on the way of life in these little communities, stuck to the canyon's walls like so many swallows' nests.

The Chinese provided no comment on their views of strange *fanquei* probing under the beds of a people without benefit of the fifth amendment or even the requirement for a search warrant. After several millenia of being shaken down by bandits, roving groups of soldiers, tax collectors, and larcenous officials, the mild intrusion of the wood oil hunters was no doubt taken philosophically.

Umsted, Jenkins, and a couple of armed sailors beat their way on foot down the river bank and around the rapids. There were the junks. They were a sad mess. Umsted darkly suspected they had been looted first, then sunk to cover the crime. The Chinese villagers, their reluctant recent hosts, must be having a quiet chuckle now, he thought. Those cagey devils always managed to have the last laugh.

The little party watched in awe as several junks shot down the fearful four-hundred-yard-long torrent of the Hsin Lung T'an while two or three more struggled up against this vast chute, where one in thirty is wrecked. The disintegrated junk and doomed men simply disappear, so say observers—sucked down into the depths, swept along by a subsurface current, everything ground up, nothing left but the memory.

With the spring rise, *Palos* headed joyously downriver to Shanghai for overhaul. Preparatory to shooting the Hsin Lung T'an, her fires were clean and hot, safeties near the popping point. A downbound ship needed every ounce of steering power. First there was the nerve-wracking approach—would the rapid be clear of junks? Then at the point of no turning back, making revolutions for fourteen knots, trembling and wallowing, the little ship hit the tongue of the rapid. It was running at twelve or

* During World War II a bomber came in for a landing at Chengtu as a platoon of Chinese troops marched across the strip. The plane's propellers sliced up half a dozen of the soldiers; more were knocked down and injured. The whole survivors simply rolled on the ground in merriment at the sight of their late compatriots, who obviously somewhere back had failed to make the proper joss.

more knots, down a hill of water that dropped eight feet in its short length. Her four rudders clanked and churned as the wheel was spun. The Chinese pilot had taken the helm himself; one could not afford the luxury even of that half second's delay in transmission of pilot thought to quartermaster reaction. Not when going *thirty miles an hour* over the ground, with quick destruction in the form of sharp rocks sometimes within reach of a long fishing pole.

Anchoring that night in a restful cove just above Wushan Gorge was misplaced confidence of a high order. By 11:00 P.M., the river had risen until the whole character of the place changed from a languid backwater to racing current. Both anchors dragged and through the night the wire hawsers out to boulders ashore popped and were replaced with monotonous regularity by a crew fighting a desperate and half-blind battle with the River Dragon.

It was with vast relief, come morning, that the bow was once more pointed downriver toward the Ye T'an. There, the uncompromising results of a slight error in pilotage were dramatically brought home. Below the rapid, a foreigner in evident distress, waving from the beach, brought *Palos* about. A sampan sculled the stranger out to the ship.

The Ye T'an dragon had snapped its jaws, he said, and down the monster's throat had gone his ship, *Pa Kiang*. She could have been saved, he thought, if the Chinese crew hadn't panicked and gone over the side.

Like a hungry horse headed for the barn, *Palos* wasted no time en route. She was soon in Kiangnan Dockyard, hideous with the din of hammers and jabbering Chinese. Like all ships undergoing the agony of overhaul, she was a dirty, uncomfortable place to live.

But there were compensations. Those silver cartwheels squirreled away during the winter made a solid, important sound when slapped down on a real stand-up bar. "Boy! Wanchee whiskey! An' nonna that Japanese slop, either. Wanchee numbah one! You savvy propah, Boy?" Scotch bottles in the Orient trade had nonrefillable stoppers, with ceramic check valves. But the ingenious Chinese, to whom time was in long supply, bored a small hole in the bottom, drained out the good liquor and replaced it with cheap synthetic, or some inferior Japanese brand.

And what a lovely, opulent sound those cartwheels made when tossed onto a cabaret dance floor to encourage some luscious, lightly clad Russian dancer to greater exertions! "Yes-shaw! Yes-shaw!" (more! more!). Other than "yellow blue vass!" (I love you), it was the only Russian most of the gunboaters knew.

Other cartwheels went for lurid embroidered velvet pillow covers for Mom, back in Minnesota or Brooklyn. Or lace handkerchiefs for baby sister, delicately and intricately worked in some mission convent school. And as has been the case since sailors first went to sea, some of the dollars slipped into black stocking tops, payment in full, C.O.D.

For those short timers about to head back Stateside, the last of the cartwheels went for tremendous black leather suitcases, richly embossed with dragons, lined in purple satin, smelling strongly of bad tanning and weighing twenty pounds empty.

Many of those same ex-short timers would be right back in Shanghai via the next transport from "Frisco," signed on for another cruise. Their blue jumper cuffs would have been casually turned back during the trip across, to make clear to the greenhorns just *who* they were. The cuff linings, naturally enough, bore the ubiquitous dragon, doubled back on himself, breathing fire in yellow, red, green, and gold embroidery.

The *Palos* left Shanghai in June. Her Old River Rats, plus a scattering of wide-eyed new hands, stepped warily to avoid still fresh paint as they regretfully watched Shanghai disappear over the fantail. The last odd overhaul jobs were taken in hand by the crew, after departure, to save time. With the summer current to buck and a pause at every whistle stop to check out missionaries or take on coal, she would, with luck, just make Ichang in time to tackle the rapids of the upper river before low water closed off travel for the winter.

The first such breather was at Kiangyin. The missionaries there proved to be in good form. Umsted also discovered that a new Chinese fort existed nearby, and being a man with a well-developed bump of curiosity, decided to exercise it.

"I got in mainly because the sentries were asleep," he wrote. Sauntering right on over to the guns, whopping big new ones, he found a spirited drill in progress. He was soon spotted, resulting in much hubbub, mixed with acute embarrassment, on the part of the Chinese. "You must leave in five minutes!" they politely informed him, which as far as he was concerned, gave him ample time to size things up. As might have been expected, the political semantics and the knotty problem of who did what to whom were pure Chinese puzzle. After six months of witnessing assorted Oriental shenanigans around Chungking, Umsted was an expert in sorting out what was what in the game of Chinese double-double take: "The soldiers belonged to a Nanking general, who was 'independent' of Peking and who was also at the same time at 'war' with the Chekiang outfit which was also 'independent' of and 'at war' with Peking."

Up the chocolate-colored river, *Palos* chuffed to dull Nanking, rice-rich Wuhu, Anking and its missions famous for drawn work, to Kiukiang for a rendezvous with flagship *Isabel,* and finally to *Little* Shanghai—Hankow—with its clubs, cabarets, and stifling July heat. En route, stop by stop, there had been the interminable haggling for coal.

The ship left Hankow with every ounce of fuel she could pack aboard: bunkers bulging, fireroom floorplates stacked high, more bags piled on deck. Then, some miles short of Shasi, they underwent the emotions of a driver crossing the George Washington bridge at rush hour who discovers his gas gauge reading empty. They had burned coal at a phenominal rate bucking the strong summer current. To turn back was no go—the next coaling station downriver was beyond their reach.

"We were going to run as far as we could, anyway," wrote Umsted, who was preparing to land, journey overland as best he could upriver to Shasi, scour the waterfront for coal, then bringing it downriver in junks.

A Japanese steamer had been following in their wake all day, plumes of black smoke pouring from her stack. Obviously she was a coal burner. Some to spare, perhaps? The Japanese and Americans were not then high in each other's affections. But in true frontier spirit, this was Indian country. Who knew whose wagon wheel would crack up next? Umsted put-putted over and brought the steamer to a halt, palavered at length from the boat and was finally invited aboard. There, a long powwow went on between the ships' officers—the passengers joining in—and wound up in reluctant Japanese agreement to provide twenty tons of coal, providing the delay would be no more than an hour.

Toward midnight, four or five hours later, with Japanese patience wearing progressively thinner at the delay, they called it quits. Eleven and a quarter tons of the powdery black stuff had been laboriously wrestled by hand out of the merchantman's bunkers into *Palos*. To Old River Rats who recall July middle river temperatures and humidity, that was more than enough. The merchantman churned on, to be reimbursed in kind from the Navy's coal pile in Ichang. There, also, *Palos* eventually arrived, black smoke pouring from her funnels as the last of the coal dust went into the furnaces. She arrived in Chungking on 11 July, one day short of an even month out of Shanghai. It had been a 1,300-mile trip for a middle-aged lady just entering her second decade, in an occupation where life expectancy was sometimes measured in months.

The unreliability of dory engines still afflicted gunboats in 1923. Shortly after Lieutenant Commander H. M. Kieffer took over *Quiros* in May, Rear Admiral Phelps, ComYangPat, instructed him to call on the commanding officers of three Chinese cruisers at Nanking. Phelps believed that if the U.S. got into a war the Chinese might be allies. Results of the 1922 Naval Arms Conference had left Japanese-American relations at a low ebb.

I anchored off the Nanking bund, recalled Kieffer, *where the current whipped by at nearly ten knots. The three cruisers lay downstream. My dory was in the skids for repairs. So I hailed a passing sampan and set out for the cruisers.*

The sampan tried to turn to come alongside the accommodation ladder of the first cruiser, but we swept down with the current toward her lower boat boom. I tossed some money to the sampan coolie and grabbed the jacob's ladder hanging from the boom as the sampan slid under it. Then up the ladder I went, white gloves, sword and all. The officer of the deck had charged back aft to meet me as I tightrope-walked the boom to come aboard.

Kieffer presented his card and Admiral Phelps's compliments, chatted awhile, accepted a thimbleful of Chinese samshu wine, and met some of the ship's officers.

Then I requested that somebody hail a sampan so I could get on with my visits, Kieffer continued. *Nothing doing. At the foot of the accommodation ladder I found a ten-oared cutter, boat cloth spread in the stern sheets. The Chinese coxwain handed me the tiller ropes. I made motions—I had just come to the river and knew no Chinese words—to the effect that the coxwain should steer. He persisted, the ultimate in courtesy, and so did I. Finally, I said to myself, 'The hell with it!' and took over. We cast off, somehow, and got the oars out, and I realized I had no means of issuing orders the crew could understand.*

We swung downstream rapidly. I put the helm over and headed for the second cruiser, shouting, 'In bows!' No result. Ten happy Chinese faces beamed at me from under their white flat hats, while their owners pulled away manfully at the oars. I shouted, 'Way enough!' Again, no result. We banged into the ladder, and the bowmen woke up and got out their boathooks, too late. We dropped down astern of the cruiser. Then ensued ten minutes of pulling against the current to finally make her accommodation ladder.

On the trip to the third cruiser, we got alongside through some sort of extrasensory perception. Then came twenty minutes of heavy pulling back home to Quiros.

If the boatswain's mate on watch expressed surprise at the manner of the skipper's return, he was well justified.

Apparently the consul general's complaint about withdrawing gunboats from river ports for target practice fell on deaf ears, because in 1923, it still was being done. Kieffer was surprised to receive orders from Admiral Phelps to take his "squadron"—*Villalobos* and *Quiros*—out to Woosung and shoot. Kieffer referred with misgivings to the commandments for *Quiros* prescribed by the Bureau of Construction and Repair. They reflected the generally feeble state of this veteran of two wars, one insurrection, and numerous lesser confrontations in China:

She shall not tow or be towed.
She shall not fire a shot.
She shall not steam where there are any heavy swells.

There was sense in what the Bureau said. "The hull's iron framing was spaced as far apart as a battleship's. The frames originally had been held together by flat, diagonal tie plates. Of these, nothing was left except irregular stars of rusted metal, all carefully red-leaded. The planking itself had shrunk so that above the waterline it was impossible to caulk it. The caulking iron and oakum simply went on through to the inside of the hull. In the bow, the cracks between the planks had been covered by strips of light metal sheet and painted over. Withal, the ship looked neat from the outside—white hull, black waterline, and spar color upperworks."

To everybody's astonishment, the practice came off well. The men who scattered below decks to check for opened seams after each shot found no more daylight showing through than before. But that was her last target practice. On 16 October 1923 the old ship became a target herself, and was sunk in the China Sea by destroyer gunfire. Feeling that disobeying the second commandment had tempted fate far enough, Kieffer willingly observed the one prohibiting towing. He left the tiresome job of retrieving the target to *Villalobos*.

Quiros had been launched at Hongkong in 1895, on Spanish account. Then, purchased by the U.S. Army along with *General Alava* for 215,000 "Mex," she was turned over to the Navy in February 1900. Her two scotch boilers and single screw drove her at eleven knots all out, to such places as her 8-foot draft would allow her to carry her two 6-pounders, two 3-pounders and two 1-pounders.

The skipper's cabin in the stern, after the fashion of Nelson's day, gave him room enough to swing a cat. But the tiny wardroom was so small that when guests were aboard, which in treaty ports was often, there was no space for the steward to pass between the backs of chairs and the built-in, sofa-like "transoms" that lined either side. So a "cheese-eye," one of the several small Chinese boys traditionally carried on board, would be stationed at the far end of the table to receive dishes passed over the heads of the diners. Between meals, the wardroom table was pressed into service as desk for the ship's yeoman or the captain, who doubled as paymaster.

Built for forty men and carrying sixty, there was no room below for all the crew. Winter and summer many pitched their cots on the forecastle, shielded in frosty weather by canvas curtains and warmed by a small stove.

Daily the Chinese merchants sampanned out with their baskets of produce, checked for quantity by the petty officer of the watch. The pharmacist's mate or doctor checked for quality, and for evidence of lurking bacteria, melons or chickens injected with water to increase weight, or eggs beyond the point of no return. Skipper-paymaster Kieffer, on to the ways of compradores, paid for the quantities delivered and not that claimed in the monthly bills. Then one time, his receipts showed that the Chinese baker had billed for three loaves of bread less than the check-in figures showed had actually been received. So Kieffer paid for the extra loaves. The next month, when the merchants filed into the little wardroom "office," they had no bills. "*You* make out bills," they said. Kieffer had been accepted as an honest man, and in bowing to the inevitable—that they could not cheat *Quiros*—they could at least profit to the extent of sparing themselves the trouble of preparing accounts.

From an initial deposit of C$20,000 from the Patrol paymaster, Kieffer doled out funds to various merchants ashore. Receipted bills went to the Patrol paymaster, who sent a check to keep the gunboat's deposit up to par. Turned into cash at a local bank and added to the sticky bundles of dirty Chinese currency in the safe, it would provide Kieffer, as was the case with all amateur paymasters, some troubled moments. At the club, in the bunk for the night, or miles away on a trip, one's anxious mind would return again and again to that same question: *Did I lock that goddam safe?* At least, amateur paymasters could be thankful they did not suffer the headaches of earlier paymasters in those parts. In 1872, an audit of the USS *Colorado*'s safe contents listed in dollar equivalents, 28,423 "Mex," 8,551.06 English gold and silver, 7,500 U.S. gold, 1,855 Spanish gold, and 5,592 in U.S. folding money.

At the end of 1923, U.S. export-import trade with China totaled US$346,699,000. For anyone speculating on the merits of maintaining the Yangtze Patrol, it and the South China Patrol (two ships) cost US$750,000 for the year, about two mills per dollar of trade, most of it concentrated in areas accessible to these patrols. In *any* insurance business that is a low, low rate.

Through untold centuries, Szechwan's only commerce with the outside world was carried by the thousands of junks which swarmed along the Yangtze. The river craft provided the support for more than a million people, who built them, manned them, supplied them, and moved them up and down the river. At the beginning of the twentieth century, when these hardy, cheerful subsistence level folk watched the first of the fire-belching *fanquei* craft struggling up the river, their reaction was curiosity mixed with amusement. The clumsy, underpowered, badly designed, "two-side walkee" steamers had to be heaved over the rapids by coolies, like any junk.

By 1910, there still was compassion left in the hearts of the steamboat men: the project to install steam winches at some of the rapids was dropped to save the trackers from starvation. But fourteen years later there was neither compassion on the part of the steamboat fraternity, nor amusement on the part of the junkmen, and for very good reason. During the week of 8–14 June 1924, thirty-two merchant ships entered or cleared Chungking; eight British, seven American, seven French, four Chinese, four Italian and two Japanese. This was a normal number for that stage of the river. What was particularly galling to the junkmen was that certain categories of cargo theretofor always tacitly considered reserved for junks—wood oil, salt, cotton—were now being preempted by steamers. The junkmen had their backs to the economic wall in a country where unemployment relief or retraining for another profession would have been roared at as hilarious fantasy. They must have cargoes, or turn bandit, or starve.

As in all Chinese professions and crafts, the junkmen were tightly organized in guilds. These powerful, monolithic "unions" could present a closed front for negotiation, strike or riot. Their demand for a monopoly of the salt cargoes generally was acceded to. Salt corroded iron holds anyway. But wood oil was different. It provided a profitable and sometimes the only downbound cargo, there being little other bulk export from Szechwan, where opium was the principal export cash crop. A further argument for the carrying of wood oil in steamers was the constant interference with the junk traffic by Chinese military or bandits—the distinction being generally unclear. Local generals complained that they received insufficient money from Peking to pay their ragtag troops; that they simply had to charge a "protection" tax to support the soldiers who rode the junks to discourage the bandits.

The junkmen were of course in cahoots with the scheme, refusing to sail without "protection," then collecting from the military for cooperation in effecting the shakedown. All this was by no means a small operation. A convoy of eight wood oil junks might carry a cargo worth US$100,000. The "protection" tax demanded was a modest US$350, but in paying and thus acquiescing in principle to an illegal Chinese demand for interior taxes on foreign-owned property, an absolutely open-ended precedent was being set that simply could not be risked.

In 1923 there had been such disorders over steamers taking on wood oil at Wanhsien that when the river rose again to steamer level in 1924, the consulates and gunboaters all recommended against steamers carrying the strongly disputed wood oil. Nevertheless, the British SS *Wanliu* decided to try it. An American named Hawley had collected a quantity of wood oil at Wanhsien during the winter and on 17 June

1924, *Wanliu* moored close aboard HMS *Cockchafer,* prepared to receive Hawley's lighters alongside. Hawley, a man of very short temper, had been warned by Lieutenant Commander I. W. Whitehorn, *Cockchafer*'s skipper, that he was making a grave mistake in trying to load the oil, but seeing that Hawley meant to go ahead anyway, Whitehorn went back to his ship and called away his landing party in anticipation of the inevitable.

Pandemonium soon broke loose ashore, complete with banners, shouted slogans, and speeches. Mass hysteria took hold. The mob commenced breaking up the lighters. Hawley, seeing his property in dire jeopardy, rushed ashore from *Wanliu* with more courage than good sense and began to lay into the mob, which held off from what must have looked to be a madman. Whitehorn, unaware that Hawley was the focus of the action, or even that he was ashore, fired a 3-inch blank, hoping to save the lighters. At this, the rioters momentarily drew back, but soon discovered the hoax and recovering their courage, went after Hawley.

Wanliu's master shouted to Whitehorn that Hawley was being beaten up ashore. Landing party to the rescue! By the time they arrived, Hawley was so badly battered that he soon died aboard *Cockchafer,* leaving her skipper with what is termed in proper British circles a very sticky wicket.

With the White Man's Burden heavy on his shoulders and foreign prestige at stake, plus something of the tradition that blood is thicker than water, Whitehorn acted with speed and firmness. There was no dallying for instructions from higher authority. Of his action, he reported:

> *I demanded that the two leading men of the junkmen's guild be brought on board H.M. ship under my command as prisoners and after I had interviewed them, to be taken ashore and shot on the site of the crime. . . . I also told the general that the funeral of the late Mr. Hawley was taking place at four-thirty that afternoon and that he and his leading officials were to follow on foot in the procession. . . . If those two . . . men . . . were not aboard . . . by 6:00 P.M. I would avenge the death of the late Mr. Hawley by bombarding the town with my six-inch guns.*

The funeral came off as planned, with General Lu, losing a yard of "face," trudging three miles on game legs. The two presumably bona fide guild leaders were sculled alongside *Cockchafer* for inspection by Whitehorn, who reported them to be evil-looking characters. But the culprits were not brought aboard the gunboat for interview as originally specified by Whitehorn. They were taken straight back to the foreshore and given the usual bullet in the neck. The delay on the program had been occasioned by the necessity to comply with Chinese law, which required a notice of execution to be posted in the city prior to the event, presumably so that a properly appreciative audience might assemble to be impressed.

Whitehorn's action was loudly applauded by American naval and diplomatic authorities in China, which tended to temper the dim view the British themselves took of this rather peremptory if not highhanded action, and may in fact have saved Whitehorn from being disciplined.

The affair caused scarcely a ripple in the United States, where *The New York Times* commented that:

One reason why the Navy Department was concerned in the Senate's failure to pass the appropriation for the new gunboats was that on Chinese rivers, Americans are dependent on the protection of British or Japanese naval craft.

This little affair by no means ended difficulties on the upper Yangtze. Nowhere in the record is there a busier and touchier prolonged period of tumult than that summer of 1924, nor one requiring more diplomacy and finesse on the part of an American gunboat skipper—in this case, Lieutenant Commander E. T. Oates, of the *Monocacy.*

The shipping trade itself suffered numerous casualties, as if the junkmen had indeed carried out their threats and put the most malevolent of devils and dragons on their trails. By mid-May, five large and all the small ships had been holed in rapids. On 21 March, the brand new SS *Chi Ping,* of the American Yangtze Rapid Steamship Company, was bulling its way up the left bank channel of the Kung Ling T'an when the pilot lost his nerve and tried to sheer out into the center channel, with the result that the ship rammed head on into a rock. She filled forward immediately, and only smart handling in beaching her in a gully on a sandy bottom between two shingle banks saved her from foundering. On 24 April, another American ship, the *Robert Dollar,* was totally wrecked, followed by three British ships. The *Monocacy* was standing by the *Robert,* to prevent looting in case the river fell, when word was received that the *Alice Dollar* had been holed upstream. Leaving *Robert* with only her funnel showing above the flood, *Monocacy* rushed up to guard *Alice* against the normally to-be-expected looting of any grounded ship.

A month later, *Alice,* patched with cement and in port, this time guarded by four U.S. bluejackets, was boarded by four armed Chinese soldiers. Paying no attention to the guards' request that they leave, one of the Chinese countered by drawing his Luger. All turned out well. The incident report said that "the guard in self defense hit this man over the head with his club, after which they left the ship without further resistance." There appeared to be so much tinder in these things that following this incident, ComYangPat decreed there would be no more armed guards without an officer.

As if riverine bandits, illegal taxes, and truculent Chinese were not enough, early May brought tidings that the Commander in Chief, Asiatic Fleet, ComYangPat, four staffers and *five ladies* would soon arrive at Ichang aboard the flagship *Isabel,* for a trip up to Chungking via American merchant ship. Not since Joe Fyffe's mother-in-law took over the cabin of his ship had there been women living aboard a U.S. man-of-war. *Isabel,* a converted yacht, was a unique exception to U.S. Navy customs.

The sighteers at least distracted attention from a new problem. Local Chinese officials had announced that on 1 June a plan would be put in effect, confirming foreign belief that in thinking up bizarre schemes to extract "squeeze," the Chinese had no peers. Balked in their attempts to collect "protection" from the wood oil junks, they blandly declared that as hostilities all over China had been brought to a close, undivided attention could be turned to pacification of river bandits. For this purpose, two Chinese gunboats were required. To build the gunboats, Chinese *and* foreign steamship companies would be allowed to subscribe to a loan, individual quota to be based on ship length. There was some charming fiction to the effect that one-half percent interest per month would be paid on the loan, plus more plausible and fairly clear innuendo that failure to ante up would result in unpleasant if not insupportable delays in such routine matters as loading cargo, obtaining clearances to sail, or hiring pilots. It took "Tommy" Oates and consul Jenkins, in concert with their perspiring British colleagues, at least a sweltering month to bat *that* one down.

On the Yangtze, 1924 was a year of exceptionally high water. At Hankow, the river rose over most of the dikes, backed up sewage and created a frightful stench, trying even to the hardened nostrils of veteran gunboaters. The turbulent river was running at over five knots. There was such violent "chow-chow" water—swirls and countercurrents—that the station ship *Elcano* had a helmsman steering at anchor. Ashore, *Elcano*'s skipper had had to put the manager of the Parisien Cafe in his place when the latter proposed that his restaurant be declared out of bounds to American sailors in uniform. "Discrimination against the American flag!" growled the skipper, who would have no part of it.

Peking, meanwhile, had certified that steamers could carry wood oil. The gunboaters warned commanding officers of river steamers against allowing members of the junkmen's guild to board their ships, on peril of their lives. The *Monocacy* tried out the new anti-mob device, "tear gas bombs," discovering that the effect on volunteers was all that could be desired. Less reassuring was a check on a slab of "bullet proof" bridge plating: "Our rifle ammunition goes through easily at 200 yards."

Balked in all their schemes to "tax" foreign commerce, the Chinese, for an astounding change, hit on something legal. They *could* "tax" *Chinese*-owned goods carried in foreign bottoms. But it was clearly understood that collection would be made at the point of loading. The word soon spread. At Hokiang, between Chungking and upriver Suifu, a local military outfit enterprisingly set up a "gate," with soldiers on both sides of the river. Their first foreign customer, the SS *Changkin*, was heavily fired on when she took the soldiers to be bandits and kept on going, and one young girl aboard was killed. Their subsequent protest was met by a counter protest from the local general. "We did *not* fire 200 shots!" he indignantly said. "We fired *300*, and we want to be reimbursed for them."

The American SS *Chi Ping* went up a week later. The "gate" was still there, but no effort was made to force it shut. The magnificent Chinese grapevine had carried the word that aboard the *Chi Ping* Lieutenant J. M. Brady and six U.S. bluejackets were ready with rifles and Lewis guns.

In all fairness, one must admit that the Chinese could scarcely afford to relax. On 26 October, the French steamer *Hsin Shu Tung* stood in and anchored abreast *Monocacy*. Then, in a few minutes her master thought better of it and moved down near the French gunboat *Doudart de Lagree*. A junkload of soldiers headed her way and soldiers took up positions ashore. *Hsin Shu Tung*'s French armed guard prevented boarding and the ship got under way. The soldiers ashore opened fire, which was returned by *Doudart*.

The whole affair turned out to be an unfortunate error. *Hsin Shu Tung,* carrying a large quantity of opium, had been assessed 16,000 taels military tax (about US$12,000). The soldiers ashore were under the impression they had been short changed, some of the tax no doubt having stuck to the fingers of the negotiator.

On 19 November, the Japanese *Teh Yang* arrived and was immediately searched by the military, who found a large quantity of counterfeit money aboard. The Japanese skipper was soon released by the intervention of the Japanese gunboat, but the money never served its purpose of debasing local currency, thus weakening the party in momentary power. As the gunboaters well understood, the reach of Chinese intelligence was long and close to all-penetrating.

One might suppose, having reviewed even these few of the many confrontations of 1924, the gunboaters would hope for a brighter 1925. Had they known what that year would bring, they would have preferred to stop the clock. In 1924, the Chinese were pitting their wits in a friendly way with the *fanquei* upstarts. The gunboats still packed a strong psychological if not actual wallop. But from then on, so far as the Chinese were concerned, the game was for keeps.

1925-1927

China in Turmoil; Rise of the Nationalists
The Battle of Wanhsien
Nanking Incident
Seige of Shanghai, 1927
The "Fourth" Marines
Of Ducks, Fish, and Wild Game

The 1924 pattern of local incidents along the Yangtze continued into the spring of 1925. The Kuomintang, or "Nationalist" party, had been attempting to consolidate its power in Canton, when a series of incidents started it on the road to Hankow and eventual national supremacy.

The train of events began with labor disorders in Japanese mills at Tsingtao. The contagion spread to Shanghai, where conditions in the factories, both foreign owned and Chinese, were not far removed from downright slavery. On 30 May 1925, some demonstrators in a protest parade in Shanghai were arrested by settlement police. In short order the station was surrounded by angry crowds demanding their instant release. A British police official gave the orders to fire, at which twelve Chinese, allegedly students, fell dead. The effect was an explosion of mass Chinese anger that brought on a general strike of near total scope. No drinks crossed the bars in the clubs, nor did wheels turn on the streets. Stoves sat cold in western kitchens where housewives struggled with unfamiliar pots and tried to find which cans contained sugar, salt, coffee, and tea; the servants and amahs had disappeared.

On 10 June, crowds rioted in Hankow, facing the foreign volunteers and a landing force from *Villalobos*. On 12 June, the USS *Paul Jones* and *Stewart* arrived and their landing forces relieved the volunteers ashore. The *Villalobos* then departed for Changsha. There was equal unrest in Kiukiang, where HMS *Gnat* upheld the White Man's burden. Complicating matters was the fact that the river was at its lowest in thirteen years. An anti-British and anti-Japanese boycott was in full swing. Eight Chinese had been killed in Hankow and more wounded.

On 18 June, in an even more serious clash at Canton, French and British troops guarding the foreign concession on the island of Shameen fired into a mob threatening to rush the bridge and killed over fifty of them. This resulted in a total boycott of British goods and a general strike which paralyzed Hongkong. Its Chinese name of Hsiangkang (fragrant harbor) was mockingly changed by the Chinese to Ch'oukang (stinking harbor), as uncollected garbage piled up. In another few months, it would become Szŭkang (dead harbor), as shipping ceased because of the walkout of all Chinese workmen on the docks, and in tugs and sampans and throughout the colony.

The chief benefit of the strike, which continued for almost a year at Hongkong, and with varying intensity at Shanghai until the autumn of 1925, fell upon the Kuomintang. It allowed them to consolidate their power in the south and prepared the Chinese peasants and workingmen psychologically to accept a nationalist, antiforeign posture which would support the KMT's march to the Yangtze Valley. It was this march which totally changed the character of the opposition the gunboats would be facing for the next seven years.

The most calamitous year for the foreigner in China since the Taiping Rebellion, 1925 ended with nearly the entire U.S. Asiatic Fleet drawn either to the Yangtze Valley or the China coast. Marine reinforcements for the ships' landing parties were sent from Guam as an afloat reserve. If 1925 was bad, 1926 would be worse, and 1927 even worse yet, when "allied" foreign guns on the Yangtze faced organized Chinese military forces in new strength and vigor and sense of purpose.

But troubles along the Yangtze and in Shanghai and Hongkong made few headlines in the United States. An officer ordered to the China station more or less said goodbye to the U.S. Navy. His knowledge of China might well be limited to a dog-eared wardroom copy of the *National Geographic,* and a ballad which alleged that the monkeys had no tails in Zamboanga, which was not in China at all. Full introduction to China was a slow process.

A transpacific crossing was a leisurely affair in the pre-air age, when one boarded a Dollar liner in San Francisco and thirty days later, via Honolulu, Yokohama, Kobe, Shanghai, and Hongkong, arrived tanned and fat at Manila.

Along with five meals a day, shuffleboard and deck tennis, young naval officers for the first time in their lives had been made aware of the real *raison d'être* for the Asiatic Fleet: among their shipmates were representatives of Standard Oil, Texaco, and National City Bank; merchants of silk, cotton, gold, wood oil, soya beans, and motor cars; an occasional diplomat, and always, in fair numbers, the ubiquitous missionaries.

Less fortunate travelers drew either the homey Navy transport *Henderson* or her bleak consort *Chaumont.* The latter was a Hog Island "double ender," her configuration calculated to confuse World War I German U-boaters as to which direction her all-out ten knots was taking her—a confusion shared by her passengers whenever a good head wind was blowing.

Some downright bad luck bachelors might have to work their way out on the old tanker *Ramapo,* or in some warship relieving a super-annuated Asiatic relic due to be broken up for scrap or sunk as a target. For a century the Far East had been the end of the line for worn-out ships.

Whatever the mode of transportation, the China-bound officer—flag officer or major command excepted—was in the dark as to ship assignment until after crossing the 180th meridian. Shortly thereafter, a radio message from the CinC Asiatic would designate his home for the next fifteen months. There would be a shift to other duty, ashore or afloat, for the second half of the cruise. Then, barring a rare extension, the time would come for the return to club-less, amah-less, whiskey-less San Diego or Norfolk, via the Pacific, or sometimes the long way around through Suez.

The 13 July 1926 entry in Lieutenant Commander Earl A. McIntyre's personal journal reflected a state of mind not uncommon in those ordered to Yangtze River gunboats. A pleasant cruise on the SS *President Madison* was drawing to a close. He had "received orders by despatch to disembark at Shanghai and assume command of a river gunboat, the USS *Villalobos*—hell of a name. Wonder what I'm in for?"

All the hell was not in the name. *Villalobos* was in for a varied year. McIntyre barely had stowed his gear in the small cabin before his museum piece of a gunboat was rattling its way upriver.

Chinkiang, the first stop en route, whose streets had been tramped by so many dignitaries over two thousand years, lyricized by emperors, described in some detail by Marco Polo, and ravaged and rebuilt a dozen times, was not the new skipper's dish of tea. He settled for a round of calls on HMS *Woodcock,* the British consul—there being no American one—the Commissioner of Customs, and a familiarization visit to that inevitable and indispensable facility in any Yangtze port, The Club.

"I'd rather live in Hoboken, Flatbush, or even Goshen," McIntyre allowed, as many others who had been to Chinkiang would readily agree.

The next whistle-stop on the itinerary, Nanking, was something even less reassuring than Chinkiang; at a temperature of 103 degrees in the shade, the neophyte River Rat straightaway composed a request for transfer to a destroyer of the "outside" fleet. "Nothing doing today," he wrote on 27 July. His entry for 28 July was, "Just as exciting as yesterday." Little did the captain know what was in store for him around just a few more bends.

At Wuhu and Kiukiang, the next ports in line, things commenced to look up. There were "nice bunds," bridge games ashore, tennis, teas, and a hint that a month's indoctrination and association with the British-oriented way of life was getting in its deadly work; lunch had become "tiffin."

Hankow, with its handsome foreign concessions, greenery, wide streets, and spotless order was an even greater improvement over gray Nanking than had been Wuhu and Kiukiang. The skipper began to take heart. The magnificent Hankow Race Club was "a great place; big club house, bar, race track, golf course, tennis, swimming pool, and everything." This was Kipling's east of Suez with a bang.

In view of the hundred-degree temperature, McIntyre certainly can be forgiven for giving the bar a whirl first, promising to check out "the other athletic activities later." The conviviality of the bar was of a high order. Three guests accompanied the skipper back aboard ship for dinner.

With the same fortitude which has marked Navy families for generations, Mrs. McIntyre bundled the two boys aboard a river steamer at Shanghai "for a hot, hectic trip, without an amah" and soon was enjoying the rounds of Hankow dinner parties and the club life, while she scoured the city for souvenirs and met the mildly exotic little league of nations which made up the foreign colony.

For Skipper McIntyre there were the added attractions of coaling ship, admiral's inspection, and reading military and international law. He was now not only judge and jury of a little empire of sixty-four souls aboard *Villalobos,* but, once upriver, a potential American consul and diplomatic representative of the United States in delicate foreign confrontations.

By the end of August 1926, the cloud no bigger than a man's hand appeared on the horizon in the form of "the Southern or Red forces . . . advancing toward Hankow." There might be a battle, McIntyre thought. And if the river continued to fall—inevitable and ultimate horror—*Villalobos* would replace *Pigeon* at Changsha.

On 23 August, *Villalobos* received from "the Rear Admiral and British Senior Naval Officer, Yangtze," a confidential dispatch:

> *In view of the advance of the Southern forces across the Hupeh border and their approach to Wuchang, I shall be glad if you will come on board HMS* Bee *at 10:00 A.M. tomorrow Tuesday, 24th August 1926, to discuss the situation with regard to combined action for the safety of the concessions should the necessity arise.*

McIntyre conferred with British Rear Admiral Cameron in his flagship, the monitor-gunboat *Bee,* on joint Changsha action "if and when," then got under way on 27 August for the trip upriver to Yochow, thence via shallow Tungting Lake and Siang River to Changsha. They passed nondescript Chinese warships at Kinkow, scruffy-looking bands of soldiers ashore, and dead soldiers floating in the river, all sombre portents of the future.

Possibly, being new on the river, McIntyre overlooked certain measures normally taken to spike the efforts of troublesome demons. Obviously they were aboard in force. "A rotten sea, wind and spray sprang up," the skipper recorded. The Changsha pilot failed to show at Chingling and the whaleboat crew had a nasty trip ashore for nothing. The ship dragged anchor all over the place, the radio antenna carried away in the gale, and when the haunted ship moored in the lee of Tien Pien Island for the night, the chain parted and she lost an anchor.

The ship had scarcely cleared the anchorage next morning when she ran hard aground on a sand ledge. This was something more than a laughing matter, even though touching bottom in the Yangtze generally was too minor a mishap to log. With the river dropping a foot a day, one could picture the ship high and dry until the following spring, with a baseball field and vegetable garden neatly laid out alongside, as had been the case with several other vessels in years gone by.

A passing tug of the British Butterfield and Swire line came to the rescue and soon everything portable was being rushed aboard her lighter to decrease the draft. Meanwhile, the USS *Pigeon,* getting out of Changsha while the getting was good, hove in sight and, together with a passing Jardine Mathieson steamer, plus the tug, was able to wrench *Villalobos* free. "It's days like this that send boys back to the farm!" wrote McIntyre, falling back into his bunk at 3:00 A.M. after the stores and ammunition had been manhandled aboard.

Possibly confused by all the diversified and fascinating opportunities for mischief, a majority of the devils which had been dogging *Villalobos* must have slipped across to one or another of the rescue vessels, as she reached Changsha on 30 August without further incident.

As soon as McIntyre had completed his round of official calls on consul and foreign warships, he dashed off a report to F. P. Lockhart, consul general in Hankow, on the state of Changsha affairs.

Lockhart's reply showed great interest in the report. Lockhart's news was all bad:

> *You have probably heard by this time of events here as of 8 September. Hankow and Hanyang are now completely under the control of the Kuomintang, and Wuchang is likely to fall at any moment. Wuchang has been subjected to terrific bombardment with heavy artillery and machine gun and rifle fire, but is still holding out. Wu Pei-fu, with several train loads of loyal troops, has retreated to a point about fifty miles north of Hankow. If he attempts to come back he will undoubtedly have a very difficult job. Practically all of the river boats are being fired upon, not only by machine gun and rifle fire, but by heavy guns as well. The Japanese convoy was fired upon this morning and the* Tuckwo *has just been fired upon seven miles below Hankow. The USS* Elcano, *the SS* I'Ling, *and the destroyers USS* Pope *and* Stewart *were all heavily fired upon near here a few days ago. The attack on the Japanese boats this morning took place between Wuchang and Hanyang.*
>
> *Order in Hankow is being maintained, but the naval units and volunteers are still on duty. All telegraph lines are now out of commission, and we are having difficulty in reaching the outside world as the wireless is not working well. Many of the Cantonese troops are now in Hankow and can be seen about the streets.*

Lockhart closed with the assurance that if "the situation should become too demoralized, I shall advise Mrs. McIntyre along the lines suggested in your letter." In other words, get scooting for the relative safety of Shanghai.

Vice Consul John Carter Vincent, in Changsha, seems to have been bypassed in much of these matters, although Lockhart expressed the wish that McIntyre let him see his letter.

Changsha, although cut off from the sea by the uproar at Hankow, turned out to be so peaceful that McIntyre, having taken over the flat of an absent professor, sent for his family.

The tale of the ensuing four months of relatively serene McIntyre family life at Changsha, in very nearly the geographical center of China, would warm the heart of any Old River Rat. There were parties at the Club and gay dinners aboard the gunboats—the British *Woodlark* and Italian *Ermanno Carlotto.* There was even champagne in honor of the birthday of the Japanese emperor, two months later to die and be succeeded by Hirohito.

With the approach of autumn, dismal weather drove people off the tennis courts and picnic grounds, indoors before the open fires which provided "central heating" and cheerful gathering places for gossip and discussion of the quaint doings of Chinese servants. And inevitably, reflecting the growing uneasiness which overhung them all, they exchanged information on the course of the fighting.

The party of the season, naturally enough in a British-oriented community, was the New Year's Eve masquerade. "All hands had a large time until a late hour, . . ." wrote McIntyre who, as a Spanish matador, escorted Mrs. McIntyre as a Chinese princess.

By early January 1927, the discussions around the open fires had become less light. The British concessions in Hankow and Kiukiang had been occupied by the Chinese with a brand of truculence not heretofore seen, infuriating most Britishers who had been entrusted, by the terms of the Chen-O'Malley Agreement, to Chinese administration under a regime of transition. Most of the foreign women and children had been evacuated to Shanghai. By 12 January 1927, affairs at Changsha had deteriorated to the point that McIntyre's family spent the night aboard *Villalobos* and the following day, along with other evacuees, boarded HMS *Woodlark,* bound for Shanghai. Anti-British feeling ashore was extremely high, partly a hangover from the Wanhsien incident but generally a manifestation of Chinese animosity toward the foreign power they felt to be the chief "colonizer" and most intransigeant of all the *fanquei* in the Yangtze Valley.

McIntyre's patience was wearing thin. The rains which had made life miserable in late summer had ceased and the river was too low for *Villalobos* to get out. "*Damn* such a country!" he wrote. "Just one damn thing after another. No British dare show themselves without a chance of being beaten up. Can't see how Britain can help declaring war. Hope she does!"

By 16 January the rains had come, but not yet in sufficient volume to lift *Villalobos* over the high spots. On 4 February 1927, Vice Consul Vincent, feeling the heat from the brush fire headed his way, asked McIntyre to radio a "situationer" to the American minister in Peking. "The Kweichow troops have gone on a rampage," he explained. General Yuan Tzu Ming, commander of two army corps of Kweichow troops stationed at Changteh (one hundred miles northwest of Changsha), had reaped the full dividends on this high-risk occupation and had been shot to death by his troops. The latter, in traditional provincial warlord army style, had thoroughly looted the city. The Changsha crowd, under nominal Kuomintang control but leftist oriented, had sent a force off toward Changteh, but they prudently stopped halfway at Yiyang to scout out the opposition.

Battles took place in China only when the contending forces were of equal or nearly equal strength. When one side outnumbered the other, it was customary for the smaller force to surrender or to join the larger. Foreigners found this philosophy to be very comic, whereas it was in fact proof of the profoundly pragmatic nature of ancient Chinese civilization. Had the ground rules not been such, the loss of life during the generally senseless civil wars from 1911 to 1926 would have been frightful.

Vincent's closing line was typical of the insouciance affected by the professional diplomat under fire: "Drop over for tea!"

At 3:00 A.M. on 28 February, McIntyre was awakened by a message from ComYangPat to proceed downriver. If getting under way at 8:00 A.M. that same day suggests an eagerness on McIntyre's part to see Changsha disappearing over the fantail, one may certainly excuse him; it was no longer the city that gunboaters remembered from palmier days, or even that McIntyre had found five months earlier.

At Lilangtan, *Villalobos* met and coaled her relief *Palos*. Then without regrets, she pointed her nose downriver at an all-out ten knots, arriving at Hankow 2 March.

There, once again, all was peaceful and serene. Flagship *Isabel,* with ComYangPat, Rear Admiral H. H. Hough aboard, had been reinforced by destroyers *Truxtun* and *Pope* from the "outside" fleet. They in turn would soon be relieved by *Pruitt* and *Hulbert.* McIntyre seriously considered sending once more for his family.

This mild euphoria was short lived. On 24 March the Chinese Nationalists savagely attacked foreign installations at Nanking. British and American warships engaged in the greatest display of naval firepower since the British expedition of 1842. On 25 March, boarding the SS *Loong Wo,* women and children once more repeated the dreary routine of evacuation of Hankow. Here at least it was carried out in relative calm; at Nanking, foreigners had been brought to safety under the most harrowing circumstances, covered by heavy naval artillery barrages.

The next day McIntyre was busy rounding up steel armor plate "to make a warship out of the old *Villalobos.*" He would have liked to have topped off bunkers, but the general strike ashore scotched that.

By 3 April, rioting and looting broke out in Hankow's Japanese concession. The Japanese sent a landing force ashore and drove back the mob, but all Japanese citizens were evacuated and most Americans went aboard ship, fearfully watching a large fire raging ashore while *Villalobos* stood by all night with steam up and guns cleared. Four Japanese destroyers arrived on 5 April, prompting Captain McIntyre to express the widely held foreign hope that "the Japs clean out the Chinese in their concession in good style."

The *Villalobos* was then sent downriver several miles to stand by the Socony Vacuum oil installation. There, the skipper chin-chinned with his opposite numbers in HMS *Teal* and *Scarab,* played poker, enjoyed "light refreshment" aboard His Majesty's far-from-dry gunboats, and in return invited their officers to dinners and their bluejackets to the movies aboard *Villalobos*.

On 21 April the big British cruisers *Vindictive* and *Carlisle* stood in and anchored off Hankow. Nanking, never of much foreign commercial interest, had been wholly abandoned to the tender mercies of the Chinese Nationalists. But Hankow was quite a different cup of merchants' tea. There was keen concern by foreigners, and particularly the British, over Wuhan—Hankow-Wuchang-Hanyang—the cities grouped at the confluence of the Yangtze and Han rivers.

Meanwhile, things were uneventful at Socony. "We fumigated all compartments below decks with sulphur to decrease the cockroach population to a minimum," wrote the skipper. Presumably for everybody but the cockroaches, things were dull all over: "The day otherwise was a total loss," he concluded.

On 27 May, relieved once again by the *Palos, Villalobos* departed Hankow and the middle river for what was to be McIntyre's last trip down the river. The cruise to Shanghai was uneventful. Clearly McIntyre had by now got the hang of making proper joss, exploding a few judiciously timed firecrackers, and saying appropriate things when passing protective Buddhas on promotories. "No luck!" he complained. "Haven't been fired on yet. The Chinks have ignored us to the point of being insulting." It wasn't McIntyre's fault, as ComYangPat was by no means equivocal in his

26 May movement order that sent *Villalobos* from Hankow to Shanghai: "If fired upon and source can be located, return and silence fire with suitable battery."

Anchoring far up the Whangpoo, unfortunately well beyond easy reach of the Shanghai Club, *Villalobos* found herself near the shipyard where six new river gunboats—*Mindanao, Luzon, Oahu, Panay, Guam,* and *Tutuila*—were belatedly being constructed for ComYangPat. Congressional delays in authorizing and appropriating funds for them had caused these handy little ships to miss the act just completing in the center ring. But there still would be enough fun for all in the fifteen years before the show closed forever.

Lieutenant Commander McIntyre was transferred to the armored cruiser *Pittsburgh* in mid-June, but not before receiving a letter which praised him and, without consent of the Navy Department, reclassified his old ship. The French minister at Peking expressed to ComYangPat his most appreciative thanks for "... directing the Commander of a destroyer which I take to be the USS *Villalobos*..." to provide asylum for French citizens at Changsha, in case of need. Those words would have made the weary old gunboat kick up her heels if she could have read them.

What McIntyre, and, in fact, what most Americans in the Yangtze Valley could not have been aware of as they sweated out those uncertain winter days of 1926–27 was that the fate of China for a generation, and perhaps of all Asia, was hanging in the balance right there in Hankow.

The trouble began eight years earlier at Versailles, where China's heady expectations of being relieved of Japanese and Western overlordship and of enjoying some of that "self-determination" promoted by President Wilson had been cynically deflated by horse-trading politicians.

Japan, on the other hand, Britain's ally for two decades, emerged smelling like a rose as the spoils were whacked up in the Far East and Pacific Ocean. It was a result the British would one day sadly rue.

Thus, China, ridden with debt, demoralized, and virtually a Japano-European colony, was worse off than she had been before the war, even though she had joined the Allies against the Central Powers, allowing the Anglo-French to recruit 19,700 coolies with their yo-yo poles as labor battalions for France. They were paid the equivalent of 19.3 US cents for a ten-hour day, seven-day week, "Chinese holidays excepted."

There was mass disillusionment over the peace settlement, naturally enough. In early May 1919, students rioted in Peking. The homes of several pro-Japanese officials were wrecked and sacked. It was indeed this disillusionment of students and intellectuals which turned Chinese thought toward the Russian revolution and eased the way for later interest in Marxist doctrine.

Meanwhile, by 1920 the Great War was well over as far as Shanghai foreign *taipans* were concerned. If any of them realized that the 1919 riots and strikes were the preliminary rumblings of the Second Chinese Revolution, nobody said so out loud. But it would be that Second Revolution which eventually sent them packing, just as the 1911 revolution had swept out the Manchus.

With the unhappy exception of the high postwar price of silver, which had forced the American dollar below the "Mex," the Yangtze Valley foreigners resumed "business as usual." The only bloodlettings were the mild skirmishes between U.S. bluejackets and Italian sailors in the Hankow bars. British and American sailors, as a result of wartime cameraderie, were for a change on good terms.

There had been, in 1918, spirited exchanges of gunfire up the Yangtze between "bandits" and gunboats, but by 1920 their echoes could scarcely be heard in the Shanghai and Hankow clubs by even the most perceptive ear. British *taipans* badger-hunted and cricketed. They enjoyed pink gin tiffins at the glass-fronted Race Course Restaurant while casually observing gentlemen jockeys gallop clockwise around the track. American *taipans* played poker and golf and listened to the satisfying clang of the cash registers. The U.S. naval officers toyed with their scotch and soda in the Palace Hotel lobby to the soothing strains of Viennese tea music. Some of the married men set up housekeeping ashore, where Missy and the kids soon made the fascinating discovery that a single word, "Boy!" instantly produced results equivalent to rubbing a magic lamp.

The erstwhile allies stood high and mighty. Germany and Russia were momentarily counted out, their share surrendered largely to Britain, partly to Japan. Even pirates prospered: "The *Palos* and *Monocacy* were active last spring," wrote the Secretary of the Navy in his 1920 *Annual Report,* "in defending vessels from river pirates and lawless elements who were holding up and looting steamers and firing on passing craft."

Yet at this time of heightening confidence in everlasting peace plus a certainty of endless profit in China, a handful of conspirators in Shanghai were helping to hatch the feeble creature which in just a quarter of a century would become a monster, sweeping the "Imperialists" from the Chinese scene. This was the CCP, the Chinese Communist Party.

Catalyst and arch pragmatist in that handful was the Soviet, Gregorii Voitinsky. The others were a mixed bag of Chinese intellectuals and disillusioned students—one could remain a "student" in old China until death did him part from academe. Voitinsky, aided by the Dutch communist Maring (Sneevliet), was there as representative of the Comintern, desperately searching for allies in the struggle to keep the new revolutionary Russia afloat on a sea dominated by capitalist enemies. The Chinese intellectuals were there partly because they were fed up with the rapacity of the warlords who had prostituted the revolution of 1911 and partly because they were desperately alarmed at the ever-broadening reach and arrogance of the foreign "Imperialists," who were merrily milking dehorned China, chief among which was Britain. Here, Voitinsky and his Chinese intellectual pals saw eye to eye. Winston Churchill's post-World War I view—that the greatest boon to humanity would have been the strangling of Bolshevism in its cradle—was increasingly reflected by Whitehall through the twenties. On 26 May 1926, Britain, the leader of the anticommunist world, broke relations with the USSR. Russia was walled in on the west by France's *cordon sanitaire* of semi-puppet allies: Poland, Czechoslovakia, Romania, and Yugoslavia. The cold fact of Russia's existence went unrecognized by the United States. In the Far East, Japanese-supported White Russians yapped at the Soviet's run-over heels and frayed coattails.

China's revolution of 1911 was what might be termed an "implosion," the collapse in a cloud of dust of a hollow stump. Sun Yat-sen, the figurehead of the movement, was a foreign-educated semi-idealist who had spent so much time abroad in exile that he had lost touch with the Chinese scene. He had no workable plans nor suitable personnel for establishing a viable "democratic" regime to replace the moribund Chinese Empire. Consequently, he soon was shunted aside by the gangster-type generals and merchant-politicians who had been developing as the real power in China ever since the Taiping Rebellion of 1850–64. What remained of Sun's influence served to coalesce a rump government centered at Canton, the original source of most of the overseas Chinese who had been the main financial contributors to Sun's movement from the beginning.

Thus, by 1923 there was an infant CCP centered at cosmopolitan Shanghai and a "nationalist" Sun-oriented provincial government at Canton, under the aegis of the "Kuomintang." The KMT theoretically was based on Sun's "Three Peoples' Principles": *Nationalism*—the freeing of China from foreign control and assimilation of minorities; *Democracy*—a guiding of the miserable masses toward self-government; *The Peoples' Livelihood*—an ephemeral, miscellaneous bundle of equalization schemes for opportunity and land ownership.

Two gentlemen scheming to rob a bank probably do not hold each other in implicit good faith. They have joined in uneasy partnership, knowing that alone neither could swing the deal. Each no doubt cherishes the hope that the job once done, the swag might somehow accrue wholly to himself, the other conveniently put out of the way, lips permanently sealed. This mild analogy should by no means be construed as applying to the CCP and KMT, but it does suggest a parallel with the situation confronting these two parties in their shared aim to eject foreigners, overcome warlords, and uplift the downtrodden.

It was at this juncture that the mortal struggle for survival between nationalist Stalin and internationalist Trotsky had shifted into high gear. The hopes for a daisy chain of revolutions in industrial Europe had irretrievably collapsed with the crushing of the Red uprising in Germany in 1923. Reduced to its simplest terms, Trotsky's ideal view of China was as a Red, brotherly, associated *socialist* state. Stalin was convinced that for the moment, Russia must turn inward, build herself up industrially and militarily and above all, get the most dangerous monkey, Britain, off Russia's back. A China built up as a strong *nationalist* state would throw out the arch enemy Britain, thus setting off a chain reaction that would engulf India and bring the British Empire crashing down. (Trotsky's theory of course was proved by events to have been right, although the ultimate means turned out to be different.)

Chiang Kai-shek came from a modest family in Chekiang, the coastal province south of Shanghai. He was at military school in Japan when the 1911 revolution broke and hurried home to become first a staff officer with a Shanghai general, then for reasons not clear, a small-time operator on the Shanghai stock exchange. Next, via Canton and employment with Sun's KMT, Chiang set out for Moscow in July 1923 for a six months' look at the Russian experiment. In May 1924, with their ward the CCP now a junior partner of the KMT, the Soviets not only set up and staffed the new Whampoa Military Academy, near Canton, but less prescient than one generally gives them credit for, tagged Chiang for director.

He eventually became head of the Cantonese government—renamed the "Nationalist" government of the KMT party—and the Cantonese troops were at last launched on the Northern Expedition, with the sacred mission of redeeming China from its exploiters, both external and internal.

That the world in general and the foreign "Imperialists" in particular felt the whole show to be communist can be appreciated. The principal political adviser peering over Chiang's shoulder was Moscow's Michael Borodin, alias "Mike" Gruzenberg, deported to Russia from the United States in 1917 as a troublemaking Bolshevik. Chief strategist and military planner was the Peasants and Workers Red Army General "Galen." As Marshal of the Soviet Union Vasilii Bleucher, he later commanded the Soviet Far Eastern Army, was liquidated in the Great Purge, then posthumously rehabilitated after Stalin's death. Numerous Russian officers were tucked in at every level. Shiploads of Russian arms came to Canton from Vladivostok, and included among the shells and ammunition was a secret weapon that would win many a battle: the silver bullet—funds for bribery.

Ahead of Chiang's forces, communist agents demoralized the military opposition. The peasantry, overtaxed, bedeviled by conscription, pillaged and put upon for a century by ever more corrupt landlords and tin-pot generals, met the Southerners with full cooperation. On 12 July 1926, Changsha was occupied by General T'ang Sheng-chih, a small-bore Hunanese warlord who had fallen to silver bullets and with a sharp eye for reality. On 22 August, Yochow, where Tungting Lake's outlet meets the Yangtze, gave up; on 6 September, Hanyang; 8 September, Hankow. Almost a month later, Wuchang, the third city of the tri-city "Wuhan," fell after a seige in front of the city's great walls in the fashion of a thousand years before.

On 29 October, there occurred an incident at Hankow which not only reflected the new Chinese temper, but makes clear to one not familiar with China of those times the infuriating condescension suffered by Chinese in their own country. There was a sign at the entrance of the small park on the Shanghai bund opposite the British Consulate General warning that the park was not to be used by Chinese nor were dogs to be introduced therein. The reason was elementary: foreigners taking the air in the evening did not want to be harassed by spitting, arguing, noisy mobs of coolies monopolizing the benches. The British proprietors also cherished the flower beds and grass, which even the best educated foreign dogs could be expected to desecrate. But the Chinese naturally darkly inferred invidious comparisons.

Shanghai was a settlement, thus its streets were open to Chinese. The Hankow enclave, as a concession, could and did bar Chinese from the bund. With the new cockiness of Nationalist occupation, on 29 October four Southern soldiers strolled out on the forbidden bund for a look.

A concession policeman recited the rules, then appended the usual *"aidah!"* very probably with some reference to their reptilian ancestry. The soldiers refused to get out and were forthwith arrested. As if by magic, the usual Chinese mob of three thousand or more screeching, rock-throwing coolies soon sprang out of the ground. Police reinforcements rushed up to dampen the enthusiasm of the crowd with fire hoses. Within an hour, the bund was quiet and deserted once more, free for unencumbered *taipan* strolling. But one tangible result was obtained; soon thereafter the British municipal council threw open the bund to "well-dressed Chinese," and before long the whole place was swarming with elated humanity. This was the crack in the dike which a scant two months later would burst open to put the $60 million British concession once more in the hands of its original owners, the Chinese.

The whirlwind advance of the Southerners to the Yangtze coincided with a tremendous expansion of labor unions and peasant associations throughout the areas influenced by or under the control of the Cantonese. *Guide Weekly,* a publication in the capital city of Changsha, claimed two million membership in peasant associations. Landlords were having a tough time collecting rents. Even worse, some of them were having to surrender their property to the tender mercy of village councils. At Hankow after its capture, more than three hundred thousand workers and shop assistants were reported to have been stuck together in a jerry-built Hupeh General Labor Union that used compulsion as well as cajolery in its organizational methods.[35]

Chiang Kai-shek, meanwhile, was well to the east, at the "front," where the going slowed appreciably once the Cantonese approached more closely the prime plum on the Yangtze tree—Shanghai. Nanking, with more symbolic than intrinsic value, could be saved for later.

By December 1926, things looked so rosy that the Canton government moved north. Comrade Borodin, the Chinese communists, and the KMT "left wingers," including Eugene Chen,* the KMT foreign minister, jumped off the bandwagon at Hankow. Sun Yat-sen's widow, Soong Ching-ling,† knowing Chiang Kai-shek well enough to have no illusions even at that early date as to the shape of things to come, parted company with her immensely wealthy and influential family in Shanghai and joined the Hankow leftists.

Vincent Sheean, at that time a visionary young American journalist already on his way to becoming a world famous author, shuttled between Shanghai and Hankow, recording for history what he saw and heard.[36]

It was Sheean's view that communists everywhere regarded Hankow as a test case for world revolutionary movements of the Trotzky internationalist flavor; that if Hankow failed, the militant world-revolutionists failed everywhere else as well.

"You could not be in Hankow a week without being aware of this," he wrote.

* Chen, born in Trinidad of mixed Chinese and Negro parentage, received an English education (London University) and in the process a bitter hatred for the white man in general and the British in particular.

† Mme. Sun, "Rosalind" Soong, educated at Wesleyan College for Women, Macon, Georgia. Returned to China at nineteen and married Sun. Thenceforth communist-leaning, and when last reported, in 1969, was in Peking, supporting Maoist policy.

"French communists, German communists, Hindoo communists, British I.L.P. people,* and numerous agitators responsible to the Comintern gave the place a fine mixed flavor of international revolt. . . . Russians in ill-defined functions appeared and disappeared."

Sheean interviewed both Chen and Borodin. The former corroborated dim foreign views as to Chen's character, which in the coolie vernacular would have been assessed by the *taipans* as being the result of a union between a turtle and a snake: "He was . . . a small, clever, venomous, faintly reptilian man, adroit and slippery in movements of his mind, combative in temper, with a kind of lethal elegance in appearance, voice and gesture."

Sheean concluded that it was evident Chen's soul had been twisted by his hatred of the (to him) pretentious white race, particularly his colonial ex-masters the British, who had blocked whatever ambitions he might have cherished, offended his sensibilities, and turned his dreams into nightmares.

Borodin was quite another breed of cat. In Sheean's words, he was "a large, calm man with the natural dignity of a lion or a panther . . . slow, resolute way of talking . . . refusal to be hurried or get excited, . . . insistence upon fundamental lines of action." In a few words, he was that rarity, a politico-historically oriented philosopher with the long view.

Definitely more attractive than all this coterie of pros from Sheean's point of view was an amateur Red, a pert, dedicated little Yankee miss, according to Sheean "slight, not very tall, with short red-gold hair and a frivolous turned-up nose." In post-January 1927 woman-starved Hankow, it is likely her name, Rayna Prohme, was noised about the bar of the Hankow Club when gunboaters and members of the strike-idled commercial fraternity gathered for the predinner whiskey and splash. "You ought to hear the way the navy people talk about 'em!" Sheean quoted Misselwitz of the *New York Times* as saying.

As for Rayna herself, she was far too busy helping edit the English language edition of *The People's Tribune,* the propaganda newspaper of the Hankow (i.e., "leftist") government, to waste time engaging in cocktail hour badinage with the capitalist likes of YangPat lieutenants.

New Year's Eve, 1926–27, at Hankow was a sort of mass *taipan* fiddling while the city smouldered. The traditionally gay bash was held at the Hankow Race Club, still opulent in spite of the closed doors of the foreign *godowns.* On hand for the celebration were several hundred of the last-ditchers among the womenfolk who had so far refused to flee beyond the seas or even to the comparative safety of Fortress Shanghai. From the many warships in the stream came a strong contingent of bright-eyed and bushy-tailed young bachelors.

Next day, some of those who had seen the New Year in too gaily worked the poisonous juices out of their systems on the golf course; the weather was near perfect. The less athletically inclined settled for the therapeutic value of the restoratives offered by American Consul General Frank Lockhart at his New Year's Day reception.

* Independent Labor Party, euphemism for Red.

A bucketful of cold water sloshed on the participants is the approved way to stop a dogfight. A heavy rain had the same effect on the Chinese; traditionally, they never fought or collected in angry mobs when the heavens opened up. Conversely, the advent of near-perfect weather following New Year's, plus the decision of the Nationalists to set aside the first three days of 1927 as a period of celebration, brewed a highly combustible mixture.

About noon on 3 January, coolies started to stroll in increasing numbers on the British concession bund. The holiday air soon became charged with tension; it was too good an opportunity for the agitators to miss. In no time the hundreds had grown to thousands. The idlers, now become a mob in a state of howling hysteria, cried vengeance against the exploiters.

British naval landing parties rushed ashore to reinforce hard-pressed concession police and hastily mobilized volunteers. It was with the most remarkable degree of control that in spite of the showers of sticks, stones, and bricks which flew into the thin foreign ranks, the orders *not* to fire were adhered to. With some small help from foot-dragging Chinese troops, a semblance of order was restored by late afternoon.

In the uneasy calm, it was recognized that the conflagration, though in fireman's terms "under control," was by no means wholly doused. Still hanging around on the concession fringes, the mob awaited with relish a rekindling of the blaze—after all, what was a holiday for?

Rear Admiral Cameron, commanding British naval forces on the Yangtze, requested ComYangPat to send help, and that officer, Rear Admiral H. H. Hough, sent twenty-five riflemen and two officers each from the *Pope* and the *Pigeon*. Their orders were carefully inscribed "to protect American interests," so that no false inferences might be drawn by an edgy Washington. As a ready reserve they were centrally positioned to back up the bone-weary, stone-battered Britishers who had already spent a bloody day manning the barricades.

But the most horrendous blow of all was yet to fall; in something of an anticlimax, Prime Minister Stanley Baldwin of Great Britain radioed instructions that under no circumstances was force to be used to protect the British concession at Hankow. The message arrived the evening of 3 January and as a consequence, all the landing forces were sent back aboard ship. The British concession and all that it contained was now, as far as foreign armed intervention was concerned, wholly at the mercy of the Chinese.

The overnight lull, 3–4 January, was due probably to a combination of satiety on the part of the riffraff, abetted by an unorthodox show of strength by Nationalist troops. But the calm period, like the passing eye of a typhoon, was both deceptive and short. By 6:00 P.M. the next day, in spite of large numbers of Chinese troops attempting to control them, mobs had overrun the entire British concession, destroying the movable barbed-wire concertinas, and dumping the barricade sandbags. In compliance with Baldwin's directive, no foreign bluejackets set foot ashore, as they could not have done so without opening fire.

The Chinese mob, like a Chinese taxi driver, goes either at full speed or stop, with naught between. By nightfall several officers walked through the deserted streets to the Hankow Club. It was as though the whole previous three days had been simply a bad dream; aside from the remnants of the barricades there was no sign of damage. The *real* wreckage was inside the club—the *taipans* in their chagrin and shame. There were no black ties that night. Gone was the stiff formality and the slightly superior British air. Americans were greeted by their first names. In the faces of those desolate men huddled for mutual support around the bar, one clearly read the message that the end of a century-old way of life had come.

There is some evidence to suggest that the Russian comrades and their Chinese henchmen were pleasantly astonished at what appeared to be a spontaneous mass action in taking over the British concession. But the curious absence of looting or damage to foreign property suggests a high degree of organization and a markedly un-Chinese mob restraint. One doubts that the agitators achieved their results without superior coordination and direction.

Kiukiang's British concession was taken over on 7 January in a repeat performance of Hankow. The British, bending with the wind and with Washington's announced readiness to join a compromise settlement, entered into the Chen-O'Malley agreements of 19 February and 2 March turning these two concessions over to Chinese jurisdiction.[37]

Whatever the facts, the "Hankow boys" pretended to have been caught wholly unawares by all these developments. But naturally enough, they were elated at what "their" mobs and "their" foreign minister (Chen) had accomplished. Chiang visited Wuhan on 10 January to assert his paramount authority by demanding a KMT Central Committee meeting at his "rightist" capital of Nanchang. The Hankow crowd, as a result of the recent events, were in a cocky mood. In fact, Borodin himself, in a moment of misplaced exhilaration at a banquet, made a nasty remark about power-seeking militarists, which must have had its obvious target, Chiang, cracking his knuckles.

Borodin and his Soviet associates by their intransigeance certainly had not fired the gun that made up Chiang's mind to break openly with the Reds; the decision had probably been made by Chiang as far back as his Moscow sojourn. But if the Hankow gang hadn't actually pulled the trigger during Chiang's visit, he probably felt they had at least laid their fingers on it; on 19 February and again on 7 March, he made speeches clearly indicating his future anti-communist, anti-Borodin course, although concurrently professing undying admiration for the USSR. Thus, by the spring of 1927, things plainly were nearing a break in KMT–CCP friendship which from the beginning had been less than beautiful.

The 2 March *North China Herald* carried an opinion by Chiang that, "It is not Russia's policy to tyrannize over us, and though her representatives have acted otherwise, insulting our every movement, I am convinced that it has naught to do with Russia but are the individual actions of these representatives."

On 11 March 1927, in Nanking, U.S. vice consul Hall Paxton informed Peking that a rupture within the Kuomintang ranks was an imminent possibility. Paxton's immediate superior, Nanking consul John K. Davis, felt himself on surer ground a week later when he telegraphed U.S. Minister John V. A. MacMurray that "although Chiang has apparently decided for the moment to subordinate his antipathy for the communists to the general welfare of the Kuomintang party, it is not believed he will give up hope of eventually eliminating the Russians and their, to him, obnoxious doctrines and domination."

In less than a month after those cautious statements, Chiang had turned savagely on the communists. Those who had not got a bullet in the back of the head were streaking it for the hills. Comrades Borodin, Chen, Mme. Sun, and pretty Miss Prohme were finding their ways secretly, deviously, and dangerously to Moscow. Stalin's infallibility had been sorely tested, and at the same time the ground had been laid for a sequence of events that has involved the United States more deeply in the Chinese scene ever since.

While stirring and portentious events took place ashore, hot action on the river was of far more concern to the U.S. Navy than any Hankow or Kiukiang concession.

On 2 September 1926, the destroyers *Stewart* and *Noa* cruised the waters off Chefoo, practicing torpedo tactics. The men were thinking of the weekend, when they would dance and watch the floor show at the Beach Cafe until midnight curfew sent them back aboard. Then the White Russian hostesses would withdraw for a minor overhaul of makeup and return to greet the young bachelor officers, who had slept through the evening to fit themselves for the midwatch of music and gaiety.

The reveries and the tactics were broken off abruptly by a priority message to return immediately to port, fuel and provision to capacity, and proceed to Hankow. In those days of stringent fuel economy, the authorization of a speed of twenty-five knots suggested something more than just a mistreated missionary.

Two days later, *Stewart* and *Pope* (which had replaced *Noa*) were charging up the Yangtze at night at twenty knots, a procedure used by warships only in real emergency.

Wuchang, defended by northerner Marshal Wu Pei-fu, was under seige by the Cantonese. Troops of opposing sides faced each other for miles along the banks. All cats look alike in the dark, and so, presumably to a Chinese soldier, do warships. The destroyers were heavily fired on. *Stewart* had holes in her stacks as big as goose eggs. Two men received flesh wounds. The fire was not returned.

Reporting in the early morning to ComYangPat, Rear Admiral Henry Hough in his flagship *Elcano,* the two destroyers found Wuchang blazing in a number of places. Wu's gunboats were drawing Cantonese fire that fell very close to the foreign ships. The river was dotted with floating corpses.

On 9 September, *Stewart* received orders to accompany a convoy. *Palos* was en route Ichang shepherding several merchant ships. Minesweeper *Pigeon* tagged along with a deckload of coal for *Palos;* one could not count on the usual scrounging for coal junks along the way. *Stewart* was to steam in company until the coal was safely transferred. The best news of all was ComYangPat's order: "If fired upon and the source of fire can be determined, you are authorized to silence such fire with suitable battery."

The elation in the *Stewart* was described later by Lieutenant K. N. Gardner:

> *What a relief! What a feeling, after days of passive submission to indiscriminate firing . . . and when we got underway the next morning we had the pleasure of running the Stars And Stripes to the fore truck for the first time.*
>
> *Just above Hanyang the fun began. Heavy firing came from the Hanyang side, and for the next fifteen minutes or more, all ships were under constant fire.*[38]

The firing by Chinese troops on the waterfront was answered with spirit by Americans using rifles and machine guns. "One member of the crew was found standing on top of Nr. 3 torpedo tube with a Colt .45 in each hand," wrote Gardner. "He was 'a two-gun man from Texas,' he said, and was going strong."

A machine gun in an old pagoda, spitting fire in *Stewart*'s direction, demanded more attention than small arms could give. As generally could be expected, the miserable 3-inch, iron-sighted excuse for an antiaircraft gun on the fantail jammed after the third shot. But there was better luck with No. three 4-inch gun, loaded with service ammunition. The first shot hit dead on the pagoda's base. There was a very satisfactory explosion, a large cloud of dust and no more pagoda. The "war" was over; firing ceased.

Aside from that action, *Stewart*'s upriver trip was uneventful. But downriver, she took a peaceable part in the last scene of one of the most remarkable encounters ever to take place on the Yangtze: the "Battle of Wanhsien." The wounded from this bloody scrap were being brought to Hankow by the SS *Kiangwo,* but she had been stopped by the same truculent outfit that had fired on the up-bound American convoy. So *Stewart* took off the wounded, hauled them to Hankow without Chinese interference, and delivered them to HMS *Hawkins,* flagship of Vice Admiral E. S. Sinclair, CinC British Far Eastern Fleet.

The action at Hanyang, for all the lively participation by the USS *Stewart,* was only a minor skirmish, terminated abruptly by a well-placed 4-inch shell, but at Wanhsien, where the pressure reached boiling point a few days sooner, the shooting sounded like a full-scale battle.

No American ships were directly involved, or even present, in the bitter Sino-British confrontation at Wanhsien, but the repercussions from it drastically affected the lives and businesses of *all* Yangtze foreigners.

There had been indiscriminate exchanges of gunfire for years between gunboats and Chinese on the Ichang-Chungking section. But this was different. This was war. Yang Sen was the prime instigator. Clever, well-educated, iron-hard Yang was the peripatetic in-and-out-and-in-again warlord of the upper Yangtze. "In" again in 1926, he was standing aloof from the Chiang Kai-shek–Wu Pei-fu hostilities at Hankow, awaiting a clearer view of the outcome, while resting secure in Szechwan.

The best and sometimes the only way to move troops along the river was by steamer, but there were not nearly enough Chinese flagships for his requirements. French and Japanese ships carried troops without serious protest. It was rumored that the Japanese had an agreement to carry Yang's men for half fare. American and British ships refused. Even in the rare instance of any payment for passage, the ships were left a filthy shambles, with smashed or looted fixtures.

Consequently, Yang Sen commenced the practice of sending a party of fifteen or twenty soldiers on board at a port of call, disguised as ordinary passengers. Once under way, they would draw their concealed weapons, rush the bridge and engine room, take over the vessel, then proceed to the loading point for troops. Pirates had been using this system for years and as a result ships' bridges and foreign officers' quarters were fenced off by iron bars and locked gates.

The British steamer *Wanliu* was so attacked on 29 August 1926. The officers, behind their barricades, managed to beat off the hijackers. *Wanliu* already was in the midst of a swarm of troop-laden sampans when the abortive takeover attempt was made. At full speed, she charged through the sampan fleet, and in dodging one, hit another and sank it. In view of the ample rescue facilities on hand and the hot fire from disappointed soldiers, the little ship kept right on going.

All that month there had been considerable bickering between Yang Sen and Lieutenant Commander L. S. Acheson, commanding HMS *Cockchafer*, over a "wharfage" tax imposed on shipping at Wanhsien and the searching of ships. There had been endless but fruitless correspondence between Yang and the consuls and shipping company representatives at Shanghai and Chungking. Acheson concurrently had promised drastic action if interference didn't stop. He had ordered British ships to remain at anchor if boarded by troops demanding transportation.

The real warm-up commenced on 27 August, when SS *Wanhsien* was boarded by a General Kow Ju-teng with a detachment of his ex-bandits, demanding a ride upriver to Foochow. Following Acheson's instructions, the captain refused to sail. General Kow refused to leave the ship, sitting tight in his cabin, playing mah-jongg, in a fine rage over the indignity offered him—a Chinese general. Tempers were high all around.

Two days later SS *Wantung* and SS *Chi Ping* arrived from Chungking. General Kow and his troops immediately took the *Chi Ping* and departed for Foochow, leaving a detachment of troops to hold SS *Wanhsien*.

Yang Sen now had his *casus belli* in the *Wanliu* incident. One sunk sampan was magnified in conventional Chinese style into *two* sampans and fifty-six men lost, plus an alleged C$85,000 in specie belonging to Yang's paymaster.

Arriving in Wanhsien on 29 August from her downriver incident, *Wanliu* was immediately boarded by Chinese troops. They would hold her, they said, until reparations had been made for the lost sampans, men, and money at Yung Yang. Her foreign officers meanwhile were having a nasty time of it.

When Acheson heard of this, he sent a party which took back *Wanliu,* disarmed the Chinese, drove them off the ship and, for final humiliation, brought back their arms. *Wanliu* lost no time getting under way and skedaddling up to Chungking.

In view of what had taken place so far, it appears that Acheson could well have garrisoned *Wanhsien* and *Wantung*. But what happened next was so absolutely without precedent that anyone might have made the same mistake.

Just after dark, Chinese soldiers in force boarded both ships, declaring they would hold them until *Wanliu* returned or the losses claimed had been paid. Things were getting out of hand to the point the British consul from Chungking and HMS *Widgeon* arrived a few days later to add their weight to the negotiations.

The final conference with Yang Sen, on 2 September, broke up in a very hostile atmosphere with nothing decided. Meanwhile an estimated twenty thousand Chinese soldiers were streaming into the city, to take up positions along the foreshore, on the surrounding peaks, and among houses dotting the slope leading down to the river. Eleven guns of various calibers were spotted in five emplacements, one within fifty yards of the shoreline.

All during Saturday, 4 September, foreigners congregated at the supposedly neutral and relatively safe American oil installation, anxiously watched every move made by the British gunboats. They knew that the merchant ship officers were held hostage under threat of death if the gunboats made a provocative move.

It was arranged with Acheson that the blue Peter at the *Cockchafer*'s masthead meant trouble; that all foreigners were to come to the American installation. From time to time, black smoke poured from the warships' funnels, suggesting that a move was about to be made, but after a few disappointments of that sort, the watchers concluded that excessive smoke did not necessarily mean war.

About 2:00 P.M. on Sunday afternoon, 5 September, the gunboats upped anchor and *Cockchafer* drifted below the SS *Wanhsien,* which lay farthest downstream. HMS *Widgeon* moved to a position about a hundred and fifty yards above the SS *Wantung,* which lay farthest upstream and just ahead of the French gunboat *Doudart De Lagree.*

About 4:15 P.M., the foreigners at the American compound were having their tea on the steps, as usual watching every move of the ships, when a queer-looking craft rounded the bend some three and a half miles below. She was painted a vivid red, with a black funnel. "A Bolshevik!" someone joked. But when she came abreast the installation, it could be seen she was the SS *Kiawo,* with a naval crew on board. She had been commandeered by the British navy at Ichang, armed with pom-poms and machine guns and manned by seamen from the cruiser *Despatch* and the gunboats *Scarab* and *Mantis.* Commander F. C. Darley was placed in command both of "HMS" *Kiawo* and of the whole operation about to begin. All told, the British counted 110 officers and men against a Chinese field army.

Signals passed between the three warships, but if the blue Peter ever broke at *Cockchafer*'s masthead, nobody saw it. Things were far too busy. *Kiawo* steamed straight alongside SS *Wanhsien*'s starboard quarter, planning to have four boarding parties enter the four after-cargo doors, get on deck, and stand off the soldiers, but they had set a trap. Fifty or so of them were sitting on deck, peacefully eating rice. As the ships touched, all the Chinese instantly disappeared. Then a bugle sounded. The landing parties leaped aboard to be met with a murderous cross fire. Beginning at 6:28 P.M., the fight lasted close to an hour, the two ships lying alongside each other in a tradition as old as Salamis.

For the first few minutes, the British suffered all the casualties, then Chinese began to fall. Commander Darley, a revolver in each hand, leaped across to *Wanhsien*'s bloody deck, leading the second boarding party, got off two shots, and was riddled by bullets. Lieutenant Ridge, gallantly holding a passageway full of armed men, had fifteen wounds when his body was recovered.

As the *Wanhsien* was boarded, furious firing broke out all over. *Kiawo*'s pom-poms and machine guns on her free side fired at the beach to counter Chinese small arms. *Widgeon* and *Cockchafer* sent shells screaming into the city, searching out artillery positions which were dropping shells in the river, some very close to *Doudart*.

A hail of fire came from riflemen ashore. Acheson barely had given the order to open fire when a better than average Chinese marksman got him through the back. For over two hours this gallant officer lay on deck, directing operations of his ship.

After close to an hour of bitter fighting, repelling Chinese attempts to board her, *Kiawo* cast off, unable to recover Darley's body. Steaming past the silent French *Doudart,* she eased under the stern of the SS *Wantung,* 600 yards upstream. Contrary to expectations, the Chinese failed to penetrate the fortified part of the ship. The officers had not been captured and, taking advantage of the confusion at *Kiawo*'s approach, they attempted escape. The mate swam safely to *Doudart*. The captain, dangling over the stern by a rope's end, was rescued by *Kiawo*. The chief engineer jumped over the side and was not seen again.

Yang Sen, meanwhile, was taking no chances with *Cockchafer*'s 6-inch guns. The gallant general was a safe twenty miles away, in the tall timber.

At dusk, the gunboats ceased firing and withdrew five miles downriver, where *Kiawo* already had anchored.

The prime object of the mission had been accomplished: the release of the detained foreign merchant officers, although one had been drowned. Unknown to Darley, the *Wanhsien*'s officers had never physically been taken, and were able to jump across from their fortified bridge to *Kiawo*.

The Chinese staff of the *Kiawo* had performed admirably, especially the "black gang," shoveling coal in temperatures of 120 degrees. The Chinese chief assistant engineer (the chief was white), shook all the next day like a leaf, and burst into tears every time anyone spoke to him. He had been at it twenty hours without relief. The strain had begun to show.

The secondary objective, recovery of the seized vessels, was not achieved. *Wanhsien* and *Wantung,* which had been shelled by the gunboats after *Kiawo*'s withdrawal, were floating shambles. The city was ablaze in four points. Of the 110 Britishers, 20 percent were casualties. Out of 7 officers, 3 were killed and 2 wounded. Four bluejackets died and 13 were wounded.

The British troubles were not yet over. The SS *Kiangwo,* carrying the casualties, was stopped twice five miles above Hankow by terrific fire from the banks. Then, on the morning of the third day, along came the U.S. convoy, including the USS *Palos, Pigeon,* and *Stewart*. The latter's rough treatment of the Chinese machine gun in the pagoda had cooled their belligerence; *Stewart* loaded the British wounded aboard and reversed her course to Hankow without interference.

On recapitulation, the firing of *Cockchafer* and *Widgeon* appeared relatively restrained, considering what they could have done had they tried. Their cannonade started several fires and knocked out some Chinese gun positions. A Chinese tinsmith, working for Standard Oil, reported to his American boss the morning after the bombardment that one shell had demolished a gun near his house and a gunner's leg had been blown through the wall. He seemed to think it was a great joke on the soldier.

The shelling had held down Chinese small arms fire, so that British casualties were far less than might have been expected. One 6-inch shell had gone howling down a long street, not hitting anything on the way, but sucking down jerry-built houses in clouds of ancient dust. "Shell no hit, Chinese man die anyway!" excitedly described a Chinese witness.

The reverberations of *Cockchafer*'s big guns were not confined to the amphitheatre of Wanhsien. They echoed all the way along the Yangtze and as far south as Changsha. At Chungking, fellow Szechwanese feeling ran high. With no British gunboat present, foreign protection devolved on the USS *Monocacy*. In December 1926, Secretary of State Kellogg received a letter of thanks from the British ambassador, saying that, "The attention of His Majesty's government had been drawn [to the fact that] . . . looting of Lungmenhao on September the 18th last, and the complete evacuation of the city [by] British subjects . . . was only prevented by the USS *Monocacy* shifting anchorage and clearing for action. . . ."

The action at Wanhsien was the turning point of foreign military influence on the Yangtze. Perhaps even more importantly, it drew worldwide attention to a situation which, before then, had either been passively accepted or beyond general knowledge. From then on, anything that happened on the Yangtze would be big news on the American front pages: the Nanking outrage, the *Panay* sinking, the Rape of Nanking, and many more. *The Nation* felt the days of the foreign gunboat were numbered, ". . . for the new China will not tolerate such intrusions of Western power into the heart of the country." *The New York World* said that foreign gunboats on the Yangtze had always been an anomaly and felt the question had arisen whether all foreign gunboats were not more dangerous as an irritant than valuable as a police force. *The New York Times* contented itself with stating the facts, without taking sides: "The business element among these American citizens . . . has always been glad of the presence of . . . warships . . . while missionary sentiment is divided."

The *Boston Herald* reflected the spirit of '76. "A state of war dominates 1,200 miles of the Yangtze . . . where between Shanghai and Wanhsien there are now about 50 foreign warships. . . . Their mission is to teach respect for the rights of all, to preserve peace, and to promote friendship." As for the friendship bit, the *Herald* would have done well to read the posters on Wanhsien's streets, offering C$100 for the head of *any* Englishman, in the absence of amplification, presumably dead or alive.

The poor missionaries were damned if they did and damned if they didn't. At a meeting of the National Christian Council, the central organization of Protestant missions, the delegates voted unanimously in favor of removal of special treaty privileges for missions and churches. But to what avail? The radicals worldwide charged that this was all a smokescreen; that the cynical missionaries were simply the "running dogs" of the "imperialistic foreign powers."

The *China Weekly Review* of 13 November 1926 got right down to basic fundamentals with the most pragmatic summation of all:

> *Wanhsien showed that the present type of gunboat used in China is becoming obsolete. A little tin gunboat on a narrow river is no match in a fight with a Chinese army equipped with modern heavy artillery. Unless the foreigners build heavier and better protected gunboats, a difficult thing to do in view of the shallowness of the rivers—the Chinese are shortly [if they are not ready now] going to drive the gunboats off the river anyway.*

During the next two years, the United States would commission six "tin gunboats" which would live out another decade and a half. But the *Weekly Review*'s thesis was right; when Chinese artillerymen learned to shoot, as did the communists by the end of World War II, time had run out, not only for "tin gunboats," but for all foreign warships on the Yangtze.

The Wanhsien affair of 1926 reminded Americans that their gunboats were getting shot at on the Yangtze, in efforts to preserve the peace of the river against the depredations of over-ambitious bandits. The "Nanking outrage" of 1927 marked the first time in history that U.S. guns spoke out in anger against organized Chinese forces along the Yangtze, and aroused Americans and the press, overseas and at home, to vociferous consideration of American interests in China.

"US WARSHIPS IN ACTION ON THE YANGTZE!" The headlines of late March 1927 were bigger and blacker than any since the USS *Maine* blew up in the harbor at Havana in 1898. World chancelleries uttered profound assessments—those of nations having interests in China expressing strong approval.

The events leading up to that climactic day of 24 March 1927 began some four or five months earlier and produced a military situation only comparable to half a dozen schizophrenics wearing boxing gloves while simultaneously playing Chinese checkers and a vigorous game of musical chairs to background music of exploding firecrackers. For a full understanding of what transpired, it is first necessary to examine the existing background and precedent for the presence of U.S. warships on the Yangtze. Among the various nations involved in China, there was an ambivalence over the use of armed force. With the exception of the destruction of the Bogue forts below Canton in 1856 by the USS *Plymouth,* ships had limited themselves to mild exchanges of small arms fire with exuberant soldiery on the Yangtze's banks. The velvet glove remained tightly buttoned over the mailed fist. Before 1840 particularly, the American government assumed toward its citizens resident in China an attitude not very dissimilar to that taken by the Chinese toward their own emigrés: let them shift for themselves.[39]

This policy suited most American merchants right down to the ground. For a century and a half they had seen their British competitors hog-tied for a year at a time by highly effective Chinese boycotts or general strikes following some display of

British military or police force. In some cases these boycotts actually helped U.S. merchants reap sizable benefits in filling the vacuum. On the other hand, one had to admit that British arms, blowing down the gates with round shot, had got foreign merchants' feet in the door and kept them there. One doubts that Commodore Perry's tactics would have opened doors in China as it did in Japan.

The humiliating surrender of British concessions in Hankow and Kiukiang had resulted in an anarchic situation of shuttered businesses, closed banks, embargoes on currency movement and ill treatment of those doughty foreigners not yet evacuated. Even the British home government had tired of turning the other cheek, and by 29 March British Foreign Minister Austen Chamberlain, as reported in the *North China Daily News* the following day, had registered a change of heart: "Britain must be assured," he told the House of Commons, "that the conditions are such that the Chinese authority established could preserve order, whenever that authority was extended, before she made any further surrender of British interests and property."

Comparisons between Hankow and Nanking were striking. Hankow was the city of sleek *taipan* fat cats, with their tennis and wood oil, bridge parties and cotton piece goods, golf and hog bristles, paper chases and duck egg albumin, masquerades and wolfram. There were cabarets, clubs galore, and orderly, well laid out, foreign-style river front settlements.

Nanking, opened formally to foreign trade thirty-eight years later than Hankow, had no foreign concessions or anything even remotely resembling such. From a foreign point of view, Nanking was a missionary monopoly, the center of all American missionary effort in China. There were more Americans in Nanking than in Changsha, Ichang, Chungking, and all points west put together. With the exception of a handful of consular personnel, oil and tobacco people, all 430 of them were missionaries. They staffed the Nanking University, lesser schools, hospitals, the Chinese YMCA and YWCA. Scattered about the city were "prayer centers," a sort of filling station for the spiritual gas tanks of wayfarers.

The American naval forces at Hankow, by virtue of limited U.S. mercantile and personnel involvement in the area, had been fringe observers and minor participants. At Nanking, the Navy had a proportionately far greater responsibility, in view of the great preponderance of U.S. nationals, and in spite of basic missionary reluctance to accept foreign military protection.

Nanking, and the country for a hundred miles south and east, was held by troops of elderly Marshal Sun Ch'uan-fang, who was given to fits of melancholy, threatening to retire and write poetry. Looking over Sun's shoulder literally and figuratively was a huge, crude six-foot-two ex-coolie, Marshal Chang Chung-ch'ang, "Warlord of Shantung," who had moved south across the Yangtze to Nanking on 18 December 1926. A veteran of the Russo-Japanese War on the Russian side, Chang was described as "a man of imposing appearance, commanding respect more through his physical than his mental qualities."

A clutch of smaller bore generals was scattered around the interior north of the Yangtze between Pukow and Tientsin, monitoring the Sun–Chang relationship like a pack of jackals. To the west, in Honan province, lay Marshal Wu Pei-fu, licking his wounds after suffering a debacle at the hands of the Southerners at Wuhan and now awaiting any decent offer.

Warlord Chang was for the moment and for convenience a junior partner of another Chang, Manchurian *Super*warlord Chang Tso-lin, famous for his White Russian harem and partiality to tiger-blood-and-gunpowder cocktails. He gave up the cocktails after discovering the nitrate in the gunpowder was an anti-aphrodisiac. Chang Tso-lin, who several years before had ousted pro-British Wu Pei-fu from Peking, was on the Japanese payroll and thus anti-Chiang Kai-shek. "The Fengtien* leaders wish to use Sun and he them," reported Consul Davis, "and both sides wish to expend the military power of the other and conserve their own."

Chang Chung-ch'ang having, camel-like, stuck his head into Sun's tent, wasted no time moving wholly inside and crowding Sun into a corner. He urged Sun to use his full strength against Chiang Kai-shek, presenting the powerful argument that Chiang was slowly but surely buying and fighting his way closer and closer to key Shanghai, source of Sun's immense opium revenues. Chang cheerfully promised to guard Sun's rear while Sun was off tangling with Chiang, something very much like leaving the fox in charge of the chicken coop. As Davis put it, Sun ". . . has climbed aboard the Fengtien horse and cannot dismount." It would have been more accurate and charitable to add that Sun's climb had been encouraged by a sharp stick.

By Christmas, 1926, Chang's Northern troops were winning friends and influencing citizens on Nanking's streets, ". . . with their conspicuous yellow arrows [on a brassard] conferring authority for summary execution, and a squad of infantry followed by four executioners, with their broadswords across their backs." A hundred and twenty White Russian mercenaries body-guarded Fengtien fatboy Chang, whose fourth and fifth concubines had been brought to Nanking by the end of December, giving the whole establishment an air of genteel permanence.

The month of January was one of classic Chinese warlord military milling around, regrouping, feeling out the opposition and testing frail loyalties. A missionary up from communist "Indian country" reported to Consul Davis that the Southerners were ominously associating their northward "liberation" march with the Taiping Rebellion of 1850–64, believing that that bloody movement had been crushed at the Yangtze only by the assistance of "imperialistic" Britain,† and were seemingly apprehensive that history would repeat itself. The recent landing of Indian troops at Shanghai and the announcement of a British division to follow tended to substantiate the Chinese fears. The missionary informer, on his part, was apprehensive too—over the propaganda being scattered around in the mission girls' colleges in Kiangsi to the effect that the girls should perform their "patriotic duty" by intimate surrender to the Southern officers.

* Fengtien was the south-central province of Manchuria, with Mukden and its mile-square arsenal as the home base of Chang Tso-lin.

† The closing phase of the Taiping collapse was materially helped along by British military action and British General Gordon, commanding anti-Taiping troops.

During this semiquiescent period of Chinese spitting on hands, rubbing rosin on shoe soles, and sizing up the opponent, the Nanking roadstead was crowded with shipping. From mid-December 1926 until the end of February, Nanking harbored in succession the U.S. destroyers *John D. Ford, Pillsbury,* and *Simpson*. The ubiquitous British had been there at one time or another with HMS *Witherington, Petersfield, Vindictive, Carlisle, Wild Swan, Wishart, Emerald, Gnat, Veteran, Caradoc,* and *Verity* —almost everything except battleships or carriers. The French were there in *Marne*, the Italians in the little *Ermanno Carlotto*, and the Japanese in *Hodero, Shinoki, Momo,* and *Katata*.

Lieutenant Commander Roy C. Smith, Jr.,[40] commanding the destroyer *Noa*, arrived there on 27 February 1927 to relieve *Simpson* as station ship, and described Nanking:

> *The city is surrounded by a high wall and is shaped something like a pear, with the northern as the stem end. This end has two gates which are about a mile . . . from the river, with the commercial suburb of Hsia Kwan between the wall and the shore. A sort of canal or lagoon, passable for small junks and sampans, runs completely around Hsia Kwan. The southern end of Nanking is the fat end of the pear and contains the Chinese city proper and its farther gate, the South Gate . . . perhaps eight miles from the river. About the lower middle of the pear are located Nanking University, the largest of the mission schools, the General Hospital, and most of the other missions and schools. The Japanese consulate is next to the University. The American and British consulates . . . are about three miles and a mile and a half respectively from the gates nearest the river.*

During the murderous reign of the Taipings, Nanking's population of a million had been cut in half. In 1927 the north end of the "pear" was largely open country. Stone bridges across ditches showed where teeming streets once had run. These wide spaces had become cultivated fields or empty wasteland—or reedbeds where masses of game birds congregated.

In 1905, Yates Stirling, Jr., on a visit to Nanking, borrowed a horse from the American consul and went to hunt pheasants in the jungle that had once been a city. Lost, he finally found an old peasant and tried to explain his predicament in sign language, as he spoke no Chinese. The old Chinaman spoke no English, but he understood horses. He took the reins out of Stirling's hands and laid them on the horse's neck and gave the animal a resounding slap on the rump. In ten minutes, the horse had delivered Stirling, safe and sound, at the consulate.

The Yangtze at Nanking could get remarkably choppy. Once ashore, there was the sticky problem of transportation—half a mile to the nearest water hole, the Yangtze Hotel; another half mile to the city gate, plus three quarters more to the club. But for those hardy souls who made the effort, the rewards in hospitality were heart-warming. The American consul, John K. Davis, was not only an exceedingly amiable fellow but one of the very ablest diplomats in China, if one can judge by the lucidity of his reports and as time proved, the correctness of his information and deductions. Smith wrote that Mrs. Davis made the consulate "a real home away from home, most attractive to all of us."

There were several other British and American families who manifested the cameraderie that mutual dependence, danger, and loneliness fostered among the professional people of all small foreign colonies in the Yangtze port cities. Such were the T. L. Macartneys of British International Export, the F. C. Jordans of British-American Tobacco, and the E. T. Hobarts of Standard Oil. Mrs. Hobart became well known to many Americans as Alice Tisdale Hobart for her book, *Oil for the Lamps of China.*

For many reasons, mostly of their own making, the missionaries formed a group apart. They turned out in force only on national holidays, to mix a little awkwardly with fellow westerners at the consulate receptions. Or, most reluctantly and at the very last minute, for evacuation when total disaster threatened.

British Consul General Bertram Giles was of a distinguished China service family. His house, the consulate nearest the river, with its hospitable hostess and two charming secretaries, was a favorite gathering place for the younger officers.

In ports like Wuhu and Kiukiang, the Standard Oil people had set aside little rooms in their compound for the use of the gunboat enlisted men, where Peking Five Star beer was on ice. But in Nanking, there was absolutely nothing for a sailor to do or a place to go. At best, he could barter for some small souvenir that might fit his limited stowage space on board, or take a few snapshots. The few antiquities to be seen were miles away, in or beyond the city proper. Rickety motor cars were not cheap to hire, even when one could be found. So *Noa*'s sailormen could only improvise, as once did Chief Quartermaster Horn and Signalman First Class Wilson. "CQM Horn and I went ashore one afternoon looking for something to eat," Wilson recalled some forty years later. "Destroyer food was not so good when you had been away from suppliers for so long. We went into a compradore's shop and found that we could get a fairly good steak broiled for us. We asked for wine, so the old boy took us down in the cellar, which stank to high Heaven, and showed us six bottles of Pommery champagne. It was dusty, covered with cobwebs, and with a date that would bring tears to your eyes if you are a drinking man. We bought the lot for a dollar mex a bottle and had the compradore put them in a tub filled with ice. Our steaks were soon ready, but we paid not too much attention to them even though they were good. We were anticipating the cooling of the next course."

The only glasses available were water tumblers, but had they been Baccarat crystal stemware, the results would have been equally predictable. Wilson and Horn discovered next day that life can be a burden, directly related in weight to the vintage of the night before.

On 7 February, Marshal Sun delivered himself of some very enlightening views to Consul Davis. Sun felt it certain that the Soviets were out to destroy British interests in China—which the British must prevent or lose India and Malaysia. Russia was the real enemy of all civilized powers, said Sun. He regretted that China had ever recognized her. On the subject of the KMT, Sun demonstrated either a shocking weakness in evaluating Chiang Kai-shek or was voicing some wishful thinking: the KMT had degenerated, he thought, until it no longer was a Chinese campaign but a forlorn hope of the Third Internationale.

Near the end of February, Sun was interviewed once more by Davis. Sun's forces had taken some severe shellackings since the last interview. Sun was convinced that the Southern forces were entirely under Soviet Russian direction. Captured field guns and small arms were all Russian supplied, he said. Artillerymen and aviators were Russian and prisoners' military orders had been signed by Soviet General "Galen." Marshal Sun looked ill, Davis thought—and small wonder.

As to the ultimate outcome, Sun felt he was reading the tea leaves with accuracy. The Russian plans would fail, he predicted, because the Chinese would awake to the fact they were being betrayed. But he was afraid that unless the other Chinese leaders and the foreign powers could be promptly aroused to the real object of the Soviet, enormous losses of Chinese life and property would have to occur.

Chang Chung-ch'ang, apparently convinced either of Sun's loyalty or of his inability to dismount from the Fengtien horse, had spent most of January and February 1927 north of the Yangtze, leaving Sun to run his front as he saw fit. Neither fears of a Sun double-cross nor the charms of the fourth and fifth concubines cut as much ice with Chang as did keeping his finger on shifty subordinates sitting astride his lifeline, the Tsin-Pu (Tsinan-Pukow) railway.

On 27 February 1927, when *Noa* anchored at Nanking, Lieutenant Commander Smith found that his arrival had coincided with a marked change in the Chinese scenario. Some of the Sun's generals, key men from Anhui—the province which straddled the Yangtze between Wuhu and Tungting Lake—had shown distinct signs of selling out to Chiang Kai-shek. Sun's front, south of Hangchow, the gate to Shanghai, had collapsed and the whole defense line had to be hastily pulled back. This startling new development led Consul Davis to sound the very first note of alarm regarding Nanking: "Provided the Anhui generals have unreservedly thrown in their lot with the Kuomintang . . . it is probable that the KMT will try to capture Nanking either by direct attack or, more likely, that they will endeavor to drive a wedge between it and Shanghai. In either event it is likely that military operations are impending which will be of very serious portent for the safety of all American citizens in Wuhu and Nanking."

Chang Chung-ch'ang's estimate of the situation evidently coincided precisely with that of Consul Davis; the marshal hotfooted it back south across the Yangtze on 23 February to find his viceroy Sun hanging on both the spiritual and military ropes. Three days earlier Sun had met a deputation of leading Nanking citizens who had urged him to keep a stiff upper lip and hold tight. No doubt with tears in his eyes as big as grapes, Sun informed them that in spite of his constant care for the welfare of the people he had met ingratitude and treachery on every side. He would hand over the whole affair to Marshal Chang when the latter turned up, retire and get on with his poetry. Then, it is alleged, the marshal broke down and had to be assisted from the room.

The arrival of Marshal Chang and a bevy of Northern generals, who straightaway moved into Sun's yamen for a two-day conference, evidently put some starch in Sun's backbone; on 26 February, the pair of marshals jumped on the train for a quick trip to Shanghai. They were back in Nanking next day, to witness the arrival of some further stiffening material in the form of an armored train (plus one more to follow), and the day after that, the arrival of the 106th Regiment of White Russian mercenaries, a hundred of its horsemen immediately being added to Chang's personal bodyguard.

Chang's earlier importation to Nanking, his "yellow arrow" boys with their executioners' broadswords strapped to their backs, were small potatoes compared to the twenty thousand Shantung troops which had swarmed across the Yangtze at Chang's heels. Many were former bandits, like Chang himself. They slapped Nanking civilians around "with disgusting brutality," according to Davis, commandeering private property without recompense. Even more painful than the slaps were the kicks the merchants got in the pocketbook, one of the tenderest parts of a Chinaman's anatomy; the Shantungers packed in bundles of paper currency, stuff even more worthless, if possible, than most Chinese provincial paper money of that day. Many of the rice shops promptly put up their shutters to escape having to accept it. The buses stopped running. Chinese business, such as it was, ground to a near halt.

On 3 March, Davis, *really* worried, wrote a memo to Skipper Smith suggesting that although he didn't anticipate any disturbance, he thought it would be a good idea to perfect without delay any plans he and the captain of the *Emerald* might have. The Chinese civilians were on the anxious seat, he said, over the possibility Northern troops might be chased out of Nanking, which would include their usual pattern of looting the city as their final beneficial act.

At this juncture the war was beginning to strike home; Smith wrote in his 17 March weekly report to ComYangPat that food was getting short, the ever faithful compradores themselves being unable to scare up anything edible either locally or from Shanghai.

In early March, Sun had made another futile effort to climb down from that Fengtien steed, reportedly having had his resignation as military governor of Kiangsu refused by Superwarlord Chang Tso-lin. Thus, lacking any means of graceful exit and in a classic Chinese example of the *double* double-cross, Sun was believed (by Davis) to be biding his time until he could join hands with his now seemingly disloyal Anhui subordinates, then *both* would turn on the Fengtien (Chang Chung-ch'ang) forces and throw them out of the Yangtze Valley. Connected with or possibly only synchronized with this Machiavellian plan, a coalition of Chihli Young Turk generals, also operating in Anhui, probably with some sort of understanding with Chiang Kai-shek, *also* hoped to fling the Fengtien crowd out of Kiangsu and Anhui.

As a result of all this hanky-panky, the Nationalists had been able to follow Davis's predictions almost to the letter; in a 6 March message, the manager of Standard Oil at Wuhu informed his Shanghai boss that, "Wuhu and the governor of the province [of Anhui] went over to the South Sunday morning very quietly."

Things were not quiet for long. The British river steamer SS *Kutwo* had collided with and sunk a Chinese troop-carrying launch. Chinese military threats to seize *Kutwo* brought the destroyer HMS *Wolsey* alongside the ship to resist by force if necessary any such attempted take-over. The USS *Preble* was ready to stand by for evacuees. Chinese feelings were further ruffled when British bluejackets ejected from an old British hulk some students staging a welcome demonstration for the Cantonese. The skipper of the *Preble* radioed that on the afternoon of 8 March, a mob of about ten thousand wrecked the Wuhu customs club and the Maritime Customs installation and hoisted the KMT flag over the remains of the club. American women and children went aboard *Preble,* and other foreign nationals took refuge in the hulk which was owned by Butterfield and Swires.

Consul Davis, in whose territory Wuhu lay, let leak some of the scorn he felt for the Cantonese. In a message to Consul General Lockhart, at Hankow, Davis asked him to protest to the "Kuomintang so-called government" over the abuse of American property at Wuhu, which included stabling horses in a mission church.

The British cruiser *Emerald,* which had been at Nanking since mid-January, at once steamed to Wuhu to back up *Wolsey* until she was relieved by *Caradoc.* Meantime, after the old Yangtze warship custom of "you scratch my back and I'll scratch yours," the British consul general at Nanking requested that in the temporary absence of all of His Majesty's ships from Nanking, the *Noa* would undertake to keep an eye on British interests, principally Jardine, Matheson steam launches at which the Northerners had been casting covetous glances.

All the concentration on Wuhu had to some extent diverted attention from a fascinating 28 February *denouement* at Nanking: the seizure by Northern Chinese naval units of the Soviet Russian steamer SS *Pamyat Lenina* (Memory of Lenin), presumably in innocent passage up the Yangtze.

Rumor had it that ammunition and some very interesting documents were found buried under coal in the bunkers. But the biggest catch of all was none other than Fanny Borodin, alias "Miss Grossberg," wife of Chiang Kai-shek's erstwhile Svengali, Michael Borodin. The latter was still at Hankow, busy exercising his manifold conspiratorial talents as one of the chief schemers of what Davis was later to designate "the Kuomintang so-called government."

Fanny was an international hot potato. Her seizure might cause the Soviets to do something unpleasant along Chang Tso-lin's feebly protected Manchurian border. She was also the subject of much hilarious speculation by the gunboaters and the foreign colony ashore, to whom "Borodin" was considered a dirty word. She had been flung to the tender mercies of Chang Chung-ch'ang's White Russian mercenaries for their personal amusement, some said hopefully. No, she had been sent north to Peking as a candidate for old tiger blood swiller Chang Tso-lin's harem, said others. Anyone who saw Fanny's picture in the February 1927 *Asia Magazine,* would not have hesitated to wash out both possibilities.

As a matter of prudence, foreign colonies up the Yangtze and the station ships of the various powers all had evacuation plans ready for use. They were rarely needed, but if such a time came, those threatened were grateful for their existence.

"While there is no shadow of danger to foreigners in Nanking," bravely began a

November 1925 plan prepared aboard USS *Pigeon,* "it is thought wise to put into the hands of Leaders of Sections the plans . . . to use in an emergency."

There was no desire to panic the civilian rank and file: "It is requested that you keep these plans under lock and key; that the contents be not divulged to those in your section, as they would cause needless questionings and alarm now." The thirteen section leaders were supplied key lists and thrown a bone of sustaining praise: "In the minds of the committee it was felt that in case of sudden emergency you could execute the plans quietly and effectively . . ." Then followed section listings of from ten to forty men, women and children, their rallying points and leaders. As a matter of interest to those not well grounded in missionary statistics, there were 14 "doctors," 10 "reverends," 61 "missuses," 79 spinsters, 112 children, and 27 just plain "misters." This did not count the Roman Catholics, who had their own private panic plan.

In terms of live bodies, American interests, for a change, were greater than those of the British at Nanking, owing to the battalion of U.S. missionaries. But here, as elsewhere on the river, by virtue of habitually having the senior ship present, the British managed to exercise their policy of Yangtze Valley preeminence by formulating the various evacuation plans. Thus, HMS *Concord*'s 1925 plan was the basis on which Captain England, of HMS *Emerald,* Lieutenant Commander Smith, of USS *Noa,* and Commander Yoshida, of HIJM Desron 24, agreed to cooperate.

Estimated foreign forces available for landing were 11 officers and 256 men, with 14 Lewis machine guns. This broke down to 7 British and 2 each Japanese and American officers; 166 British (2 seaman and 1 Royal Marine platoons) enlisted, 40 American and 50 Japanese bluejackets.

Once on shore, all troops would proceed to the New Gate in the Nanking wall. If the gate was open, march through. If closed, blow it down! Hold the gate with two Lewis gun sections of the Royal Marines. Here the remaining Britishers would go to their consulate to act as escort from there back to New Gate. One officer, twenty men and two Lewis guns of the U.S. force would occupy Socony Hill. The rest of the Americans would go to the U.S. consulate as escort. One Japanese officer and twenty men would proceed to the Japanese consulate for escort duties and the remainder stay at the New Gate in reserve. The rest of the Lewis gunners and half a platoon of British stokers were to be posted close together in a position to command the bund and large open space opposite the landing place.

The plan was basically all ready during March 1927 but it was set up with the assumption the opposition would be mobs of unorganized coolies who would turn tail and disappear at flank speed if somebody so much as dropped a light bulb. What turned up instead were tens of thousands of professional Cantonese troops, fresh from a victorious march of a thousand miles, heady with success and powerfully propagandized into hating the foreigner.

By mid-March, the merry-go-round had gone half a turn at Nanking. Marshal Sun Ch'uan-fang had at long last managed to dismount; he and his troops were headed home to Kiangsu province. There, they would be in a convenient bargaining position, either to defect south to Nationalist Chiang, or threaten the left flank of the Northerners and chief tormenter, Chang Chung-ch'ang. No doubt with this in mind, Chang had prudently withdrawn his headquarters to Hsuchow-fu, two hundred miles north of Nanking, turning over immediate command of the front of General Chu Yu-pu.

The warship crews had watched with interest as General Chu's men built railroad tracks down to the foreshores and with the usual magnificent Chinese ability to improvise, plus thousands of strong backs, had chivvied two complete armored trains south across the Yangtze. Besides the trains, tens of thousands of well-equipped Northern troops had been ferried over to Nanking. It looked like they meant to put up a real scrap to save the city.

As late as 18 March, Consul Davis felt the Shantungese were holding, or even gaining, ground. The front was forty miles away. Students at Southeastern University caused some disturbance by tossing a small bomb, but the Northerners soon put a stop to *that* foolishness by occupying the campus with a thousand troops and decapitating two of the more troublesome lads.

On 19 March, Davis commenced to revise his thinking. Rumor now had the Southerners only sixteen miles away—so close, in fact, that Davis had alerted a mission in the southern part of the city to put a watch on the roof of their tall tower to cock an ear for gunfire.

The next day things were confused all over. But Davis had not given up hope and had even suggested to the Legation in Peking that they cool off the newshawks, whose alarmist messages were panicking Stateside relatives of the local missionaries.

By 21 March, it began to look like the newshawks were not so far off the beam after all; even from the American consulate itself star shells could be seen. One of the armored trains, manned by White Russians, had been spooked on the way to Chinkiang and was rattling back toward Nanking. At Chinkiang itself, Northern troops were leaving—scrambling across the Yangtze, toward that good old north-pointing Grand Canal.

Actually, the Cantonese had made rapid advances on all sides and by the afternoon of 21 March a complete breakdown of the Northerners' front was evident. Consequently, the British and American consuls decided to evacuate their nationals and informed the Japanese of their plans to commence at six-thirty the following morning.

Monday afternoon, 21 March, Ensign Woodward Phelps and ten men landed from *Noa* and went to the American consulate. It was felt provocative for this small party openly to carry arms, so five .45 caliber automatic pistols were hidden under coats. The rest of the "artillery" was available at the consulate—two dozen .30 caliber Springfield rifles. The wisdom of not carrying arms soon became evident; as Phelps and party, in three motor cars driven by American civilians, passed through New Gate, they saw fifty armed British Marines halted outside. There, the frustrated British had spent half the night in an unsuccessful attempt to palaver their way past the Chinese gate detachment, then had gone back aboard ship. The next day, nineteen British Marines, a couple of signalmen, and two British officers insinuated themselves into the city individually, and reached the Britsh consulate.

The evacuation began as scheduled and by noon, 102 were aboard *Noa* and 73 aboard the U.S. destroyer *Preston,* which had arrived that morning from Wuhu.

That same morning, John D. Wilson, a *Noa* signalman, had been established in the Standard Oil house on Socony Hill, where a rooftop platform gave him visual communication with the American consulate and the ships. On *Preston*'s arrival, two of her signalman, D. Taylor and H. S. Warren, joined Wilson. Their platform was an ideal target for snipers on the land side, and all three men were later awarded the Navy Cross for their heroic performance under fire.

The next day was relatively calm ashore, although many Northern troops passed to and fro near the consulate. With complete evacuation of those who wished to go, the various guards "stood down" and took it easy, although at night heavy rifle and machine gun fire was heard to the south.

Wednesday, 23 March, was quiet enough in the forenoon, but the firing was coming closer. Smith of the *Noa* and G. B. Ashe of the *Preston* made the rounds ashore and found all peaceful at the consulate and Socony Hill. Nevertheless, with the disorganized state of the Northern soldiery, Ensign Phelps felt they might go on a looting binge before the final pull-out. Farthest from their thoughts was the possibility the victorious Cantonese would harm either foreigners or their property. Consequently, to improve their position against disorganized mobs, he drove down to the dock and picked up a Lewis machine gun, wrapped it in canvas, and drove back to the consulate without interference.

By midafternoon, Northern troops began pouring in disorganized torrents through the city, in full retreat to the waterfront. Traveling fast, General Chu Yu-pu lost no time making himself scarce on the south bank. No sooner did he and bodyguard board a ferry than it got under way in a fusillade of shots between the bodyguard and enraged soldiers left on the bank. It was estimated that seventy thousand retreating Northerners were ferried across the river during that hectic afternoon and night.

Phelps watched the soldiers and walking wounded pour by all afternoon in what looked like a general rout. At 7:00 P.M. he posted sentries, with orders to fire the first shot in the air, the next to kill. It was a sleepless night for all. Unidentified Chinese tried three times to climb over the wall of the consulate compound, but were frightened off. A Northern Chinese officer, wounded and robbed at the back gate, was dragged inside by the Americans to have his wounds dressed, then put out once more, crying piteously for help until he died. It appeared a heartless act, but had the Cantonese found the Americans harboring the man, the results easily could have been disaster for the whole small group. Three Chinese were executed outside the gate early in the morning—whether KMT sympathizers or deserters no one knew.

Against a background of rifle and machine gun fire, the consulate received word about 8:00 A.M. that the Japanese consul had been killed and his consulate looted. Miraculously enough, the telephones were still operating and the electric power still on. About the same time, scattered groups of Cantonese began appearing, and Consul Davis engaged several officers in conversation. That big American flag over the gate was not necessary, they said. The place would be quite safe. On Davis's advice, Phelps stowed his guns out of sight to avoid fracturing this delicate aura of peace.

By 9:45 A.M. things looked far less rosy; in spite of the nineteen Marines, the office of the British consulate general had been looted and the consul general himself reported killed. The missions and schools, where many of the men and some of the women with their children had refused evacuation, were having, if possible, an even rougher time of it. Obviously, much of what was picked up at the consulate was distorted by the prevailing excitement. A month later, when all reports were in, Admiral C. S. Williams, CinC Asiatic Fleet, in his letter to the Chief of Naval Operations, described the ordeal:

> *. . . The British, American and Japanese Consulates were apparently singled out for attack. Nationalist soldiers in the uniform of the 6th Nationalist Army under the command of General Cheng Chien, and also apparently in accordance with a prearranged plan, descended upon all foreign buildings and homes and committed all manner of outrages. Men, women and children were insulted, threatened, shot at, robbed and stripped of their clothing. The Japanese Consul, ill in bed, was deliberately shot at twice. Two members of his staff were wounded. Two British civilians [Dr. L. S. Smith, port doctor; Mr. Frederick Huber, harbormaster] were killed during the attack on the British Consulate.*
>
> *The faculty of the Nanking University and other missionaries were subjected to the greatest indignities. Dr. J. E. Williams, vice president of Nanking University, was deliberately murdered, Miss Anna E. Moffat . . . of the . . . Presbyterian Mission was purposely shot twice through the body by a Nationalist soldier without cause or provocation. . . .*

At the American consulate, reports from the Japanese and British consulates, which were closer to the Cantonese point of entry than the American, indicated that it was advisable to move; their turn was only a matter of time. Davis and Phelps decided to head for Hobart's, on Socony Hill. They broke out arms, left packs and bedding rolls, and set out about 10:00 A.M. with Consul and Mrs. Davis, two young children, and five missionaries who belatedly had decided to join them. A coolie carried two suitcases. The missionaries, after considerable grumbling, consented to lug the machine gun, broken down into several more portable and less warlike looking parts. Over all floated the Stars and Stripes on a pole carried by a missionary—a miniature version of Washington crossing a dry land Delaware.

They had covered only about two hundred yards of the mile and a quarter of open, hilly country when they met an armed soldier. The consul had explained who and what they were and the soldier affably waved them on. They had gone fifty yards when the same fellow, whose aim was as erratic as his disposition, fired on them. From that time on, it was a game of hare-and-hounds, with two or three soldiers in the rear of the Americans potting away, their accuracy fortunately no better than the first fellow's.

For reasons either of fright, fatigue, or wariness of being tagged a partisan, the missionaries toting the machine gun surreptitiously dropped most of the parts, including six pans of ammunition. Meanwhile the coolie had been shot and the Davis baggage liberated. They had covered almost three-quarters of the distance when the law of averages gave a Chinese soldier a hit. Fireman third class Ray D. Plumley, the only U.S. naval casualty in the whole "Nanking war," was wounded in the back. This was too much; the sailors fired and wounded or killed two of their tormentors. The remainder took the hint and abandoned the pursuit, and in a few more minutes the party was safe at Hobart's. Or were they?

At Hobart's, there were about thirty men and Mrs. Hobart. A few prospective looters had turned up, but Mr. Hobart had outpointed them in debate and they had left emptyhanded. With the arms once more stowed out of sight of any inflammable and covetous chance Chinese visitor, all hands settled down to watchful waiting. About 1:00 P.M. the pattern became clear; other houses in the area were entered and thoroughly looted. Soon, soldiers in twos and threes were seen in the near vicinity. Those coming to the Hobart house were met at the gate by Hobart, Davis, or vice consul Hall Paxton, all of whose lives constantly were in danger, with rifles at their chests and no-nonsense expressions on the faces of the brutish-looking coolie soldiers. Meanwhile, more and more Chinese were knocking on the gate. Money was running low to buy them off. Groups of fifteen to thirty were commencing to clot up. It became clear that loot was not the sole object: "We are Bolshevists!" one of the soldiers told the consul. "We are proud of being Bolshevists, and we are going to act like Bolshevists." Clearly, time was running out.

In a final, ominous encounter, Consul Davis gave the soldiers the last of their money, rings, watches, and spare clothing. There was nothing left to buy off any more. "It's no use. Get your arms," he said.

Somewhere along the line, there had been time to clean the cosmoline from the consular Springfields. "We emptied many quarts of [Hobart's] scotch whiskey into a bowl and cleaned the bolts with it," Signalman Wilson remembered forty years later, lamenting that ". . . the whiskey had a slightly oily taste after the job was done." To reach his perch on the roof, Wilson had to climb out a bathroom window. One of the larger missionaries had elected to armor plate himself by lying in the dry tub, hindering use of the window. Wilson just turned on the shower and he got right out.

Not only were Signalmen Wilson, Taylor, and Warren exceedingly brave, they were essential. Without them it is doubtful that the trick could have been turned at Socony Hill. From the time Wilson took up his post on 21 March until the final evacuation, he handled practically all communications between the foreigners ashore and the ships in the river, until he was joined by Taylor and Warren from *Preston* on 24 March. During that last day, snipers' bullets continually whistled by the platform from which they worked. Chief Quartermaster Charles W. Horn did most of the signaling on the *Noa* and was continually exposed to sniping.

The signalmen, although under fire the entire time, maintained a log in regular Navy style, and its stark phraseology was more eloquent than flowing prose:

24 March

Noa from *Socony*	We hear heavy rifle and machine gun fire in city but everything quiet here. 0200

Noa from *Socony*	We do not know for sure just how much of city is occupied by Cantonese but there seem to be large bodies of soldiers still entering city. 0215
Noa from *Socony*	Everything still OK up here. Still in communication with American Consulate. 0300
Noa from *Socony*	From Consul. Southerners surrounding city from Hansi gate on west to gate on east. Chu Yu-pu started to Pukow and is on a car there. He is attempting to make a stand but has no hope at present of any reinforcements from the north. 0750
Noa from *Socony*	From Consul. The Southern troops in city came to American Church Mission. Entered house to look for Northern soldiers. Broke open all trunks of Americans and robbed Chinese. American Christian Mission reports same treatment. 0817
Noa from *Socony*	Consul advises if Southern troops enter consulate he will disarm guard and conceal arms. Will do same here. Apparently Southern troops advancing rapidly through city and firing near consulate. 0905
Emerald from *Socony*	Party of officers from British consulate have been cut off from consulate and taken refuge on Socony Hill. 0945
Emerald from *Socony*	Unable to approach British consulate account of looting. The only information we can give you. 1005
Emerald from *Socony*	British consulate has been evacuated. All consular officers are now on Socony Hill. 1115

[Four messages here inform that consular personnel from both consulates, plus Phelps and American guard now on Socony Hill as of 1150.]

Emerald from *Socony*	Regret to report that British consul was killed while disarming looters. . . . 1205
Emerald from *Socony*	Further reports British consul wounded. Please keep a glass on us. 1215
Emerald from *Socony*	American consul is getting in touch with Cantonese officials. No immediate danger. Do not open fire until requested. . . . 1226
Socony from *Emerald*	*Emerald* is preparing to fire with six inch shrapnel in area beyond Socony Hill. 1310
Preston from *Noa*	I am preparing to use my main battery if it becomes necessary. 1330

Socony from *Emerald*	To American consul. I suggest for your consideration that you should demand an interview for me as Senior Naval Officer Present with Cantonese commander. I am prepared to land with or without escort but perhaps an escort should also be demanded in view of their total inability to control the situation. 1337
Noa from *Preston*	Have located a sniper who has been firing at us. Request permission to silence him. 1342. [Note: Permission was granted.]
Noa, Preston, Emerald from *Socony*	Commence firing. 1452
Noa, Preston, Emerald from *Socony*	Do not fire. 1453
Noa, Preston, Emerald from *Socony*	We are being attacked. Open fire. SOS SOS SOS. 1453
	[Note in log: During this period all hell broke loose.]
Emerald from *Socony*	Send armed guard to foot of hill. We are being held up. 1525
Noa from *Socony*	We are trying to go over the wall to the south end. Please send help. 1537
Noa from *Emerald*	Send relief immediately. 1550
Noa from *Preston*	May I have permission to put a shell in that black hulk. 1610 [Note: OK]
Socony from *Noa*	Come over wall and come down toward beach and meet landing party. 1615
Noa from *Socony*	We are clear. 1650

Mrs. Hobart, in that last horror-filled half hour before the barrage, had thoughts of her baby speared on a Cantonese bayonet, herself a momentary plaything, her husband and the other men shot like fish in a barrel, as she knew others already had been. Cocked guns were held at Mr. Hobart's chest as he tried to placate the ever more demanding and arrogant troops. They had been told by some of the pro-Kuomintang servants that there were many more foreigners hidden in the house. The clock had only a few minutes or even seconds to run. A fierce, wild joy welled up in her as the crash of exploding shells shook the house. Better to be blown up in a sort of *Gotterdämmerung,* all together, rather than suffer the indignities expected at the hands of those cruel-faced soldiers. Let them *all* be blown up!

The sailors, at last released from restraint, joyfully pumped the bolts of their Springfields, uttering triumphant oaths that were music to the ear as they hurried the fleeing Chinese on their way, running in their characteristic terror of artillery.

In his official report, Ensign Phelps wrote:

It wasn't three minutes after our first signal to commence firing that the first salvo landed. The gunfire was most effective and undoubtedly saved all our lives. Missionairies in the interior who were being looted stated that all looting ceased when the bombardment commenced. At Socony Hill the soldiers ran like scared rabbits, but not until we had killed three or four of them ourselves. When the soldiers left, we made three ropes out of sheets, in order to get over the wall at the base of the hill. . . . Armed men were lowered first over the sixty-foot wall . . . then came the women and children followed by some unarmed missionaries. The last to leave were the signalmen. . . .

We still had about a mile of open country and one canal to cross before reaching the shore . . . forty-eight men, two women, and two children.

Captain Hugh T. England, Royal Navy, commanding HMS *Emerald,* was a Britisher of the old school, ready and in fact too eager to land half his ship's company in the face of immense odds. More important, in Lieutenant Commander Roy C. Smith of the *Noa,* the United States had a man with guts, initiative, cooperative attitude and good, common sense. Generations of such officers had served with honor through many wars and had shown the Flowery Flag with pride over all the world's seas. The action in which Smith participated at Nanking was a model of effectiveness—force judiciously applied, masterfully executed, and precisely right in timing and degree. The modesty of the action in no way lessens a claim to naval immortality for his order to Lieutenant Benjamin Franklin Staud:

Well, I'll either get a court martial or a medal out of this. Let her go, Bennie!

During the next ten minutes, Bennie's gunners sent nineteen flat-nosed, high-explosive projectiles howling shoreward from *Noa*'s 4-inch guns. The *Preston* contributed fifteen more. Downstream, HMS *Emerald*'s 6-inch 100-pounders added a touch of *basso profundo* to the orchestration. All told, she fired seventy-six rounds.

Upstream lay three Japanese destroyers. Lacking either Japanese authority to shoot or Anglo-Saxon initiative to act on their own, they maintained discreet silence. But at each Anglo-American salvo the Japanese crews, massed on the fantails of their ships for a better view, gave vent to a thunderous "*Banzai!*" The *Noa* delayed opening fire after the SOS from Socony in order to give senior ship *Emerald* the chance to shoot first. But *Emerald* had trained in her guns after the negation of the original Socony request to open fire. So after an agonizing wait of perhaps half a minute, the two U.S. ships let go anyway, regardless of seagoing etiquette. Meanwhile, HMS *Wolsey,* boiling downriver full speed from Wuhu, arrived on the scene just as *Noa* blasted away. "Well, there those blooming Yanks have got the wind up again!" cried her skipper. Then *Emerald* let go and it was *Wolsey*'s turn to sweat, frantically trying to get a target designation. Meanwhile, she improved her time by spraying shoreside snipers with her machine guns.

Snipers had been potting away all morning at the ships and at the small boats traveling between the ships. This exasperating fusillade had kept people under cover and had made many hits, but miraculously had killed only one man, in *Emerald*. The word to open fire with the main battery included the machine guns and riflemen. With intense satisfaction and vast relief the men opened up on the pestiferous snipers. The rattle and pop of small arms erupted from both *Noa* and *Preston*. Civilians and soldiers who minutes before had swarmed over the foreshore simply melted into the ground and the sniping ceased forthwith. Three Nationalist gunboats that had been shelling Pukow likewise turned tail and disappeared upriver in clouds of black smoke.

In the days of windjammers, muzzle-loading cannon, and Marine marksmen in the fighting tops, warships had carried half a dozen or so youngsters in their early teens who did odd jobs, scrambled about in the rigging, or got into spaces too small for a grown man to work. In time of battle they became "powder monkeys," carrying charges from magazines to guns. The *Noa* at Nanking was unique in having on board a powder monkey—the skipper's fourteen-year-old son, Roy C. Smith III, half a century and more after the custom generally had disappeared into history.

Young Roy, aboard "informally," had helped build the signal platform at Socony Hill. Then he had come back aboard in time to carry Lewis gun pans to the machine gunners in action against the snipers.*

Meanwhile, things on the north bank were in no less a state of confusion than at and around Nanking. At Pukow, SS *Pamyat Lenina,* scuttled by the retreating Northerners, was slowly slipping beneath the surface.

Apropos of *Pamyat Lenina,* Fanny Borodin had, by easy stages, finally wound up in Peking, after a lengthy stay at Tsinan, Chang Chung-ch'ang's home base. There, she probably had constituted highly negotiable trade goods for Chang Chung-ch'ang in his dealings with boss Chang Tso-lin.

Concerning the latter, Vincent Sheean, who was in Peking at the time and on good terms with several Soviet embassy staffers, wrote that "the old bandit was dissuaded from strangling her, and after a great many delays and legal difficulties she was brought to trial in July."[41]

The trial and its aftermath were typically Chinese. The trial was called wholly unexpectedly at some unearthly early hour in the morning, well before old Chang had torn himself from the embrace of one or another of his forty-seven concubines. The hearing was short; Fanny was acquitted. By the time Chang had collected his forenoon wits, both judge and Fanny were "Ningpo more far"—disappeared from Peking. The judge shortly thereafter surfaced in Japan, a country he no doubt felt fitted his peaceful nature and desire to engage in uninterrupted contemplation.

Chang Tso-lin, naturally enough, had been fit to be tied, shaking Peking upside

* Young Smith was acting in best Smith family tradition. His grandfather, Lieutenant Roy C. Smith, as skipper of the *Villalobos,* had sailed the Yangtze from July 1909 to June 1910 and set a unique record of steaming a total of 7,299 miles in that time. Powder monkey Smith's great-grandfather was Rear Admiral W. T. Sampson, the hero of Santiago. Smith III was a member of the Naval Academy class of 1934, and his son, Roy C. Smith IV, graduated from there in 1960.

down in an effort to find the culprits. Then, about a month or more after the "trial," Sheean was approached by one of his Soviet embassy pals with a proposal that Sheean, with the help of a passport to be supplied by a sympathetic American lady in Peking, help smuggle Fanny Borodin out of Peking and out of China. This bizarre plot, as far as Sheean was concerned, fell through. But perhaps it was only a cover plan—Fanny soon turned up safe and sound in Moscow to join her husband there.

With the refugees from Socony Hill on board the *Emerald, Noa,* and *Preston,* it was time to think about the remaining foreigners, some ninety American men, forty women, and twenty children, deep in "enemy" country. England and Smith proposed to remedy this situation by showing the mailed fist; they composed an ultimatum, *demanding* that the commanding general come on board before 11:00 P.M., 24 March, to negotiate regarding the day's unpleasantness. Furthermore, all foreigners must be brought to the bund by 10:00 A.M. next day, or . . . "we intend to take immediate steps . . . treating Hsia Kwan and . . . Nanking as a military area."

At 6:30 P.M., with some of the dust settled, Rear Admiral Henry H. Hough, ComYangPat, arrived from Hankow, adding his flyweight flagship *Isabel* to the growing "allied" fleet, and setting to rest Smith's anxiety by heartily approving all that had been done.

The bold demands of the ultimatum to the Chinese general had produced a somewhat insolent and evasive reply, pointing out that he was too busy fighting a war to be visiting foreign warships. Early on 25 March, Admiral Hough bounced the ultimatum right back, with improvements. He had a message, he said, from Chiang Kai-shek requesting foreign warships to hold their fire, while he, Chiang, would guarantee the safety of foreigners and would be in Nanking next day to see to it. The new ultimatum conceded the Chinese general's point that he was fighting a war and thus agreed to settle for *any* major general coming aboard to powwow. And as if the mention of Chiang were not enough, there was an added clincher—if there was no compliance by noon, such action as deemed necessary would be taken.

As the afternoon wore on, without the appearance of general or refugees, General Cheng Chien, commanding the 6th Nationalist Army, was warned that unless U.S. and British nationals were on the foreshore by late afternoon, Nanking would be declared a military area and the American and British ships present—USS *Isabel, Noa, Preston,* and HMS *Emerald* and *Wolsey*—would shell the military yamen and all the salient military points in and about Nanking. Reference to the yamen worked; the refugees started appearing on the bund by late afternoon and by 8:30 that night, all were safely on board.

Although the list of prospective bombardment units did not include the Japanese destroyers, there is no doubt Squadron Commander Yoshida would have been happy to join. He and four or five sailors, all unarmed, had bravely gone out to the Japanese consulate and were shaken to the core by what they saw and heard. His nationals had suffered gross indignities and many injuries. To save all at the consulate from general massacre, the little naval unit there of nine bluejackets under Lieutenant Araki had not resisted. One of Yoshida's beach guards was killed by a sniper while Yoshida was checking the consulate. At an "allied" conference the morning of 26 March, debating the probable course of events, Yoshida interrupted from time to time in his sketchy English: "I wish to fire! I will shoot!" he cried, unfortunately a little *ex post facto* to the main event.

As a strange sequel, Lieutenant Araki committed suicide on 29 March, at Shanghai, where he had gone to report to the Japanese admiral on events at Nanking.

"In order to ensure the safety of the Japanese residents at Nanking," he wrote in his suicide note, "I endured what we could not tolerate. The lives of the Japanese refugees could be saved, but I am ashamed that the honor of the Imperial Japanese Navy was disgraced by the Southern soldiers." According to the *North China Daily News* for 30 March, "The crack of the suicide shot rang out simultaneously with the notes of the bugle announcing morning colors."

In a grim if not eerie corollary to the Nanking epic, all three of the junior principals met the same fate—suicide. On 12 September 1939, Benjamin Franklin Staud, the *Noa*'s gunnery officer—newly promoted to lieutenant commander—killed himself. Two days later Lieutenant (junior grade) Woodward Phelps died in the same way.

So ended the "Nanking outrage." As for Lieutenant Commander Smith's remark that he would "either get a medal or a court martial" when he told Bennie to "let her go," records show that he got neither. But the opinion of the official and unofficial world, other than China and the USSR, was near unanimous; the action had been wholly justified.

The Shanghai foreign press was editorially warm in praise of the American action, calling attention to the cooperative good feeling that had long existed between the British and U.S. Naval forces on the Yangtze. Some "letters to the editor" contributors were less diplomatic, voicing satisfaction that the United States had at last stopped talking out of both sides of its mouth and had done something. One signed "Americano" summarized the new Yankee cockiness:

> *The sooner the Chinese get the idea out of their heads that they could by any chance use force to drive the foreigners from China, the better it will be for all concerned. If they want force, all they have to do is keep on as they are now doing and they will get all they are looking for.*[42]

Smith's immediate operational boss, Rear Admiral Hough, reported to the CNO on 11 April concerning ". . . the excellent, cool-headed judgement and efficiency displayed by Lieutenant Commander Smith. . . ." Also, he considered that "the conduct of Ensign Phelps and the men under his command was particularly commendable."

Admiral C. S. Williams, the CinC, Asiatic Fleet, expressed to the CNO on 22 April his "unqualified approval." The U.S. consul at Nanking, John Davis, wrote to ComYangPat on 5 April that he ". . . found Lieutenant Commander Smith invariably willing to cooperate in a most whole-hearted and able manner and his hard common sense and good judgement were of great value . . . and [his] willingness to take this firm action unquestionably saved a situation which would otherwise certainly have resulted in unprecedented disaster."

At Shanghai, G. E. Gauss, the consul general, reported to the Secretary of State on 8 April that:

The American officials, naval officers and naval personnel . . . conducted themselves in a manner which reflects only the highest credit and glory on the flag under which they served.

The commercial gentlemen felt good about it all, too, both personally and in their pocketbooks. The assistant manager of Standard Oil in Shanghai, Harry F. Emerall, wrote Admiral Williams that the services rendered during the Nanking incident, the evacuation of Changsha, Ichang, and Chungking, and the convoying of their vessels had placed the company and staff under deep obligations . . . "not only protected the lives of the staff, but saved us from greater financial loss."

Minister MacMurray wanted to know the names of the bluejackets with Phelps, especially the signalmen. In a telegram to the Secretary of State on 26 April, he conveyed his admiration for Smith "for his moral courage."

Best of all, an 18 June letter from Washington closed with, "The Department commends you [Smith] for your timely action . . . so excellently executed." The painstaking squiggle underneath was that of Curtis D. Wilbur, Secretary of the Navy.

President Calvin Coolidge said nothing, as was his wont. But as "Silent Cal" was born to the Down East tradition of "Don't tread on me," one suspects the saturnine old gentleman was pleased.

Most heartwarming letters of all were from those who escaped being stripped naked, or worse, being looted of that which some spinsters held more precious than life itself. In an account of one unlucky participant in such an incident, sworn to before the U.S. consul at Shanghai on 3 June 1927 and printed in a pamphlet by the *North China Daily News,* titled "A Bolshevized China—the World's Greatest Peril," a missionary lady was well on the road to chalking up some sort of Far East record, but on the ninth violation by Cantonese soldiery she swooned away and thus lost official count.

"No words are adequate to express all the gratitude we feel," summed up a resolution "unanimously approved" by all those taken aboard *Noa,* and addressed to Skipper Smith. Some appended a postscript: "We, citizens of Norway, also guests of the *Noa,* heartily concur in these sentiments." And one a month later, from Tokyo, ". . . wanted to express in writing," to Skipper Smith, that she had ". . . been thinking not only of the service in protecting our lives, but also of the kindness and courtesy shown us . . . aboard *Noa*. We are indeed proud of our United States Navy!"

At 3:00 A.M. on 25 March 1927, the USS *Preston,* commanded by Lieutenant Commander George Bancroft Ashe, left the little allied fleet anchored at Nanking and fell in astern of the refugee-laden SS *Kungwo*. The mud was barely hosed off her anchor chain when firing broke out from the Nanking side and two bullets hit the destroyer. A burst or two of *Preston* machine gun fire soon silenced it.

Some three hours later, the convoy was again shot at between Silver Island and Hsing Shan fort—small arms fire aimed first at the merchantman, then the destroyer.

"At the same time," Ashe stated in his report to ComYangPat the next day, "and even before we could reply with our machine guns . . . the three-inch gun in the fort opened fire at point blank range."

The Chinese gun barked worse than it bit. The first shell splashed about fifty yards ahead of the ship. A second went through the fire control platform and a third splashed astern.

After the second shot, *Preston* fired three rounds from her forward 4-inch gun, plus a few bursts of machine gun fire which silenced the fort. Nobody aboard the convoy suffered a scratch. Chinese accuracy was so incredibly poor that one would think they *must* be shooting blanks. But it was this singular inability of the Chinese to hit anything with either small arms or artillery that allowed tin-plate foreign warships to keep the river. When Mao Tse-tung's Red army reached the Yangtze in 1946–47, accuracy was greatly improved, as several large British warships would discover.

There were more forts downriver, armed with much heavier artillery than 3-inch guns. So Ashe, with the safety of *Kungwo* and refugees uppermost in his mind, arranged to catch up to and travel in company with HMS *Cricket* and her convoy SS *Wenchow,* also refugee-loaded. *Cricket* was a 625-ton monitor, with one whopping 6-inch widowmaker forward and another aft, behind shields, plus the usual bristling array of shielded gunboat machine guns.

Joining up fifty-two miles below Chinkiang, the convoys ran into more heavy fire from both banks, some two miles above Kiangyin Point. *Cricket*'s machine gun and pom-pom fire proved to be effective treatment and the Chinese firing stopped.

For the *Preston,* this was to be a mere practice run. Passing *Kungwo* to *Cricket* below Kiangyin, the destroyer turned upstream and headed back for Nanking, where she found herself once more in the refugee business. With about seventy men, women, and children on board, she got under way for Shanghai on 27 March. That trip was still vivid in Ashe's memory, forty years later:

Among them was a Miss Moffat, who had been shot through the stomach. Found in a gutter somewhere, she was brought aboard the Preston *just before sailing time. We made her as comfortable as possible on a cot in the chart house.*

On our way downriver there was sporadic firing at the ship—from many places along the shore line. At one point, where the river made a large bend and the channel was close to the bank, we knew that a 3-inch field piece was in place about the middle of the elbow.

It previously had fired on a river steamer, killing, I believe, two men. Before arriving at the bend, I sounded general quarters, manned all guns and trained them shoreward. The field piece soon came in sight, manned and trained on the river. Several hundred Chinese armed with rifles popped up from behind the levees. Evidently our 4-inch guns, trained on the shore, were more than they expected; not a shot was fired and you can bet I breathed a sigh of relief.

On arriving at Woosung, we found men-of-war—some twenty-eight of them, we later found out—strung out along the river all the way from the Standard Oil com-

pound to Shanghai. As we passed these warships—British, French, Italian, Spanish, Dutch—each one in turn, including one U.S. gunboat—manned the rail!

And *that* was gunboat duty for a destroyer on the Yangtze in 1927.

Highlighted, one might add, by the fact that as *Preston* and the Chinese confronted each other with their 4-inch guns and 3-inch field piece, respectively, across a mere fifty yards, some of the Chinese considered discretion the better part of valor and saluted the ship as she passed. That, and the fact that Miss Moffat recovered.

The course of events during 1926 might have sent the *yang kweidzah* (ocean devils) running at Hankow and Kuikiang, but 1927 opened with a different trend and proved Shanghai to be an entirely different matter. Not only had British and American overall views hardened in the short interim between Hankow (early January) and Nanking (late March), but they now had the military wherewithal to make "Fortress Shanghai" a meaningful term, and more force was on the way. Nowhere in the prior history of western civilization could one find a comparable situation.

At the end of March, destitute evacuees from Nanking, Wuhu, Kiukiang, and Chungking streamed into Shanghai. The British river steamer *Suiwo* arrived on 29 March, carrying 320 Hankow evacuees in cabin space for 18. Their worldly possessions, accumulated in some cases during a lifetime in China, had been left behind to be looted or burned. Some of the men wore coolie rags, their own clothes having been "liberated" by covetous soldiery. Women had their hair hanging down simply because they had fled their homes without time to save a handful of hairpins.

What the refugees saw at Shanghai must have lifted their hearts, although the *North China Daily News* of 28 March 1927 may have gone a little overboard in its ecstatic headlines: *"FROM WOOSUNG TO THE UPPER RIVER,"* the *News* screamed, *"SUCH AN ARRAY OF MEN-OF-WAR AS NEVER SEEN BEFORE!"* The *News* then somewhat inaccurately listed twenty-eight warships, totaling over 109,000 tons. Actually there were more, they forgot the destroyers. The U.S. had four cruisers, four destroyers, an oiler, a transport, a minesweeper, and a gunboat. Britain had five cruisers, an aircraft carrier (actually operating off the Yangtze's mouth), a destroyer, and two gunboats. Japan had six cruisers, seven destroyers, and three gunboats. Two cruisers, a sloop, and a gunboat flew the Italian flag. Two cruisers and a sloop represented France, one cruiser each for Spain and Holland, and a sloop for Portugal—total, forty-seven warships.

Over this "fleet," superior to most of the world's navies, floated the flags of *eight admirals:* three U.S., three British, one French, and one Japanese. And for a pleasant change, that of four-star Admiral C. S. Williams, U.S. Navy, was senior. At that time, unless one had pressing business topside, he carefully avoided being trapped in the open at 8:00 A.M. Proper observance of the ceremony of morning colors meant standing at attention and saluting for twenty minutes while the *Pittsburgh*'s band raced through the national anthems of nine nations.

The strength ashore was equally impressive: 9,800 British, 1,500 Americans, 1,500 Japanese, 50 Italians, 400 French, plus the mobilized Shanghai Volunteer Corps of about 1,500 more. With these, added to the strength aboard the ships, foreign manpower stood not too far below the 40,000 Shanghai foreigners, including 3,000 Americans, they were there to defend.

The situation faced by this relatively powerful and well-coordinated force was described by American Consul General Gauss in a 22 March telegram:

The Southern forces . . . are just outside Shanghai . . . mobs having taken arms and attacked the native police stations. . . . General strike called for noon of March 21. . . . Disturbances and firing reported at North Railway Station. Workmen have very generally donned red arm bands and are quarrelling amongst themselves. The Municipal Council has mobilized both volunteers and the police . . . and declared a state of emergency to exist. They have also requested the assistance of the Foreign Naval Forces for internal defense of the Settlement. I have given my approval . . . and American Marines are now landing. The Japanese have also landed forces.

Gauss had been generally correct in his assessment except for one highly significant misconception; the workers were not "quarreling amongst themselves." Their communist-led union organizations were establishing themselves in a position of total power in the Chinese city. The 21 March general strike was, in effect, a declaration of war on the Northern garrison troops, the municipality itself and its police. For a day and a half, armed workers slashed their way through disorganized police and Northern soldiers, all taken completely by surprise. The bitterest battle took place at the North Railway Station, where the Russian mercenaries of General Sun, in the armored train which had been ferried across the Yangtze at Nanking, fired some of the last shots of the "war" before fleeing to refuge in the International Settlement. By nightfall of the second day of the uprising, all of Shanghai except the French and International Settlements was in the hands of the workers. Such was the position when Chiang Kai-shek landed on the bund from upriver on 26 March, the native city already having been handed over to his troops.

During the next three weeks, the workers in their red arm bands coexisted in fraternal amity with their Cantonese "brothers," while the foreign defenders of the settlements looked on uneasily from behind their barbed wire concertinas and blockhouses. "Down with imperialism! Exterminate the feudal forces!" screamed the posters in huge black letters during the last of March and the first days of April. Moscow's *Pravda* and France's communist mouthpiece *L'Humanité* extolled the virtues of Chiang's Nationalists and his "new order in China."

Then, at dawn on 12 April, a siren wailed on a Nationalist gunboat off Nantao. A bugle sounded at Chiang's headquarters. With stunning surprise and magnificent organization, Chiang's soldiers and detachments of gangsters recruited from the underworld struck savagely at the workers in their union halls and on the streets. Red arm bands merged indistinguishably with the red that flowed in the path of flashing executioners' broadswords and spitting Lugers. In a few days it was all over. Once secure in his own right on the Yangtze and at China's financial heart, Shanghai, Chiang had cast off the uneasy marriage of convenience with the communists as he would have a leprous cloak. The interminable war between the Nationalists and communists had begun.

Concurrent with the climax at Shanghai, the upper river was in ferment, with communist bands in control of the Chungking area. Conditions had never really been anywhere near normal since the incident at Wanhsien the previous year. On 24 March, the American flag was torn down from the consulate at Chungking. The consul and vice consul took refuge aboard *Monocacy* and American residents concentrated on the waterfront, while brigands burned and looted their properties in the hills on the south bank. All foreign firms closed. The British Asiatic Petroleum Company and Standard Oil carefully removed vital parts from the oil can fabricating machinery so that their stocks could not so easily be confiscated. But they had failed to agree on *what* vital parts, so that the canny Chinese soon were able to assemble the necessary machinery by combining parts from both. Needless to say, when the owners eventually returned after the troubles had subsided, there was no more oil. Nor, equally needless to say, money to pay for it.

During the evacuation of all foreigners which took place, Chinese pilots working for the British had decamped, in fear of their lives from retaliation. There seemed only one solution for the British gunboat *Mantis,* lying at Chungking; blow her up to prevent her falling into Chinese communist hands. But Captain Tornroth, veteran American captain of Yangtze Rapid ships and former contract pilot for the U.S. Navy, volunteered to take her downriver by himself—the only case on record where a foreigner, unassisted by a Chinese pilot, successfully brought a steamer down the gorges and rapids of the upper river.

In Shanghai, the Nationalist marriage of convenience with the communists, so rudely broken, was at once replaced by an equally pragmatic romance with the Chinese bankers. "Loans" immediately were forthcoming to put the new government on a solid basis. All the former talk about relinquishment of concessions and "extrality" (extraterritoriality) went *pianissimo.* The foreign *taipans* began to breathe more freely. By May 1927, the last of the Yangtze Valley communists, in the form of the rump "government" of Eugene Chen at Hankow, had discreetly, surreptitiously and with speed unusual in China, departed for more salubrious surroundings, such as Moscow. Comrade Borodin, Rayna Prohme, and Mme. Sun Yat-sen no longer trod the streets of Hankow, where the Union Jack no longer flew over the late British concession.

Also, by May 1927, Shanghai had shaken down into business (almost) as usual, meaning that the tennis games, horse races, paper chases, the French Club, and so forth were all operating again. The Shanghai Volunteer Corps was back in mufti, even though no traffic moved up the Yangtze and trade with the interior was at a standstill. The once more peaceful weekend warriors had little to do in their offices.

As for the Chinese military, affairs were slipping this way and that like a pony on a wet path. Some military actions bordered on pure buffoonery, but the humor thereof could hardly be appreciated by the sweating, starving peasantry and the river folk.

An idea of Chinese naval warfare of those days can be gained from the contents of a message intercepted from the Chinese cruiser *Hai-Chi* on 18 May. That venerable craft, of 4,400 tons and with 8-inch guns, was the largest Chinese warship and still loyal to the Peking government.

The *Hai-Chi* had first broadcast a warning that "all merchant ships and the foreign warships should leave the Woosung fort approaches to avoid the danger of gunfire. If not, our fleet is not responsible for any accident occurred." Then she got into the meat of the thing. Throwing down the gauntlet in "I can lick any man in the joint" style, she announced to *Hai-Yung, Ying-Swei, Hai-Chow, Tung-Chi, Yung-Kien, Yung-Fung,* and *Kiang-Nan* that:

> *We are sorry to relate that since we first fight you at Wusung on March 27th we have arrived here three times without seeing any of your ships. It is not a good plan for you to hide inside the upper and narrow part of the river. We just destroyed your Wusung fort but that is not our job. Now we are waiting at the mouth of the river hopes you come to there soon and then try a sea battle with us for the unity of Navy.**

The message was signed, "Admiral Shen Song-lich, of R.C.N." [Republic of China Navy]. Obviously, the admiral won the day; many Tsingtao summer sojourners of the mid-thirties remember Admiral Shen, commander of the Tsingtao squadron, *Hai-Chi, Chao-Ho,* and a smaller ship, who was also mayor of the city.

Years later, at a Tsingtao party at which a number of U.S. Navy families were present, the mayor, after the fashion of Chinese officials, wore a loose white gown, practically indistinguishable from that of a servant. A young Navy wife with a cigarette but no light, and not knowing the mayor from any other Chinese, gave a little tug at his gown. "Boy! Go catchee match!" she said. The mayor toddled off, brought a match, bowed, lighted her cigarette and then, with an ingratiating smile, backed away. The lady was blissfully unaware, either of her *gaffe* or of this gesture of exquisite courtesy on the part of an old Oriental gentleman in not embarrassing her.

When the *Hai-Chi* was sparkling new, Admiral Robley Evans, the CinC Asiatic Fleet in 1902, met her and later described her in *An Admiral's Log* as "the cleanest thing in the shape of a warship that I ever saw." When the *Oregon* ran on to the rocks between Chefoo and the Taku bar, during the Boxer troubles in 1900, the *Hai-Chi* went at once to her assistance, secured alongside of her, and did all in her power to aid her in getting afloat. The Russians had their eye on the Chinese navy, and hoped, no doubt, to secure several of their ships by capture before the war was over. While the *Hai-Chi* was employed in aiding the *Oregon,* a Russian cruiser hove in sight and

* English was the only common spoken language between Cantonese and Northern Chinese naval officers, many of whom had been trained in England. The written language was identical all over China, but to transmit it by wireless or telegraph, it was necessary to enter a code book listing each ideograph by number. This transmitted number would then be decoded on the receiving end into "plain Chinese."

remained at a distance cruising about. Captain Sah, convinced that she meant to capture him, reported the matter to the captain of the *Oregon* and asked his advice, which was promptly given. . . . "Hoist the American flag at your foremost-head, and let us see who will attack you!" This was done, and the Russian ship went on her way.

In the summer of 1927, Admiral Mark Bristol assumed command of the Asiatic Fleet. For the previous eight years, he had served under the State Department as High Commissioner in Turkey, where his adroit diplomacy had built up a firm Turko-American friendship. On arrival in China, he decried the lack of social contact with the Chinese and proposed to change the approach to one of co-equality. The Yangtze was where this policy could most profitably be applied, he thought. Stop shooting. Make friends. Intermingle.

The admiral, like most newcomers to China, found it difficult to believe that fundamentally, the Chinese did not like Americans. But why should they? They had been perennially put upon by outsiders—Mongols, Manchus, Japanese pirates, and then the "Red Hairs"—who humiliated them and debauched them with opium. Sir Robert Hart, Inspector General of the Chinese Maritime Customs for close to half a century, probably knew Chinese officialdom better than any Westerner ever born, and wrote ". . . it is apparently beyond dispute that, however friendly individuals may have appeared or been, general intercourse has all along been simply tolerated and never welcome . . . ," justifying this with, "Just as one can paralyse the body or corrupt the soul of a human being, so too is it possible to outrage the spirit and antagonise the nature of a people."[43] That, the Americans had managed to accomplish with signal success. Western businessmen pressed trade at the point of a gun to a degree unique in world commercial intercourse. Missionaries in their thousands, zealous and well intentioned, none the less manifestly felt and acted superior to the "heathen Chinee," to the indignation of Chinese who felt that what they considered culturally inferior people were attempting to foist on them inappropriate morals and dogma.

In spite of Admiral Bristol's lofty ideals, Sir Robert Hart's pragmatic estimate represented reality. In essence, things continued as before; perhaps even deteriorated, as the future would soon prove.

Closing out the tumultuous, bloody year of change, 1927, the Nationalists' shift of partners from Hankow Reds to Shanghai bankers was paralleled by their leader; in December, Chiang Kai-shek "retired" his wife of many years and mother of two children and in a Christian ceremony married beauteous thirty-four-year-old Mei-ling, daughter of rich, influential Charley Soong, publisher of Bible tracts and KMT propaganda.

No account of Shanghai could be complete without some mention of what a local annual publication on Shanghai's foreign protectors of those days called "the best known and perhaps the most popular of the foreign military forces," the Fourth Marines.

The Fourth Regiment, U.S. Marines, was the seaward anchor of the Yangtze Patrol during the period which might be called the Patrol's heyday: 1927 to its flaming end in 1942 at Corregidor.

The Fourth Marines were not old as military organizations go, having come into existence on 15 April 1914 as a result of the trouble looming between the United States and Mexico. Between then and their arrival in Shanghai on 24 February 1927, the Fourth Marines had served in a sample of almost everything. In mid-1914, they waited off the west coast of Mexico for three months, prepared for a landing that never came. On 21 June 1916, they landed in Santo Domingo and there remained until August 1924. Then action shifted to the United States, where they revived a tradition of the old west in riding shotgun on the mail trains that had until then been plagued by frequent robberies.

The transport *Chaumont,* on a record twenty-one-day run from San Diego, reached Shanghai with the Fourth Marines—then two battalions—on 21 March 1927. Shanghai was to be the scene of almost constant turmoil for the next decade and a half. Beginning on that blustery March day when the regiment first hit the streets of Shanghai to stay, they were to make for themselves over the years a reputation for smartness, toughness, and downright indispensability that Shanghai will never forget.

In September 1932, following the Japanese attack on Chinese territory surrounding Shanghai, the newly organized second battalion of the Fourth arrived in Shanghai aboard the transport *Henderson.* This was the first time since Santo Domingo that all three battalions of the Fourth had served together. The regiment remained thus until December 1934, when the third battalion returned to the United States.

Other than a rising scale of banditry on the street, relative tranquility reigned until August 1937, when almost overnight the city became the theater of modern, large-scale warfare between Japanese and Chinese. Fourth Marines once more manned the barricades, sentry posts, and dugouts that protected the International Settlement from a spillover of the fighting which blazed around the city.

One can be sure, as a basic premise, that in the old tradition of Leathernecks versus sailors, the Fourth was happy to provide sparring partners for those of the "outside" fleet men who happened to stray too far up Bubbling Well Road into the Leatherneck sphere of influence and what they considered "their" cabarets. Gunboat sailors not only knew better, but had their own favorite, hotly defended hangouts anyway.

Under the proper stimulus, Marines were never slow in tangling with men of the various other foreign detachments. A very satisfactory state of belligerency could be established by a leading question or a facetious remark concerning a Seaforth Highlander's kilt. A typical sporting event of this nature occurred one day in uptown Shanghai, wherein the American Marines and their British opposite numbers had become enthused over the advent of a payday coupled with a Saturday afternoon, and with more luck than good management found themselves in the same drinking establishment, on their joint border. The result was easily predictable, and soon proved the need of intervention. In the tradition of the Texas Rangers—send only one man for one riot—the Settlement police dispatched two Sikh constables to restore order. These tall, immensely dignified fellows, their beards swept back under their turbans, had a system: Whenever a battler downed his opponent, one or the other of the Sikhs would gently push him in the direction of someone similarly disengaged. In a fairly short time, there were only two combatants left on their feet, both groggy. These, the two Sikhs had no trouble subduing and sending on their way. Not a Sikh whisker had been displaced nor a club swung. No knives had flashed, no shots rung out. The Chinese "boys" climbed out from under the tables and from behind the bar. Honors mutually secure, those badly damaged withdrew to lick their wounds. Those in better shape amiably closed ranks and called for beer: "Aye! Thot was a bloody good scrap, Yank. Come and hahve one on me!"

In retrospect, it appears that in the persons of those two Sikhs there stood the British Empire in microcosm. This was the old strategy of balance of power, wherein a small country with modest resources but with brilliant diplomacy, had pushed now one, now another perhaps bigger power, so that the outcome cooled to a near standoff, with Britain always intact and profiting. But in World War I, Britain had become the blundering world policeman in a brawl too big to handle.

Since the first impressions of China were usually of centuries-old cities filled with multitudes of people, one tends to picture the entire land as teeming with people, and with every square foot of soil cultivated and inhabited. Under such conditions, one would expect wild game to be scarce, both from lack of proper environment and from the tendency of perpetually hungry citizenry to manifest a deep interest in anything edible.[44]

Somewhat surprisingly, such impressions and expectations are both erroneous. There were swamplands along the rivers and canals in South and Central China and heavily wooded areas in the uplands, all favorable to the flourishing of wild game. Deer, bear, and a multitude of lesser food and fur animals roamed Manchuria. Clouds of migratory water fowl at times made one think he was back in the Chesapeake Bay.

In 1914, the USS *Elcano* was heading downriver when, as her skipper, Lieutenant R. A. Davis, put it:

> *. . . my eyes sprung open in amazement. Where the river was not supposed to be dry, there was a large expanse of bright, shiny mud. I swung over where there was deeper water and continued to look at the mud-flat. Suddenly, the whole thing rose off the river and went flying away. Ducks! There must have been over an acre of them, possibly two acres. Not wishing to lose another such chance to give my crew a duck dinner, I had the gunner's mate load several 3-pounder cases with bird shot. Never had the chance to use them.*

The Chinese themselves had very few firearms suitable for hunting, but nevertheless game could be bought almost anywhere. In both quantity and variety it came at prices competitive with the barnyard variety.

In some sections of the country, pheasant was cheaper than chicken; after all, domestic birds not only had to be closely chaperoned against theft, but after a fashion, even fed. Wild pheasants required only time and patience, both commodities in long supply in peasant China. Countrymen made ingenious traps of barbed sticks ringed in a small circle, barbs inward, a few grains of bait in the center. A pheasant, one of the stupidest of birds, would peck at the grain and, when he tried to pull out his head, would become impaled on the barbs.

More affluent hunters stalked their prey with homemade guns constructed of a length of wire-wrapped pipe tied to a makeshift stock. A matchlock firing device lowered a smouldering piece of punk to the touch-hole. This ignited a charge of powder mixed from charcoal, bird droppings and sulphur. After a short pause, a mushy "bang," or rather more accurately, a sort of "whooosh" then followed. If the barrel didn't burst, a handful of short bits of cut-up wire or nails scattered in the general direction of the target. Long shots were the exception. A hunter spent most of his time in cautious stalking.

The inclusion in Navy menus of such exotic fare as venison and pheasant was not as well received as one might suppose. "Sir!" complained a spokesman for the crew one day in Hankow. "Them goddam pheasants is bustin' everybody's teeth from the shot! Couldn't we have more just plain chicken for a change?"

The big lakes, such as Poyang and Tungting, were favorite stopping-off places for thousands of migrating wild ducks. These were "fly-fly" ducks in Chinese pidgin, not the tame, barnyard type "walkee-walkee" ducks. The Chinese managed to catch a few of these wily birds by hiding under a tiny raft of bamboo and floating down amongst them. The seemingly harmless bundle of flotsam would slowly approach a covey of blissfully unsuspicious ducks. Soon, a frantic strangled "quaaaack" would be heard as a duck was yanked underwater by the legs.

The process was scarcely what one might term either convenient or wholesale. There was another method which a sportsman might hesitate to boast of in his club, but which certainly hauled in the birds. It must be remembered that the ducks were no fools; they knew exactly the point at which an approaching boat or human became dangerous. The wholesale method took advantage of that hole in the ducks' education, the long range possibilities of a 3-pounder saluting gun loaded with a sailor's white hatful of BB shot.

Gunboats with a saluting battery sometimes bagged as many as fifty birds with one blast. This was enough to feed the wardroom officers and those of the crew who fancied wild duck, with some left over to trade for beer with the sampans which clustered astern after the gunboat anchored in the evening. In return for the feathers, feet and innards, the local sampan colony would be happy to do the plucking and dressing, as ex-farm boys in the crew were canny enough not to reveal their knowledge of such special skills.

The rice bird was a small creature not much larger than a sparrow. Its small size and relative rarity made it a favorite delicacy for epicures, but not the sort of thing ordinarily fed sailormen. The unique method of preparing it for the skillet was learned only through investigation of another interesting aspect of Chinese life, the lottery. The National government had instituted a state lottery, the proceeds of which were supposed to be expended on road building. A first prize of 500,000 Chinese dollars, then equivalent to over US$100,000, was large enough bait to interest foreign ticket buyers.

"What are the chances of winning?" a gunboat officer asked a Chinese friend.

"You know lice bird?" the Chinese answered. "You savvy how Chinese man fix lice bird for eating? No savvy? Bird very small! No can take out insides same like chicken. Put lice stlaw down throat of bird, then blow. Insides come out back end of bird. Some day one lice bird he no like this. He get very mad! He blow back and blow insides out of Chinese man. You have same chance win lottery!"

For a race as clever as the Chinese in exacting tribute from the unwary foreigner, it was to be expected that the hunting field could be counted on for some special sort of exploitation. Thus, one should not have been surprised at the feigned injury. One common stunt was for a Chinese to lie in wait in the path of a foreign hunting party, rise unexpectedly from cover and cry out in anguish that the shot just fired had winged him. As proof, between howls for *cumshaw,* a few shot holes in his rear end would be uncovered for inspection. This grievous damage, the wounded man would cry out in his pain, could be assuaged by fifty *yuan,* applied to the palm of his hand.

The tenderfoot was inclined to pay off, to avoid unpleasantness with the local police, who could be counted on to rise up immediately out of the ground and back the injured party. The more experienced hunter would take the bitterly protesting victim back to the gunboat where the doctor would pick out a few of the shot. Snippets of telephone wire, the standard Chinese sporting load, proved fraud. The fellow had had a confederate pepper him from a safe distance. In the rare event that the shot turned out to be the Abercrombie and Fitch type, a deal could generally be arrived at somewhat below the original demand, with the police receiving a small present for their trouble and cooperative attitude.

The average American not only knew very little about the proper preparation of game for eating; he generally failed to appreciate the results produced by experts. A case in point was a brace of pheasant offered as a dinner by a German whose wife, a handsome northern Chinese, had arranged their house in a modified Chinese style. One rather expected to find in it the odors generally met in a Chinese house: the faint aroma of burnt peanut oil, the pungence of dried fish, wood smoke from the charcoal cooking fire, and a sort of damp mustiness which accompanies underheated houses in damp climates. But the predinner odor was so overpowering that a limburger cheese factory would have been positively fragrant by comparison. The stench seemed to cause no embarrassment to the host and hostess. During the martinis and ancient eggs, they discussed with animation the details of the preparation of the delicacy about to be laid on the table. The source of the odor, it was soon discovered, was the pheasant which, in accordance with the best European tradition, had been allowed to hang by their necks until they dropped free.

The European predilection for aged game was well recognized by those stores in Shanghai which catered to foreigners and the cooks who fed them. The wishes of the master were law to one such cook, Hu, and Hu demonstrated his understanding of the strange tastes of barbarians by serving at least once a week, venison, wild duck, or some such selected from the tons of game which, in season, arrived from Manchuria by ship. On these occasions, the character of the evening's entree was strongly apparent as soon as one opened the front door.

Hu, personally, had no taste for game; one could be sure he would not share the master's ration on such occasions. With the exception of lacquered duck and ancient eggs, the Chinese do not enthuse over long-killed or gamey food. Gunboat officers upriver sometimes capitalized on this trait in sending parcels via Chinese mail from Hongkong to Hankow. The inclusion of a small piece of ripe limburger in the package was a fair guarantee that it would not, in the usual manner, lie around the post office a week or two, but would immediately be speeded on its way.

Caveat emptor was the watchword in Chinese food markets as well as other emporiums. To sustain customer confidence in the better markets, the carcasses of animals and birds still retained a strip of the original fur or feathers in order to verify their ancestry. Those rare European housewives who did their own marketing or checked the coolie's basket on his return home with the goods were thus credited with the ability to differentiate between cat and rabbit fur, moose hair and caribou skin, or pheasant and gamecock tails.

Chinese fishing methods shared some of the bizarre aspects of their ways of capturing game, but of the two a far greater national effort was concentrated on the pursuit of the finny tribe. One got the impression that for every fish in China there were at least two Chinese scheming how to net it. Everything from a one-inch minnow on up was acceptable. There was no Fish and Game Commission to set a bag or size limit. Fishermen cast their nets along the river banks from dawn to dark, more often than not drawing them in totally empty. Occasionally a school would be hit and to the accompaniment of much loud and happy comment a haul of a half dozen or so glittering, jumping little things no larger than a sardine would be dumped in the sampan's bilges. Shrimp no larger than a fingernail, a couple of hundred to the pound, were not unusual. Nothing edible ever really had a chance to grow up in a Chinese river.

In Chungking and Ichang, otters sometimes were used in conjunction with net casting. These squirming, sinuous little beasts went over the side of a fisherman's cockleshell skiff, a leash around the neck tight enough to keep any chance fish from going down. From time to time this would be loosened and a small fishy reward offered to maintain the otter's enthusiasm.

A slightly more Machiavellian type of fishing was confined to periods of bright moonlight. A low freeboard sampan would moor, athwart one of the smaller streams, its downstream side coated with whitewash. A version of the Chinese one-stringed fiddle played a crucial part in the drama. The instrument in this case had been altered by an extension which reached into the water. When the fiddle bow was sawed across the strings, not only were the usual hideous sounds produced, but the vibrations were transmitted through the wooden extension into the water, at a spot downstream from the sampan. The fish, whose appreciation of harmony could only be more discriminating than a Chinaman's, apparently had one burning desire—to put as much distance as possible with minimum delay between them and the source of the vibrations. Rushing upstream, they would see shining in the moonlight the white sampan flank, evaluate it as a minor obstacle to their escape, and jump over it—into the boat. It was as simple as that. And unlike the hook and line method, none of the victims ever escaped to tip off their brethren on this joker in the Chinese Department of Dirty Tricks.

Carp, generally considered in the United States as fit only for feeding other fish, were a great delicacy in China and were raised in stagnant pools that would have had any less hardy fish floating belly-up in short order. Gathering the crop was simplicity itself; a quantity of mashed-up outer hulls of nuts would be dumped into the water at harvest time and very soon the carp would surface in a dazed condition, ready for easy netting.

The great fish mystery concerned the rice paddy eels. In some parts of China, the rice paddies marched up the sides of what were almost small mountains, tier on terraced tier. In the wet season, the higher overflowed into the lower and so on down to the bottom. In the dry season, the water disappeared and the earth sometimes became almost as hard as stone. Yet, in the spring, after the young rice shoots were a pleasant light green with a foot or more of water around their stalks, old men sat on the banks and caught eels. It is well known that eels journey to the open sea to spawn. By what magic did they get into those mountain rice paddies?

A good many foreigners hunted because it offered a way to escape the boring routine of office, club, cocktail parties, and evenings of bowling or pinpool. But the average Britisher, particularly the gunboat officers, hunted with a passion that amounted almost to a religion. Along with the notes on moorings, currents and local conditions, British journals, logs and diaries generally contained copious memos on the character and quality of the hunting.

A page from the journal of Lieutenant Douglas Claris, Royal Navy, who served in HMS *Thistle* in 1911–12, listed the variety and quantity of game available, and the obvious reverence with which this sport was held in His Majesty's wardrooms. Over the month of January, the menus of *Thistle*'s wardroom and crew's messtables were varied by the addition of twelve snipe, seven partridge, ten pheasant, four wild ducks, five hare, four geese, two pigeon, and two plover.

Not all gunboat officers were members of this dedicated breed that underwent considerable hardships tramping through broken fields and dried rice paddies in pursuit of wild meat. But one could enjoy long horseback rides in the low mountains back of the moorings at Chungking, and partly out of youthful bravado and partly out of prudence in being ready for some of the meaner native dogs which liked to worry the ponies, one often carried a small Colt automatic, dangling in full view.

"You are a couple of bloody fools to wear those guns!" an old-timer up the river told two young gunboat officers at Chungking. "The commies don't give a damn about *you*—but they want firearms. One fine day a commie will spring out of the ground and pinch your bloody pop guns and very likely do you in if you are foolish enough to resist. Those blokes play for keeps. They bloody well spit on the gunboats and on Chiang's chaps too!"

Nevertheless, the gunboaters strapped on their guns and rode away into the hills. The only real annoyance on such expeditions were the swarms of small boys who tagged along and fought to retrieve the empty shell cases, still smoking hot as they flipped out of the guns during target practice. Nothing ever went to waste in China.

1923-1931

New Gunboats on the River
Life "At Sea" on the Yangtze
Changsha Tales
Bandit Shootups
Captain Baker Up for Ransom
Light Tales of Shanghai and Upriver

When six new American gunboats reached the Yangtze all at once, ComYangPat and his gunboat sailors walked with a springier step. The Shanghai American Chamber of Commerce and the U.S. consuls along the river rejoiced. The missionaries didn't say so, but obviously felt better. ComYangPat considered that he no longer rated "with but after"—the Navy's differentiation between almost-but-not-quite equals. He now stood solidly "with" the Royal Navy's Rear Admiral, Yangtze. In fact, so clearly "with," that by the last quarter of the thirties, the Royal Navy felt the pressure to the point of exercising their traditional Yangtze Valley one-upmanship by assigning a vice admiral to the job, with a rear admiral on roving patrol.

All the new U.S. gunboat power arrived not a moment too soon. For the third time, the wheel had come full turn. In the early days, it had been the *fanquei* warships against Imperial Peking. Then, following the Taiping rebellion, bandits proliferated on the Yangtze until the Imperials gradually reimposed order. With the crack-up of the Imperials in the 1911 revolution, disorder prevailed on the river during the nearly two decades the provincial warlords were riding high.

Chiang Kai-shek's Nationalists reestablished central control and order in 1927. But his "Times of Trouble" soon brought back riverine disorder in the form of bandit communists. As bandits and gunboats all worked the same beat, they were bound to meet.

With few exceptions, every CinC Asiatic since Rear Admiral Bell in 1866 had clamored for new, suitable, shallow-draft gunboats. Instead, the Navy had sent out a succession of misfits and worn-out crocks. The *Palos* and *Monocacy* opened U.S. naval operations on the upper Yangtze, but they were obsolete even as they slid down the launching ways in 1914.

The 1923 disorders on the Yangtze brought a loud chorus of complaints from the Shanghai American Chamber of Commerce and powerful U.S. church societies anxious about their missionary brethren. On 16 October of that year, Secretary of State

Charles Evans Hughes wrote a strong letter to the Secretary of the Navy urging him to expedite some new Yangtze gunboats. The United States was losing too much prestige, he said, by having to depend on the more modern ships of other powers.

A year later, on 19 December 1924, Congress at last authorized six new gunboats, but cagily refrained from appropriating the necessary funds. Four more months passed before the money was forthcoming. The CinC Asiatic wasted no time. Ever since 1919, the staff had been writing and rewriting specifications, even calling on the "Old Man of the River," Captain Cornell Plant, Inspector of the Upper Yangtze, for advice. On 30 September 1925, bids were opened for final consideration. To meet "the peculiar conditions of navigation on the Yangtze Kiang," the CinC wanted a 150-foot, 385-ton, twin-screw, tunnel-stern ship of 4½-foot draft, with two diesel engines to give 15 knots, and bunker capacity for 1,000 miles at 10 knots. One never knew where the next filling station might be open on a troubled river. The steel, engines, principal auxiliaries and armament—two 3-inch, high-angle guns and eight machine guns—would be sent out from the United States.

The contracts were let to the Kiangnan Dock and Engineering Works of Shanghai, and the keels of all six boats were laid in 1926.

On the raw, blustery morning of 28 December 1927, the USS *Guam,* first of the "new six," was commissioned.[45] Except for triple-expansion steam engines in place of diesels, she followed specifications almost exactly.

Lieutenant Commander R. K. Awtrey read his orders to take command in the presence of Admiral Mark Bristol, CinC Asiatic, Rear Admiral Yates Stirling, ComYangPat, and a host of officer and civilian guests. Chinese yard workmen sculled themselves around the ship in a fleet of sampans, half-hidden in a pall of smoke from firecrackers. This racket was guaranteed to frighten away the hordes of devils sure to be skulking about, waiting to move on board and make mischief. The workmen themselves, emboldened by the hideous noise of gongs, rattles, and crackers, added to the general din by shouting highly obscene suggestions as to what the devils could go do next.

The *Guam* was christened with Whangpoo River water by the young daughter of Commander Bryson Bruce, superintending constructor at Shanghai. Years later, she recalled that he "was of the old school, and felt the launching of a ship was its maiden voyage, the honors to be done only by a virgin," adding that these were not abundant in Shanghai.

As for the river water, it was felt slightly blasphemous to spatter champagne over a warship representing bone-dry America.

Perhaps the unorthodox christening affected the *Guam*'s fate. She would indeed become the "Chameleon," unique in naval history, destined in her lifetime to sail under four different flags and five different names.

As with people, ships have distinct personalities. They can be perverse, or capricious, or dependable, or even brave. Some ships are held in such affection that one would like to caress their well-rounded stern sheets. Others, cranky and mean, arouse only a basic impulse to kick them where it hurts. The big question was, what would the *Guam* be like? Would she be lovable, a bitch, or the comfortable old shoe sort? The gunboaters who sailed her found she had a little of all those qualities.

Admiral Stirling could scarcely wait to give the new ship a whirl, and went aboard on 19 January 1928 for a real shakedown—thirteen hundred miles to Chungking. Five days later, he was able to proudly show her off at Hankow to six British warships, three Japanese, three French, one Italian, and *five American: Palos, Monocacy, Penguin, Villalobos,* and *Helena*. If ships can feel jealousy, indeed these aging queens of the Yangtze must have, at the sight of their spanking new colleague. With four line officers and a doctor aboard there would always be enough talent available in even the most remote port for a bridge game, a professional appendectomy, or tennis at the club, plus enough conversational potentialities to keep the wardroom from going "Asiatic."

The fifty sailors in the crew felt they were on a yachting cruise. In the fireroom, oil flowed to the burners at the twist of a valve to keep a constant full head of steam in the boilers through the longest pull up a rapid. There was no filthy coal to shovel, no clinkers to break up, no ashes to hoist or superannuated machinery to coddle. On deck, five Chinese "boatmen," at the equivalent of US$5 a month each, did all the dirty work. Every man had a springy pipe-frame bunk in the airy, screened deckhouse above the waterline. Second helpings of eggs and pancakes, shined shoes and pressed pants were taken care of by the mess coolies.

Even the most unpromising garbage never reached *Guam*'s spotless slop chute. Not so much as an empty egg shell went unwanted in that hungry land. Vegetable peelings, coffee grounds, plate scrapings, and more choice items were segregated with minute care by the Chinese mess cooks, who passed this nectar tenderly over the fantail into the eager hands of the sampan fraternity. In return the sampan coolies scrubbed the ship's sides to gleaming whiteness twice a day.

Working her way slowly up the middle river, giving ComYangPat a chance to take everything in, *Guam* was initiated into the business by shepherding a convoy. On 1 February 1928, Standard Oil vessels *Mei Lu,* towing *Mei Ying,* and *Mei Foo,* towing *Mei Hung,* started upriver through the "bad lands." In a couple of hours, *Mei Foo* ran aground but *Guam* pulled her off. In two more hours, *Guam* was at general quarters; *Mei Hung* had been fired on from the beach. Two days later the convoy again exchanged fire with some "sportsmen" on the banks. Foreign skippers of the merchant ships carried high-powered hunting rifles on the bridge and cheerfully joined the armed guards in potting at the opposition ashore. It was just like the christening ceremony all over again, except the firecrackers were not blanks.

Like chicks popping from a clutch of eggs, the rest of the "new six" were out of the nest by mid-July 1928. *Tutuila* joined the *Guam* at Ichang, after *Guam* had made a round trip up the gorges to Chungking. Then the two took a convoy upriver, where at Chungking Admiral Stirling shifted his flag to "*Tutu*."

Like the *Guam, Tutuila* also had had a slightly unorthodox christening. Her sponsor, fifteen-year-old Beverly Pollard, was well-indoctrinated in the ritual of the Roman Catholic faith, and the solemnity of it all must have been too much for her. She hauled off with the bottle (champagne this time), and smashed it across the ship's bow. Then, blowing her carefully rehearsed lines completely, she shouted, "I christen thee USS *Tutuila,* in the name of the Father, the Son, and the Holy Ghost!" This clearly unpremeditated piece of powerful joss must have been appreciated at command level, as *Tutuila* was the only one of the six to survive World War II in allied hands, and one of the pair which escaped total destruction.

The standard American theory, that if it's *bigger,* it has to be *better*—was nowhere better disproved than by the other four new gunboats. The *Oahu* and *Panay,* next down the ways, were thirty feet longer and eighty tons heavier. The last two, *Luzon* and *Mindanao,* were *forty-eight* feet longer and almost *two hundred* tons heavier. All shiphandlers know that in any tight maneuver, the closer the bow is to the stern, the better. For demanding tight maneuvers, the upper Yangtze has no equal. By 1933, the gunboaters would have concluded that the planners had gone wrong more than once.

In August 1928, *Mindanao* shook down with a trip up to Chungking. Only in the summer and early autumn months was there water enough to accommodate her clumsy oversize. Her commodious flag quarters, almost half as big as *Monocacy*'s entire deckhouse, were occupied by Admiral Mark Bristol, CinC Asiatic, personally finding out what was what up the Yangtze. His program of winning friends among the Chinese, as he had in Turkey, had failed on the river above Hankow, for reasons made evident in the *Guam*'s first upriver chore—convoying merchantmen.

On 28 January 1928, the American SS *I'Ping* reached Ichang from downriver. Belonging to the Yangtze Rapid Steamship Company, she was typical of the ships that sailed the upper river, and no beauty. Unlike the low freeboard gunboats, her slab sides rose high above the water, to provide cargo space and to make it impossible for uninvited guests to jump on board from sampans. Six transverse watertight bulkheads would localize flooding when she hit the inevitable rock. The triple-expansion steam engines gave her 2100 horsepower. She was steered by three rudders. Twelve first-class passengers could be accommodated in neat little cabins, their muddy river bathwater cleared by powdered alum. There was a comfortable lounge, a dining saloon with food worthy of the Palace Hotel, and a busy bar. The whole topside was a large promenade where the wonders of the upper river could be enjoyed in full panorama. Below, there was space for 325 tons of cargo and steerage passengers in numbers kept a secret by the Chinese staff; the captain neither knew nor cared. The bridge was a little citadel of armor plate and steel bar gates, hopefully proof against bullets from ashore and bandits among the passengers.

With 128 tons of cargo and 31 tons of coal on board, *I'Ping* would have to wait until the spring rise, perhaps April, to make it up the rapids to Chungking. Meanwhile the captain could look forward to a comfortable two or three months playing pool, loafing, and reading, while checking from time to time the quality of the whiskey at the Customs Club.

His hopes were soon dashed by a peremptory directive from the local Chinese general to hurry downriver to Kulaopei to pick up some of Yang Sen's troops, retreating before the Nationalists out of Hankow. At her destination the following day, *I'Ping* was soon swarming with two thousand lousy, starving, undisciplined soldiers which she took to Ichang. A second trip was ordered and an even greater state of confusion found at Kulaopei. A number of sampans carrying troops got swamped in the panic of loading and many of these screaming, disorganized wretches were drowned. Loaded to the main deck scuppers, *I'Ping* once more chugged back to Ichang, tried unsuccessfully to lighten ship by offloading cargo into sampans—the soldiers refused to work—was forced to take more soldiers aboard, and received preemptory orders to get under way upriver instantly.

On the morning of 1 February, approaching Kung Ling rapid, *I'Ping* struck a submerged rock and water began to pour into the hold. She was beached, unloaded, and given the usual jury repair patch of cement hurried in its hardening by the addition of soda. The whole operation was speeded by a gentle hint on the part of the Chinese commander that unless she was ready to proceed by noon two days later, the captain and crew would be shot.

The ascent of the next two rapids required unloading the troops below and reloading above, during which indiscriminate firing between those ashore and those aboard was routine. To fully appreciate the feat of clawing up these rapids, one must remember that the river was at the lowest of low water. The current sometimes ran at twelve or more knots. In the wickedest rapid of all, the Hsin T'an, fully a thousand men heaved and struggled at bamboo ropes and steel hawsers, literally lifting the ship eight feet in her own length as she foamed and slithered up the rushing tongue of water.

The *I'Ping* arrived at Kweichow on 5 February and found the city in a complete state of confusion. Soldiers ashore demanded to be taken aboard. Those aboard insisted the ship proceed immediately upriver. All concerned punctuated their arguments with indiscriminate rifle fire, and if anybody was in command at any echelon, there was certainly no evidence of it.

Thus, with even more troops jammed aboard, *I'Ping* approached another stiff rapid, the Hsin Lung T'an, finally disembarked the troops after prolonged quarreling and firing amongst them, heaved the rapid, reembarked troops, and after a week-long sleepless, near-foodless nightmare for the ship's officers and crew, reached Wanhsien on the evening of 6 February.

"You should have seen the vessel after the men had gone," a witness reported. "You would have thought that we had been carrying hogs instead of men. It was covered with filth from one end to the other, so deep that we had to shovel it off. Beds were destroyed, mattresses ruined, tables and chairs broken, and all fixtures gone. The ship needed a complete refitting."

In South China, the Navy maintained as feeble and unsuitable a collection of craft as it had on the Yangtze before the arrival of the "new six." The *Wilmington* and *Helena* served in rotation as flagships until *Wilmington* sailed out of Hongkong on 20 May 1922 for the last time, headed for Cavite and the United States. Their 9-foot draft limited their penetration inland to Canton, a day's voyage squeezing over bars with little to spare. The hinterland, such of it as was reachable on a 6½-foot draft, was left to the *Pampanga,* an 1888 ex-Spanish twin of the *Samar.*

Finally, the Navy decided to equalize conditions, so on 5 October 1928, *Guam* set out from Shanghai, headed south. She island-hopped down the coast, convoyed by the minesweeper *Penguin,* and on 14 October proudly steamed into Hongkong. Anchored by night and steaming by day, she only once had to turn back, to Amoy, to await better weather.

Shortly before *Guam* departed for Hongkong, her skipper noted in his journal that American interests had just bought the Shanghai Electric Company for 81,000,000 taels, roughly US$50,000,000—another large package for the U.S. Navy watchdog.

On 1 January 1929, the foreign concessions in Hankow took the final step. Surrendered almost exactly two years previously to become "special administrative districts" in transition to Chinese control, the reversion to China was now complete. The governing body with foreign representation was abolished. Taxes collected were pooled with those of the native city. The rate inexorably would go up and maintenance down. Clearly, Chiang and his Nationalists were feeling their oats.

This period was also one of rising Chinese business influence and independence. Chinese banks and firms were dealing directly with foreign countries and by-passing the local *taipan* representatives. Chinese Christians were assuming greater control of the church in China. Greatest change of all was the adoption of a protective tariff on 1 February 1929 which replaced the five percent ad valorem that had short-changed China and tied her hands financially since the mid-nineteenth century.

The *Monocacy* and *Palos* were in a class by themselves—certainly far below the "new six," but not as decrepit and unsuitable as the Spanish prizes. Perhaps a face-lifting would help, and in 1929 it was proposed that either diesels or oil-burning boilers be installed. As far back as 1926 the sad state of the *Palos* had been reported to the CinC. Between 14 June and 15 July of that year, optimum cruising time for the upper river, she made three unsuccessful attempts to climb up to Chungking. Her boilers were of faulty design, ComYangPat's letters repeated for the umteenth time. Combustion took place halfway up the smokestack which spouted flames after each scoop of coal. There was no ash hoist; when fireroom hatches were opened to manhandle the buckets of clinkers on deck, the forced draft was killed and the fires lost their zip. The ship had to spar moor every three hours, to clean fires and give the sweat-drenched stokers a blow. There was no armor on the main battery. There was not enough space for the crew, some of whom had to sleep and eat on deck, regardless of weather. But, as always, nothing came of ComYangPat's appeal.

By 1929, Chiang and his Nationalists had consolidated their control of central and north China to the point that the U.S. Marine brigade at Tientsin had been withdrawn, leaving a 500-man legation guard at Peking and 1,200 at Shanghai, plus the permanent unit—one battalion—of the 15th Infantry at Tientsin. It was a year of mild sniping up the Yangtze, rather than the former organized animosity. The *Tutuila* had been the target of some potshots while approaching Ichang, which prompted her skipper, Lieutenant Commander S. D. Truesdell, to complain to the local general.

"They are only country boys. They mean no harm—just plain, youthful exuberance," explained the general. Truesdell admitted that he had the same problem. His crew included many new recruits, and just the other day he had caught some of them fooling around with the 3-inch gun on the fantail. They had a live round in it and it was pointed at the general's yamen. The yamen was white and conspicuous, Truesdell explained. It made a good target. Thereafter, when *Tutuila* steamed past Ichang, there were the same troops, picking lice out of their nondescript uniforms and sunning or bathing themselves on the foreshore as before. But there were no potshots.

There were many variations of the fringe nuisance game. One little band had established a "toll gate" on the middle river on the authority of an old Krupp field piece. Their knotty fire control problems had been solved by boresighting at a whitewashed rock on the opposite bank. But the project failed to prosper; when a gate-crashing vessel interposed between gun and rock, as seen down the gun bore, a shell was slammed into the breech and fired and a miss astern invariably was rung up. The vessel would proceed untaxed and unharmed, to the chagrin and mystification of a gunner who, one suspects, had never heard of the merits of leading a moving target. The average gunboat skipper felt genuinely reluctant to discourage such harmless enterprise.

Events elsewhere on the river had commenced to "normalize." The U.S. consulate general at Nanking, so brutally violated in 1927, had been reopened. The original American proposal had been that for the flag raising, the Chinese furnish a band and an honor guard, with high-ranking representatives of the Nationalist government, army, and navy on hand. The Tiger forts would then fire a 21-gun salute, returned by an American gunboat. *Then,* provided that meanwhile the U.S. government had already recognized the Nationalists, or was prepared to do so, the gunboat would fire twenty-one guns with the new "sun" flag at the main, to be answered by the fort, gun for gun. But the Chinese refused to "humiliate" themselves, and soon Minister MacMurray came down from Peking to announce that Washington had ordered the consulate general opened without Chinese participation. A sailor guard from *Luzon* stood by while the colors climbed the pole, to the accompaniment of a 21-gun salute from the gunboat. *Then, Luzon* fired a *second* 21-gun salute with the Nationalist flag at the main, returned by the Tiger forts, gun for gun. Thus, by diplomatic pirouetting of a type sometimes only dimly understood or appreciated by gunboat sailormen more used to direct means, the deed was done. The "Nats" were in.

Presumably, all hatchets were buried, and nervous and restless Chiang Kai-shek was entertained at dinner aboard *Luzon* by ComYangPat, Admiral Stirling. Also present were several who would make headlines for years to come: the one-time crown prince to Sun Yat-sen, Sun Fo; moneybags T. V. Soong; double turncoat Wang Ching-wei, and others of lesser renown.

In late April 1929, *Mindanao* lay in Shanghai's Kiangnan Dock and Engineering Works preparing for the tricky passage to Hongkong. "The great danger to be avoided is having water enter the fire and enginerooms," wrote Lieutenant Commander A. W. Ashbrook in his diary. He was building bulkheads around the fireroom blower intakes and engineroom doors to prevent this.

Upriver from Kiangnan, the Chinese aviation station had just received four new German-built float planes. They were furnishing daily entertainment for the gunboaters, leading Ashbrook to observe that "it seems as though the Chinese pilots handle planes about the same as Chinese chauffeurs do automobiles, that is, they plan on coming as close to objects as they possibly can without hitting them. They appear to take delight in just missing sampans or making sampans get clear of their way." This method apparently had certain drawbacks, as Ashbrook noted that one of these daredevils crashed about a thousand yards from *Mindanao*. "Rushed our motor sampan to the rescue," he wrote, "but a sampan had already pulled the pilot out of the water. The plane will probably be a total loss."

On 30 April, the tanker USS *Ramapo* filled *Mindanao* to 10,000 gallons less than full capacity, leaving the two forward tanks dry to put her nose up and her tail down and thus give a better bite for her propellers.

Departing Shanghai 2 May, *Mindanao* sailed by day and anchored by night, hugging the rugged China coast. On 18 May, belying the apprehensions of Skipper Ashbrook, "Fat *Minda*," as her crewmen affectionately called her, tied up to Hongkong's man-of-war buoys.

In justice to truth and the history of the Navy in China, it would be derelict to overlook some of the ancient craft inherited from Spain after the Spanish-American War. No matter what their faults, they had become like old friends who did the best they could with what they had.

Built of wrought iron in Spain in 1885, *Elcano* (or as the Spanish called her, *El Cano*, the Canal) displaced 620 tons. For a ship her size—165-foot length and 26-foot beam—she packed probably the most formidable armament of any ship in the U.S. Navy: four 4-inch rifles and four 6-pounders. Lieutenant (junior grade) Roger E. Nelson, during his time aboard, never saw any of them fired: "We had no ammunition allowance for target practice and no orders to conduct any." Actually, the 4-inchers were museum pieces, with iron sights and a complicated three-movement breech block. There was no loading machine nor dummy ammunition, so that the gun crews remained innocent of any experience in actually slinging a shell into this venerable artillery.

Ships with distinct personalities generally had nicknames, and the *Elcano* was no exception. Each of her propellers was driven by a two-cylinder compound reciprocating engine. Each engine had a pump which circulated cooling seawater through the condenser. At each stroke of the engine, a stroke of the pump squirted water well above the surface of the river, which led to her nickname, the "Yangtze Sprinkler." There was no vacuum at all until the main engines started to turn over, and not much then. An additional unusual feature, suggesting a railway locomotive as a remote ancestor, was a steam jet in the smokepipe to induce draft, which sent a white plume spouting up at each revolution of the engines.

As the two Scotch boilers lay side by side, when one was drained for repairs a heavy list of the ship made life difficult aboard until the boiler was refilled and trim once more restored. Nelson later wrote that:

It took a long time to raise steam, then, the optimum 125 pounds showing on the gauge, the ship was ready to get under way. By the time the anchor engine had done its work, the pressure would be down to a hundred pounds and as soon as the main engines began to turn, the steam would drop still further. We then had a choice; make turns enough to stem the current and have the pressure continue to drop until water came over into the engines and blew off a cylinder head, or slow the engines to allow pressure to be maintained and so have insufficient power to move upstream. The Captain would try to maintain a balance. But in the end, he'd have to shout, "Let go the anchor!" The pilot would scream that we couldn't anchor there. But with no more steam, the anchors had to go down, and then, likely as not, they'd drag a bit before they caught and held.

Dragging anchors usually meant the two would have come together and locked flukes. So then began the repeated raising and lowering until they shook loose, by which time the pressure would be gone again.

The steering engine, a small steam affair, was if nothing else, in a position for easy checking. If the conning officer on the birdge under way stepped two paces backward from the open bridge rail he got a burned posterior. Sharing the bridge with the steering engine was a wheel, a voice tube to the engines, the two engine room telegraphs and a magnetic compass. The fire control equipment was there too—a voice tube to the 4-inch guns and a single pole, single throw open knife switch to the buzzers.

The captain's quarters were fairly commodious for such a small ship. They were aft, and because of the hull conformation, shaped like the end of a bath tub. The skipper enjoyed the luxury of a tub and flush toilet, but lesser officers used a small steel cubicle on deck, barely large enough to squeeze into, which housed shower and archaic sanitary facilities.

Just forward of the captain's cabin was wardroom country, with two small staterooms on each side of the ship opening directly into the very small rectangular wardroom itself. Up through the center of the latter went the mainmast, bearing a set of deer antlers as decoration. [The ship originally had been rigged for auxiliary sail.]

The rooms were so small that Nelson had to step into the wardroom to pull on his trousers. Ventilation was sketchy. Each stateroom had a screen-covered porthole so near the waterline that not infrequently the breeze, if any, was generously adulterated with dirty Yangtze water. To the benefit of the multitude of bugs, Chinese boatmen, struggling upriver in their junks, were constantly knocking out the screens by using the portholes as a convenient boathook hold.

There was no pantry or galley for the wardroom. A messboy carried food along the deck and down a ladder from the crew's galley, dodging flies en route. The menu was largely wild duck, wild pigeon, and pheasant, with very little red meat. The fact that this gourmet fare invariably came on toast in no way mitigated the monotony of it. With no ice machine and no cold storage, life was hand to mouth. Or more properly, compradore to cook, day by day.

In Ichang, Wednesday dinner was baked bean night, a welcome variation from wild game on toast. The treat was jointly much looked forward to by ship's officers and the routine two guests—all the extras the wardroom could seat—from the foreign colony of some sixty missionaries, businessmen, and customs officials.

For one hour each evening, electricity provided by a gasoline-driven generator coursed through the otherwise dead circuits. This powered the radio for the daily contact with "outside," the only source of news. It also allowed the ancient light fixtures to cast a feeble glow and churned a mild breeze from one or two electric fans. "When the generator stopped, we used candles," Nelson recalled. "Each officer carried one to his room. But there weren't enough candlesticks to go around. Being the junior, I had to put mine on a saucer, and by that light I did my reading and wrote my letters in my little room."

Through the long winter, the crew kept busy on the rifle range—a flat piece of river bed uncovered by low water. It was sometimes cold and raw on the windswept shingle bank. Still, fortified by a whopping breakfast, the men enjoyed the competition and the picnic atmosphere of a day-long stay in the open. They also noted the Chinese distaste for cheese; even hungry beggar boys clustered around the men eating their basket lunch passed up the sandwiches of rubbery standard Navy cheese.

For mild recreation ashore, there was the inevitable baseball, abetted by a little place rented to serve as a sort of club. The chief commissary steward, a fellow enormously fat in the tradition of that rate, was in charge of the "club." On being refused more beer one fine night, an aggrieved seaman, among other uncomplimentary things, called the chief "a pot-bellied son of a bitch." His honor thus sullied, the chief reported the man for a court martial offense. The trial required every officer on the ship—the captain as recorder, the doctor as counsel for the defense, and the three remaining officers as the court. All the witnesses clammed up, and it could be proved only that the chief had been called "pot-bellied." He was that all right, so they had to acquit.

In 1925, C. R. Cox Company, Ltd., operated six American flag steamers, three pairs of sisters, on the upper Yangtze, and Nelson made a number of armed-guard trips on the smaller *Chi Chuen*.

When the *Chi Chuen* was shot at on the downriver passage, and Cox had received a threat she would be sunk by gunfire if she attempted to sail upriver again before handing over the proper "squeeze" for alleged opium violations, Nelson and guard were told off to protect her.

The armored bridge shutters were pierced by 4-inch square peepholes, too small for practical use as fireports. So Nelson emplaced his three Lewis machine guns on top of the pilothouse. They were well exposed, but depending on the usual execrable Chinese marksmanship, he felt he had gained an advantage by having an open field of fire.

The strategy paid off. "When we came abreast of Paotung," Nelson recalled, "there was a sort of seawall. We had to cruise fairly close to shore in a swift current, so that the ship moved very slowly ahead. On a level space, a couple of companies of soldiers were drilling. When we hove in sight they lined up on the bank in two ranks, the front rank kneeling, rifles pointed at *Chi Chuen*."

Belly-flopped there on the pilothouse, creeping by this silent line of potential executioners, Nelson and his two best marksmen carefully set their sights and pulled back the cocking bolts. If not eyeball to eyeball, it was certainly muzzle to muzzle.

As the ship inched by, the theory was proved once again that time is not an absolute yardstick; some five minutes are longer than others.

The powerful all-pervading stench of opium, constantly being smoked all over the ship, was one of Nelson's strong recollections of river boats. Another was the "pidgin" cargo, the illegal freight—tiny packages or large bales brought aboard by crew members for private profit—to the point that the ship's draft was increased beyond the safety mark. There was little the lonely captain—sometimes the only non-Chinese aboard—could do in such cases. A call for help from the gunboat would engender bad blood in the crew, little more than a band of cutthroats at best. But sometimes there was no choice. The guard would be called, then have to stay with the ship for the full trip, to make sure the pidgin cargo didn't surreptitiously return, or the black-tempered crew make trouble en route.

What with all this demand for armed guards, *Elcano,* the perennial Ichang station ship, had to be overmanned in order to provide them. Consequently, when less than the usual number of guards were off the ship, there were more men aboard than the below-decks hammock hooks could accommodate. The overflow had to make-do topside, flemished down on deck, screened from the sometimes nippy river winds by canvas curtains and awnings.

On one occasion, far more than the usual four or five rode shotgun when the *I'Ling* had been accused of swamping a sampan with its wash. As might confidently be expected in any such case, this was no ordinary sampan; it was carrying the payroll of the local ragtag army.

When the offending merchantman dropped anchor at Kweichow, her skipper, D. B. Hawley, a Naval Academy graduate who had resigned, was promptly taken into custody by the Chinese and hauled off into the city. There he would stay, and the ship would be held as well, proclaimed the Chinese, until that payroll had been made good. Lieutenant Commander Rush S. Fay, the *Elcano*'s skipper, rose to the occasion. Commandeering another Cox steamer, and putting forty of his crew aboard, he set off to the rescue.

Nelson was in the party, and had his first sight of the upper river beyond the Ichang bend. He found it impressive:

We emerged from the Windbox Gorge at dusk the next day to a scene that seemed to my young eyes like some sort of inferno. On a flat space at the end of the gorge, kettles boiled over many fires. The flickering flames, the smoke, the steam rising from the kettles, the vaguely seen figures moving around, the towering walls on each side of us, and the near darkness of nightfall gave the thing an eerie quality. Actually, it was nothing more dreadful than the thrifty Chinese boiling brine for salt.

As we anchored off the walled city of Kweichow, a detachment of soldiers came down to the foreshore and drew up facing us. Was it shoot or talk? Captain Fay decided to see what he could do with talk. He got himself into full dress uniform, medals and all, and very courageously I thought, went ashore alone to beard the lion. How he did it, I don't know, but in four hours he was back and with him was Hawley.

The latter was none the worse for wear, but very much relieved. All he wanted, he said, was a stiff drink and a bath in lysol. The next day, with Hawley once more restored aboard his ship and with part of our guard, we sailed together back to Ichang.

Nelson, as junior officer aboard the *Elcano,* generally was tapped for the official boarding calls on other men-of-war. The British, he found, and as most Old River Rats discovered, always made one welcome, plying the visitor with horse's necks, each insisting that one be had with him in turn. "Then they had their fun, pouring me into my motor sampan to return to *Elcano* with my little boarding book." The French served wine and were very proper and polite. Nelson spoke French quite well, unlike most of his American colleagues, and could exchange small talk with them. "But the Japanese never consented to anything more than the required formalities, and damned little of that. I was never permitted more than a step inboard from the gangway, and none but the required questions concerning rank of commander and so forth were permitted."

To the average civilian, the formality and inevitability of official calls might seem to have been right out of Gilbert and Sullivan. For these rituals, one wore a knee-length frock coat embellished with grapefruit-sized gold-fringed "swabs" (epaulettes) which extended the wearer's beam several inches. Down the pants legs ran "railroad tracks," wide gold stripes. A gold-encrusted belt held up a four-foot sword, which commanders and above sometimes affected to carry by the grip. It was also a tradition that a commodore rated having one button of his fly undone, and that anyone who had rounded Cape Horn could cast loose the top button of his jacket. But the Panama Canal had made Cape-Horners few and far between, and when zippers replaced certain buttons these privileges fell into disuse.

Topping off this magnificent display was a cocked hat, projecting fore and aft some eight inches, decorated with cockade and gilt trim. Clawing up the vertical ladder and through the scuttle hatch of a British destroyer wardroom while so attired called for finesse of a high order, especially after three or four glasses of hospitality.

And hospitality was inevitable and overpowering. A bottle opened for an official guest was "on the King, God bless him!" The remainder of the bottle was taken in quick hand at no cost to the wardroom after the guest had fought his way out and climbed into his boat alongside. If the official guest could be cajoled into sampling first a scotch and splash, then a brandy, then topping it off with perchance a Pimm's Cup No. 2, each out of a virgin bottle, then *three* practically full bottles of His Majesty's best fell free into the waiting hands of his faithful officers.

It was not uncommon to receive a signal from a neighboring British gunboat about 9:00 A.M. on a Sunday morning, requesting that a boarding officer be sent as soon as possible for official dispatches. Gunboat officers, in full regalia, performed these errands of mercy on numerous occasions, probably qualifying for some sort of minor

British life-saving decoration. The preceding Saturday evening's activities would have been late and hairy for the British. Their wardroom bar opened by strictly observed rules at 11:00 A.M. But if an official guest were aboard, this burdensome schedule no longer obtained. In his honor, a horse's neck (brandy and ginger ale, with a twist of lemon peel), favorite restorative in such emergencies, would be instantly brought forth all around and the "official despatches" considered as having been handed over. The boarding officer would then ease on back home, while his grateful hosts could with fair equanimity manage to sit out the Sunday forenoon religious services droned through by the skipper, at which attendance was obligatory for all hands.

Religious services in the Royal Navy, while not necessarily inspiring, could be memorable. A well-known and well-liked British skipper was conducting the Sunday service aboard his gunboat when officers from the *Tutuila* arrived on board for the usual Sunday curry tiffin a little too early, and assembled on the forecastle, enjoying a gimlet in anticipation of the end of the sermon. It was a hot, cloudless day. Flies droned around the sailors' heads as they sat on deck, fidgeting and thinking of the beer awaiting them in the canteen ashore. "Look hyah, you blahdy bahstahds!" boomed the preacher-skipper to his congregation. "If you don't stop that damned moving abaht, I'll read the whole blahdy thing ovah agayn!"

It was not on that same Sunday, nor even in that same port, where there occurred one of those typical examples of British low-key humor which fairly fractured the collective funny bone of the River Rats. The British Rear Admiral, Yangtze, was staging a welcoming party for his colleague and American counterpart, a newly arrived admiral ordered to duty as ComYangPat. The new admiral had a charming daughter, but she was a teen-ager—younger than the British considered eligible for formal social affairs. The British invitation, accordingly, was to Admiral and Mrs. ComYangPat only. The new ComYangPat, whose daughter was used to being included in such affairs back home, thought that perhaps a broad hint was indicated, and signaled to his British host: "I have a daughter." The answer was immediate and memorable: "Congratulations!"

Duty on board the *Elcano* proved to Nelson that it was not necessary to ride a Cox steamer as armed guard to enjoy action. If one waited around Ichang long enough the action eventually sought one out. This, *Elcano* confirmed one morning during the process of sweeping away weekend cobwebs.

Part of the routine was the periodic "sighting of anchors" to insure that they had not been silted in to such a depth that the anchor engine could not break them out again.

With the anchors picked up and redropped one day, *Elcano* found herself in line abreast, thirty yards outboard of the Japanese *HIJMS Seta,* who in turn was thirty yards outboard of HMS *Scarab*. The latter was closest to the beach—thirty yards from the post office pontoon.

In the afternoon a freshet was reported on the way downriver. A rising river meant the ships would yaw and be too close together for safety, so about 6:00 P.M. *Elcano* prepared to move. While heaving in the anchor chains, she took a rank sheer to starboard, bringing her across *Seta*'s bow. The sheer had broken out the newly laid anchors, which commenced to drag. Ringing up full speed, *Elcano* tried to clear *Seta* ahead. She cut the latter's port anchor chain, tore out her windlass, and wound up both *Seta*'s anchor chains in her churning screws.

Both ships then dragged down on *Scarab,* holing her and *Elcano* above the waterline. Like drowning rats clinging to each other for mutual support, all three gunboats next dragged down on the pontoons moored to the shore and there, for the moment, they all managed to hold on. With the river rising and current increasing, wires were got out in every direction to hold what they had. In the darkness and confusion the beset gunboaters could see that if the current hit the starboard (land side) bow of *Scarab,* the whole circus would carry away and sweep downriver, taking with it everything in its path.

Then the heavens opened up. For two days all hands worked watch-and-watch around the clock, drenched to the skin in pelting rain, trying to clear the chain anchor cable fouled in *Elcano*'s screws. No ship could move until this was done. Finally, on the second night, one of *Seta*'s chains was picked up by grapnel and brought aboard *Elcano.* Nelson had vivid memories of the scene that followed.

> *We finally convinced the captain of the* Seta *that we had to cut his starboard chain. I can see him sitting there on his capstan, stubbornly refusing to permit it and sucking wind through his teeth, saying, "I am sinking an idea." Actually, he never produced one. After we cut his chain with a hacksaw we passed the anchor end of it to him. He shackled up his port chain to his starboard anchor and hove it in. We still had the other half of his port chain wrapped around our props.*

Thus freed from the riverine Laocoon group and with at least one anchor and chain to hold her, *Seta* moved out into the stream. It was one down and two to go. But life had become an increasing misery. *Elcano* coaled ship in the midst of it all, and the dust mixed with the incessant rain to make a black slush that gave hawsers, deck, and hands a slimy coating. The rain poured off the awnings, into shoes and down necks. Nothing was dry. Nothing was clean. No one slept.

On Thursday, they at last grappled up *Seta*'s second chain and anchor, and on the following day *Elcano* was swung out by tugs and moved downriver to the Standard Oil pontoon, almost a week after that fatal Monday. There, she received her starboard anchor and chain from *Scarab,* who had freed herself and dropped alongside.

Divers finally unwound the chain from *Elcano*'s screws and a week later she moved out to anchor in the stream, well below the city, to put things in order. Upriver, the usual dozens of ships lay at anchor.

A couple of days later, the midnight calm was shattered by hoarse cries from the quartermaster: "Turn out! All hands on deck! There's a ship coming down on us!" The Frenchman *SS Kikin* was fast approaching, headed broadside, directly for the ram that protruded well forward of *Elcano*'s above-water stem. If she became impaled on the ram, *Kikin* was a good bet to sink.

On the *Elcano,* they awaited the collision, veering chain desperately so that the shock might not break out her anchors. The ram caught the *Kikin* a bit aft of amidships. She swung around the bow, hung up on the billboard, then came alongside with a crash of rending wood and metal.

Men sleeping topside on *Elcano* had trained out the 'midships 4-inch gun to make more room for hammocks. *Kikin* brought up against it, sending a piece of the shattered training rack flying through the air to hit a man on deck, the only human casualty of the night's work. Holed and leaking, the *Kikin* was hauled off by a tug before she could sink alongside.

The *Elcano*'s existence was so near typical of all the ex-Spanish gunboats that one might almost substitute *Villalobos, Quiros,* or *Samar* in any account of her life. Leaving Ichang for the last time in November 1927, *Elcano* steamed to Shanghai to end her career in support of her successors, the "new six," by acting as receiving ship for their assembling crews.

On 30 June 1928, her blythe new consorts manned and gone to "sea" and her own usefulness ended, *Elcano* was decommissioned. She and the *Villalobos* were towed to sea and sunk in "special target practice" on 9 October 1928. The skipper of the towing ship noted that *Elcano* kept creeping up on them during the night as they lay to, awaiting daybreak. It might have been the weight of the bight of towing chain, but he had an eerie sensation that the old ship simply was lonesome and wanted to snuggle up to a friend for comfort in her last hours.

For a ship, forty years was a long life. The two old veterans had survived floods, rocks, shoals, dragging anchors, Chinese artillery, and other hazards of life on the Yangtze. When they finally went down, it was at sea, under the guns of their own Navy.

Since the dawn of the Christian era, Changsha has found itself in the path of armies surging across China. Lying astride the river-canal-portage system connecting the major commercial hubs of Hankow in central China and Canton in the south, it was a way station for the murderous Taipings in the 1850s, for Chiang Kai-shek in 1926–27, and between times, for groups of bandits, communists, or troublesome rabble under one local warlord or another.

Three attacks by the Japanese in October 1939, another a year later, and a heavy, determined drive in January 1942, were all beaten back by the Nationalists, justifying Changsha's title of "The City of the Iron Gates," earned when it withstood an eighty-day seige by the Taipings.

Proud and independent Changsha, with its twelve gates and high walls, was one of the last cities in China to hold out against demands of foreign missionaries and merchants for entrance. Opened as a treaty port in 1903, it was still almost impossible for foreigners to rent property inside the walls as late as 1920, so that most foreign establishments were situated on an island in the Siang River, which flows past Changsha, thence into broad, shallow Tungting Lake fifty miles below the city.

During the spring and summer floods, Tungting Lake may rise as much as thirty feet above winter level. But if any gunboat drawing more than three feet was caught at Changsha between November and February, the crew could plan on spending several months playing acey-ducey, or becoming experts on grass linen, white brass and pewter teapots, bamboo articles, and firecrackers, for all of which Changsha was well known, while they waited for high water again.

The railway to Changsha from Wuchang was opened in 1918, but from then on, disorder along the line made it far safer to stock up in advance on such locally unobtainable and fundamental items of gunboat existence as fuel oil, tobacco, coffee, toilet paper, and beer. The *Monocacy* and *Palos,* although converted to coal, could burn wood in an emergency, and thus to a certain extent were able to "steam off the country."

In May 1910, Lieutenant Claris, of HMS *Thistle,* wrote in his journal that, "More antiforeign feeling could hardly be found in any other treaty port and there is nearly always a gunboat at or near it. . . . Serious riots occurred in April 1910 and much property was lost. All the hulks were burned and finally before a gunboat could get up all the foreigners had to escape in a river steamer and anchor ten miles below the city."

Changsha's reputation, in July 1930, had in no way improved over that of 1910. On 26 July, rumors spread through the city that "bandit-communists" were approaching. There were no longer any walls to stop them. The walls had been leveled some years before and the stones used for paving. The new invaders, like those of the past millenium, were busy looting, burning, and murdering. The three foreign gunboats present, USS *Palos,* HMS *Teal,* and HIJMS *Kotoga,* had alerted their nationals ashore and informed them of the signal to be given in case evacuation was decided on.

By the next day, the waterfront was a jabbering, swarming mass of thousands of Chinese attempting to beg, buy, or bully their way aboard the many tugs and small craft lying in the river between the island and the mainland. With nightfall, the shooting ashore increased in intensity and at 10:00 P.M. sirens of the three warships sounded the signal for the foreigners to abandon their homes and property to the mercy of the mob and evacuate the city.

The morning of 28 July, with fourteen Americans aboard, the *Palos* put the burning, exploding city astern and dropped ten miles down the river to join the Standard Oil tugs and the other gunboats.

Looting, burning, and shooting continued, with rumors brought out by intrepid runners that the communists were demanding a million Chinese dollars ransom for the city and that they planned to cross to the island in the river, the foreign settlement where British, Japanese, and American business interests were located. Thus, on the morning of 29 July, *Palos* turned her refugees over to the big British gunboat *Aphis* and returned to within a mile of the city for a better look.

There the *Palos* observed one of those typical cases of Chinese ingenuity and ability to survive. The ship's compradore, Fu Chang, who always dressed to the nines and gave every indication of being a rich man, had overnight become Fu Chang the poor man. No longer was he the dandy, arriving in a fancy sampan; now he dressed in blue coolie cloth, went barefoot, wore a grass hat, and rowed himself out in a decrepit wreck of a sampan with a ragged tarpaulin covering the vegetables, eggs, and chickens that kept the crew off iron rations.

Fu Chang also provided good information on the basis of which *Palos* got under way the morning of 31 July and eased upriver to the city. HMS *Teal,* upriver the evening before, had been the object of a few scattered shot and had retired ten miles downriver. The mud was barely hosed off the anchor chain before the fun started. "About twelve shots struck the *Palos* before she returned fire," a crew member* wrote.

By then, the ship was at general quarters, and each man was at his battle station well supplied with arms and ammunition. Both three-inch guns were swung into action and the Palos *became a death-dealing piece of machinery. . . . After the elapse of fifteen minutes, the* Palos *had stood well up into the city and was running very close to the shore . . . from one hundred to one hundred and fifty feet . . . and the little ship was being badly battered by deadly gunfire—many shells penetrating and some passing completely through the ship, through the parts not protected by bulletproof steel. Every man on the ship was doing his part, every man had a certain duty to perform, many were fighting with little or no protection from the communist gunfire—the gunners of the three-inch battery had only a narrow shield to protect them and the machine gunners were little better off. . . .*

Captain Tisdale had left the bridge and was on our little front porch where the forward three-inch gun is located, giving orders, directing the fire. . . .

At 7:45 the Palos *had completed its port run and had completely passed the city and to the head of the island, then she turned downriver and headed once more into the fray.*

Dr. Farnum, a lady missionary who had elected to stay with the ship, was drafted to tend the first casualty, a man wounded in the chest. Meanwhile, *Palos* barreled on downstream, small arms popping and blasts from her main battery jarring the little ship from stem to stern.

A Yangtze gunboat had very little fire control equipment other than the human eyeball and larynx. One peered down the iron sights of cannon or rifle and pulled the firing mechanism when the skipper shouted, "Open fire!" With the exception of the method of charging the ordnance from the breech rather than via the muzzle, things had changed little since 1812. Hence, when the city had been left astern and worthwhile targets thinned out, the exuberance of the crew was such that it took some time for the "cease fire" order to percolate aft over the crackle of machine guns and Springfields.

Proceeding to the old anchorage downriver to lick her wounds, *Palos* found her joss had been good. There were over a hundred bullet holes in her sides and stacks and one bullet had passed through the radio room window, missing operator and equipment by a whisker. But nobody had been killed and only one man lightly wounded. She had fired 67 rounds of 3-inch and 2,000 rounds of .30 caliber.

That afternoon brought reinforcements—two Chinese gunboats and the Italian gunboat *Ermanno Carlotto*—which began lobbing shells into the city. HMS *Aphis* reappeared on 2 August and was promptly fired on from the shore. *Aphis* was something special in river boats. Built during World War I for river service in Europe or the Middle East, she carried two 6-inch guns in turrets and was in effect a small, 600-ton monitor. After a few well-placed shots from those whopping guns, the Chinese realized the grave tactical error of provoking so formidable a foe and their firing

* Believed to be, but not verified as, W. W. Bradley, radioman third class.

promptly ceased. That final action ended the "war" as far as the gunboats were concerned. Whatever the cause—boredom, exhaustion of supplies, payment of sufficient ransom, or just plain lack of purpose—the communists shortly thereafter commenced withdrawal. And as the military, like nature, abhors a vacuum, the Nationalist troops began drifting back across the river and picking up the pieces, their courage increasing as the enemy forces decreased.

On 5 August, the international fleet plus the two Chinese gunboats moved up to the city, which had been put under martial law by the once more cocky Nationalists. Four days later martial law had been lifted, the city had begun to sort itself out, and those of the foreign missionaries who through conscientious objection had refused evacuation and had gone to earth, once more surfaced and announced their unexpected survival. If nothing else, the foreign gunboats had saved the international installations, had rescued from grievous danger a number of foreign men, women, and children, and had perpetuated for at least a while longer the belief that the *fanquei* teeth had not yet wholly been drawn.

Palos remained in the somewhat less than salubrious environs of Changsha until 15 November, when she headed back to Shanghai. A crew member, anticipating their arrival, added with typical Yangtze caution ". . . if we ever get there."

Her shallow draft made *Palos* an ideal candidate for duty in turbulent Changsha, but she was running low on practically everything from fuel to blank forms. So the *Guam,* carrying stores and a deckload of coal in bags, set out from Hankow for a rendezvous below Changsha. Off Yochow, at the entrance of Tungting Lake, all hands had just tucked away a special Fourth of July dinner when the general alarm sounded for battle stations. It was an appropriate date for fireworks, and fireworks there were.

Chinese troops had been sighted ashore. Ten minutes later the first shots rang out, at which *Guam* closed the beach like a terrier after a nest of rats. In another couple of minutes, two shots hit the ship. This was an indignity not to be passed over; *Guam* opened up with the port machine gun battery. Then she reversed course and let them have it with the starboard Lewis guns. The sporting nature of the exchange soon took a more serious turn; Seaman Samuel Elkin was hit in the chest and five minutes later was dead. At this, both 3-inch main battery guns commenced to bark. The ship pressed in closer toward the point from which the greatest concentration of fire was coming. By 1:44 P.M., the celebration of the Glorious Fourth was over as far as the Chinese were concerned. It had taken 34 rounds of 3-inch and 1,200 .30 caliber to silence them, but not before they had hit the ship 40 times.

The British joined in complete unanimity of spirit in celebrating the national holiday. On passing the unfriendly point near Yochow, following *Guam's* skirmish, flagship HMS *Bee* opened up with her big 6-inch, while Rear Admiral McLean, Rear Admiral Yangtze, looked on approvingly.

The *Guam* participated in several more dust-ups in 1930. On 12 December, she silenced communist fire with six 3-inch, high explosive shells, and the next day did it again with a mere four rounds, in both cases accompanied by some 250 rounds of machine gun fire. These were routine repeats of several similar actions earlier in the year, one only 198 miles above Hankow. In three years the *Guam* had easily paid back her purchase price and keep.

Changsha was by no means the only disturbed area. The Yangtze had its share of the action. Most of the dirty work was the irresponsible activity of bandits or communists—the labels were interchangeable. As a result, *Guam, Tutuila,* and *Panay* were on more or less constant roving patrol along the upper and middle river, convoying merchantmen, furnishing armed guards, and exhibiting a brand of truculence when fired on that let the "bandits" know the gunboats had a bite and meant to use it. Gunboats were everywhere. In mid-March 1930, *Guam* shared the Chungking anchorage with a British, Japanese, and French gunboat. *Tutuila* and a Britisher were at Wanhsien. At Ichang, *Panay* lay moored with another little gunboat squadron—HMS *Gnat,* HIJMS *Hodzu,* and the Frenchman *La Grandiere.* There were the usual two or three Chinese gunboats, presumably maintaining themselves by taking in each other's washing, but otherwise contributing nothing to peace and order on the river.

Here, for the historical record and nostalgic Old River Rats, are the names that put Americans in the front lines of their first undeclared war: *I'Ling, I'Ping, Chi Ping, Chi Ta, Mei Lu, Mei Foo, Mei Ming, Chi Nan, I'Fung, I'Chang,* and *Chi Chuen.* They had melodious Chinese names, Chinese crews, and operated sometimes with typical Chinese confusion—but with an American skipper and American guts, enterprise, tenacity, and, at the stern, the badge of American ownership and right to gunboat protection, the American flag.

It was in providing protection that one of the hottest firefights in YangPat's history took place. The *Chi Ping* was on her eighty-fourth voyage, Chungking to Ichang, with 155 tons of tobacco, wood oil, hemp, and native medicines. The cosy passenger cabins were occupied by six *Panay* sailors, Lieutenant (junior grade) Cameron McRae Winslow, and two British subjects, the Nathans. Oiled and ready in the cabins or by the gun slits in the armor plate set up abaft the bridge and on the armored bridge itself were a Lewis machine gun, two Thompson sub-machine guns, two Springfield rifles, two sawed-off shot guns, and two Colt automatic pistols.

The first action came soon after a fog-delayed 10:00 A.M. departure from Chungking. On a long stretch of wide shingle beach, about two hundred soldiers were goose-stepping around in the usual Chinese parody of a drill. Then, with absolutely no warning, it developed that this was no drill; the troops suddenly wheeled toward the ship and let go with a volley. The Lewis gun behaved more or less predictably and straightaway jammed. But after 145 rounds from the Thompsons and a rifle, the battle was over, the soldiers on the run, and *Chi Ping* had acquired a dozen more bullet holes for her collection.

The message must have gotten through to Fowchow, several miles farther downriver. A hundred soldiers stood in double rank, rifles pointed from the hip. But none raised their weapons and none fired.

Chi Ping moored for the night 246 miles above Ichang, where she was joined by *Chi Chuen* and *Chi Nan,* also en route downriver. The armed guards of all three rallied around for beer and to chin-chin over what trouble, if any, might lie ahead.

For once, the unlucky thirteenth of March was devoid of traditional ill omen or luck. The *Chi Ping* left her slower consorts astern and charged on downriver, dodging junks and watching the scenery flash by at close to twenty knots over the ground. She stopped momentarily at Wanhsien and anchored for the night at Kweifu, at the head of the great gorges. Under way at dawn the next day, the *Chi Ping* again soon outdistanced the other two. At 11:00 A.M., she met the Chinese *Wan An,* headed upriver. There was a real hot spot a few miles farther on, they warned—watch out! The river was too narrow to turn and go back. There was only one way—through the shooting gallery, if such it turned out to be.

At mileage 48 above Ichang, *Chi Ping* met a sight to chill the blood. For as far as the eye could see, the north bank was swarming with grey-clad shapes. For two solid miles, the gauntlet stretched along the river; the range to a passing ship was from 50 to 300 yards.

The ship was barely abeam of the first troops when about a hundred Chinese, rifles at hip, opened fire. Immediately, there was the staccato rat-tat of the Lewis, a sound like ripping cloth as the two Thompsons spewed out .45 caliber slugs, all punctuated by the intermittent zaps of the two Springfields, firing fast as the bolts could be pumped. The Chinese very soon thought better of the standing position and were on their bellies. The fire from their rifles rippled back and forth along the interminable line. From time to time, a machine gun spat out a burst. Then there was a lull. The Americans took stock and looked to their ammunition supply. Bullets had spattered the armor plate, some passing through at joints. The Chinese steersman, standing without murmur, stoically twirling the wheel on Captain Opperman's cool commands, had a bullet in his leg and the end of one finger shot away.

The pause was short. At mileage 46, a fieldpiece shell screamed across the bow and exploded in the opposite bank. A second shell whizzed directly overhead amidships. The third and last fell close astern.

At mileage 44, a point of rock jutted out from the north bank. This, the downbound ship must directly head for. Then just short of collision, a turn was made, leaving the rock twenty feet abeam. A final turn put the rock dead astern. The rock was covered with soldiers. Out in front, waving a red flag, stood an officer. Captain Opperman's reply to this truculent signal to heave-to was in effect a thumb at nose; he held down the whistle cord and the ship went shrieking straight on down as fast as all-out speed would carry her.

The flag waver jumped back and the soldiers crouched low. With *Chi Ping* abeam, several scattered shots rang out from ashore. Lieutenant Winslow's men instantly opened up with everything they had. The return volley from ashore was tremendous. Then, with her stern showing to the Chinese, the vessel was being raked and the guard's fire masked. Winslow was hit. The ship was riddled with 250 bullet holes in the side, a few through the bridge "bullet-proof steel" shields, and several through the "armor plate" screens farther aft. But it was the end of the long, running battle. The grey-clad mass grew small in the distance. *I'Ling,* coming upstream, was warned of the cannon, two thousand yards up on the hillside, out of reach of the guard's small arms. Her guard already had silenced some firing at mileage 39. So *I'Ling,* in company with *Chi Ping,* returned once more to Ichang.

It was now time to be concerned about *Chi Chuen* and *Chi Nan,* following *Chi Ping* an hour or two later through the same hellfire canyon. Was it one of those strange anomalies one found so often in China? Or was it Winslow's potent sting? The two other ships soon arrived unscathed and un-shot-at. Some soldiers on the north bank had only grinned at them.

Winslow, later awarded the Navy Cross for the heroic defense his little band had put up, had received a nasty thigh wound, although he had not mentioned it until a careful check of other possible injuries had been made. Two Chinese passengers had been hit in addition to the helmsman. There was a mild sequel to the affair: Sitting in a Hankow cabaret during his recuperating period, Winslow, like nearly all hands present, was admiring Helen Webb, one of the most exotic Eurasian damsels ever to embellish an east-of-Suez cabaret.

Exhibiting some of his battle-proved determination, he rose to his feet. "To hell with these crutches!" he said, throwing them aside once and for all. "Let's dance!"

Panay cheerfully shared the 1930 exchanges of fire with "communists" ashore. On 7 October, between Shanghai and Hankow, she blasted a couple of troublesome trench mortars located behind a dike. Seven high explosive 3-inch shells and 450 machine gun slugs did the trick nicely in two firing runs. On 12 November, 201 miles above Hankow, smoke from a cannon shot provided an intriguing possibility, but even though she slowed to one-third speed to avoid a miss, she sighted no target nor any splash and had to proceed disappointed. At the perennial trouble spot, Temple Hill, 235 miles above Hankow, a cocky trench mortar started crumping away in *Panay*'s direction, distance 400 yards. After five 3-inch rounds, one a direct hit, the mortar joined a milennium or more of Chinese history of frequent frustration.

British and American gunboats were never reluctant to use the mailed fist. The Italians, with little commercial interest at stake, were generally unaggressive. Records are sparse on Japanese actions of this period, but if their fist didn't strike, they at least didn't mind shaking it; in mid-April 1930, Japan's new 5,000-ton cruiser *Naka* and fifteen Japanese destroyers visited Hankow in the greatest show of force there since the 1927 Nationalist takeover.

As for the French, things in Szechwan were falling their way. The current local Chungking warlord had bought five French warplanes and hired two French officers as flight instructors. Unless forced into it, or spurred by Gallic honor, their policy was to sit tight. In a slight deviation from this policy they eliminated the old Krupp field piece that never hit passing merchant ships because the gunner did not appreciate the merits of leading a moving target. The tolerant attitude of Anglo-American gunboaters toward such harmless sport had allowed it to carry on. But the French gunboat *Balny* apparently harbored no such whimsical sentimentality. In four attacks between November and December she finally managed to silence the old Krupp forever.

About the same time some Red "bandits" had committed the grave indiscretion of kidnapping a French priest for ransom, holding him some distance inland in a small village. The *Balny* put her landing force ashore, marched them to the village, liberated the prisoner, and all returned to the ship without interference from the partisans.

Unacquainted with foreign uniforms, they had taken the Frenchmen in their red pom-pommed sailor flat hats to be Soviet Russians. Naturally enough, the pragmatic Frenchmen felt that if it made the partisans happy to rub elbows with their ideological kinsmen, why disabuse them?

Irrational as it may seem, in view of trouble on the river, 1930 was a banner year for Yangtze mercantile traffic.

In that year, not less than 12,291 craft, of an agregate 8,056,111 tons, entered and cleared at Hankow. At Shasi, 1,614 craft entered and cleared.[46] Of the latter, only 96 were flying Chinese colors, but it would be anybody's guess how many more had somehow finagled that magic little piece of *fanquei* bunting that gave a fair degree of protection from rapacious tax collectors and bandit interference.

By 1930, 39 steamers and 26 motor vessels were operating between Ichang and Chungking. There were 879 clearances at Chungking: 274 British, 240 Chinese, 168 American, 129 Japanese, and 86 French. Indeed, things had come a long way since Archibald Little's appropriately named *Pioneer* opened the twentieth century on the river.

Since the days when Ichang was the end of the road for steamers, Ichang coolies had insisted on the right of transshipping cargo there, even if it meant taking cargo out of the ship and replacing it in the same vessel. During the high water season of 1931, the Yangtze Rapid Steamship Company decided to stop this foolishness once and for all by running two of their largest boats directly to Chungking. In retaliation the Chinese accused the *I'Ping*, one of the ships, of swamping a junk, although actually she was thirty miles away at the time and a Chinese vessel was guilty. Unfortunately for *I'Ping* and the Americans, the Chinese vessel belonged to some local warlords and so was "un-sueable." The next nearest ship thus became the villain. Seventy thousand Chinese dollars were demanded. The magistrate was in a predicament. Something must be salvaged for Chinese "face." To drive home the point, the junkman's guild had placed the body of a drowned crewman in the yard of the magistrate's yamen, where by tradition it must remain until adjudication of the case. In the summer heat, the corpse soon began to make its presence apparent.

There being no U.S. consul at Ichang, the commanding officer of the station ship, *Tutuila*, served as acting consul, so W. E. Brown found himself the mediator. The tedious negotiations, hurried along by strong recollections of the departed junkman drifting in the windows, finally resulted in a compromise: Four thousand Chinese dollars would be paid for the junk, plus another thousand for its cargo, which included the junkmaster's life savings of three hundred silver dollars. The drowned crewmen were evaluated at $100 apiece. A mother-in-law came higher, at $300.

As Brown summarized it,[47] justice was served:

The magistrate told the junk owner he had got him $5,500 for loss of junk and cargo and had better take it and like it. If he heard any more about it, he'd throw them in jail. The result pleased everybody. The junk owner received a fair price for his loss, though he probably had to turn a large portion of it over to the association. The association leaders got their squeeze and maintained face. The Yangtze Rapid

Steamship Company was let off with about one-fourth of what they thought they would have to pay. The magistrate was rid of one very decomposed corpse in his courtyard. The consul general at Hankow could check at least one item off the list of innumerable settlements to be made in his vast territory. The commander was most surprisingly relieved to find out how simply the whole affair had come out. And justice—as understood since time immemorial in China—had been done.

If Old China Hands thought that the end of 1931—the Year of the Monkey and one of ill omen in China—would bring about an improvement in the situation, they were in for a sad awakening. The Secretary of the Navy's *Annual Report* for 1932 summarized conditions as the Yangtze gunboaters saw them from Shanghai to Chungking:

China, at the close of the year, was more completely disorganized than at any time since the Revolution of 1911. The Nationalist government now controls only the area adjacent to the lower Yangtze Valley; Manchuria is uncertain; Canton again estranged; and the country, in the face of the havoc wrought by floods and warfare, is called upon to support the largest number of troops that has ever been under arms in China.

In January, the Japanese Naval Landing Force occupied large portions of what once had been the American settlement, Honkew, but had for years become more and more the center of Japanese mercantile and residential penetration in Shanghai. The Japanese forces were soon in bloody struggle with the Chinese 19th Route Army, under General Tsai Ting-kai, who received only token assistance from Nanking, where Tsai and his army were under dark suspicion of being communists in Nationalist clothing and a threat to the regime. In fact, Chiang sent two of his crack divisions to the area to keep an eye on the 19th while it was locked in a death struggle with the Japanese. They took little or no part in the fighting.

The 19th Route Army had been brought north at the end of 1931 when control of the Nanking government was transferred to the anti-Chiang Kai-shek clique, mainly Cantonese. Troops were quartered in the vicinity of Shanghai and along the Shanghai-Nanking Railway, to insure the safety of the southern politicians. When the new government found itself unable to function and Chiang returned to power, he intended to remove the 19th, and if necessary to do this, disarm the troops. Then came the clash with the Japanese in January 1932 and the 19th became national heroes, remaining at Shanghai until its flank was turned. The army then went south to Fukien, where it was broken up in early 1934 after the Nationalists captured Amoy (10 January) and Foochow (13 January) from the communists, with whom the 19th had been cooperative.

The Japanese were not ones to delay action. On 26 January, they warned the Chinese boycott associations to call off their strike by the next day or face the consequences. Fully aware that Chinese inertia would make compliance impossible, the Japanese were prepared to commence military operations the next day and did so. On the same day, with Shanghai's International Settlement on the edge of a full-fledged battlefield, the municipal council declared a state of emergency.

The Americans had anticipated the crisis and beefed up the Fourth Marines from 1,247 officers and men to 1,694 by reinforcements from Manila. Also from Manila came the 31st U.S. Infantry, 1,056 strong, shivering in tropical weight uniforms and such overcoats as could be scrounged on short notice. Marching up Nanking road behind the regimental band, colors flying in the chill midwinter wind, they represented the first U.S. Army troops to "invade" China since the Boxer Rebellion in 1900, although there had been a battalion-sized unit of the 15th U.S. Infantry at Tientsin for many years, as part of the international garrison guaranteeing access to Peking.

The USS *Barker* had a grandstand seat, alongside the Texaco Oil compound opposite Woosung Creek, a few miles up the Whangpoo from the forts at the point and about eight miles below Shanghai proper. Before dawn each day, her commanding officer, Lieutenant Commander James K. Davis, would climb a 125-foot water tower for a bird's-eye view of the battle between the Japanese and the Cantonese 19th Route Army troops on opposite sides of the creek. Then he would descend and report to the CinC, Asiatic Fleet, whose flagship was moored off the bund opposite the Shanghai Club.

This vantage point was so well located that when war correspondent Floyd Gibbons and comedian Will Rogers arrived in Shanghai, Gibbons moved aboard the *Barker* to live, after safely parking his private supply of scotch ashore in the Texaco compound club.

"Gibbons, who was really a great war correspondent, slept peacefully through everything, every morning," according to Davis.

Later, at the Texaco compound club, I would brief him on battlefront goings-on. Then he would go to the telephone with his notes and give his story in rather cryptic style to his secretary in the Cathay Hotel in Shanghai. He would tell the secretary that there had been some spirited gunfire in such and such a place. Then she would transform it into something like: "Well, boys and girls, there was a hot time going by the old Shanghai racetrack today. Your scribe saw the Jap troops really take it on the chin! . . ." and so forth and so on. . . .

Broadcast listeners of that time will remember that in his radio delivery Gibbons had the staccato quality of a machine gun. "The secretary, deeply schooled in the technique, would expand and decorate the phrasing into Gibbons' style, then pour it into the cable office."

Skipper Davis, on his lofty perch, saw one of those anomalies that could happen only in China:

There was a handful—three, as I recall—of Japanese destroyers which would put in time every morning running up the Whangpoo, shelling the Woosung forts. They would steam upstream beyond Texaco, then reverse course and give the forts a lacing on the way out into the Yangtze. This would go on for several round trips.

On one such occasion, the first run of the day, a Nationalist Chinese gunboat headed down the Whangpoo and passed the Japanese destroyers directly opposite

Davis. The Japanese had added a new wrinkle—lobbing a few salvoes in the Cantonese along Woosung Creek in addition to punishing the already half-destroyed forts. Suddenly their firing ceased. They trained guns amidships. Bugles blared "Attention!" As the Chinese gunboat passed abeam, Chinese and Japanese alike rendered formal passing honors, then the Chinese craft proceeded out to sea. The Japanese then reversed course, headed downriver and soon were shelling the forts again.

Perhaps in recognition of this Chinese whimsy, but more probably following the ancient Navy Department formula of sending the oldest ships available to China, the Asiatic Fleet was soon buttressed by the venerable *Rochester,* a 9,000-ton Spanish-American War relic, and the 1,200-ton gunboats *Asheville* and *Sacremento*—the latter a coal burner. All had come from the "Banana Fleet," the Special Service Squadron operating off Central America. The *Rochester* had 8-inch guns, the others only 4-inch guns, but all had stacks as high as those in a powerhouse, which made them quite impressive to the Chinese, who measured the power of a warship, not by guns or armor, but by the impressiveness of her smokestack array. The venerable French cruiser *Waldeck-Rousseau,* with six stacks in two groups—on the horizon she looked like a pair of ships overlapping—was by Chinese standards the world's most powerful warship. As she carried coops of live fowl and other animals on her fantail, ships downwind could verify that her power was not confined to guns and stacks.

Conditions were particularly bad along the middle and upper Yangtze during low water season—from mid-October to April. In the middle Yangtze, in order to find sufficient water, ships had to stick close to the outboard sweep of a bend where bandits on the high banks could unexpectedly shoot down on them. Even the gunboats found it hard to reply effectively as there was no room in the channel to turn and go back for a firing run. In fact, the spread of banditry and communism became so acute that on 23 January 1932 the American consul general at Hankow recommended an increase in the Yangtze Patrol.

Kidnapping for ransom had become a way of life throughout the harried republic. The American missionary J. W. Vinson was captured at Yangchiachi, Kiangsu, on 1 November 1931 and killed two days later. The Reverend Burton Nelson, kidnapped in 1930, was still awaiting a sweetening of the captors' pot.

When Captain Charles Baker of the Yangtze Rapid Steamship Company was kidnapped on 16 January 1932, the case set a pattern in negotiations of a later day between Oriental communists and their western adversaries. Captain Baker was skippering a motor barge up the middle Yangtze in lowest low water. Navigation markers had been removed by the "bandits" and in spite of taking soundings continually with bamboo pole as was the custom, the barge ran aground off Low Point, some two hundred and fifty miles above Hankow. In jig time the bandits had picked the barge clean of cargo and had Baker and six Chinese crewmen on the auction block. Their demands were nothing if not ambitious: "We have taken the Captain and ammunition from Ship Number Two while grounded at Chinghow," they wrote. "Several days ago we sent a letter to American gunboat to which we have received no answer. We will give you one week to have proper answer. We want two hundred thousand dollars, one thousand boxes of rifle ammunition, and one thousand boxes pistol ammunition."

Their communique was noted in the log of the *Oahu,* which left Mopanshih (mileage 122 above Hankow) on 23 January, after having returned the two bandit messengers to Chenglingki. The *Panay* was ordered in April to guard the grounded *I'Ping,* which belonged to the Yangtze Rapid Steamship Company, with instructions to handle Baker's case if any more developments took place.

The letter was accompanied by one from Captain Baker to the American consul general. He was living under the most deplorable conditions, he reported, and had been interviewed "by an agent of the Soviet government," who confirmed the terms of the ransom. Baker suggested a quick and favorable answer, lacking which, he warned, "I am to die."

In typical Chinese fashion, the bargaining then began. "Mr. Baker poor man his friends will pay five thousand dollars if and when delivered and where he will be delivered," came the radioed reply.

Then began an incredible exchange of letters which was to continue for months. On 7 February Baker, in his sixties, wrote that the price tag read C$50,000. "For God's sake, see what you can do for me," he pleaded. "I have five thousand dollars of my own. The YR Company will probably help. I can work out my debt to them but don't let me die this awful death. . . . The cold is intense. I cannot last long under these conditions."

Complications appeared. The bandits claimed that Nationalist troops had winged some of their comrades and that as a result Baker's ransom had increased to "$200,000 in [Chinese] silver, 1,000 cases bullets, 1,000 rifles, 10,000 bags rice." There was an additional shopping list of army stores and supplies.

The YRS Company next moved in, represented by an American, a Mr. Case, and a Chinese from Shanghai with a "firm offer of $3,000 silver," pointing out that Baker, no spring chicken, had outlived his usefulness in the business world. This was a pragmatic approach any Chinaman could be expected to appreciate.

The reply, from the head of the local Pailochee branch of the bandits, indicated that these boys were no simple country bumpkins. "In reply to yours of the 8th instant," they elegantly began, "we learn you have arranged $3,000 as ransom for your friend Captain Baker." Then with a mild touch of irony, they laid their next trump on the table: "Many thanks for your kindness, but we think you are not honest enough. The Captain himself has stated that he has a sum of $5,000 deposited in the company and has promised to pay $5,000 for the loss which we have suffered while the Nationalist Army was attacking us." The letter continued at some length, dwelling on the advantages of honesty all around, to which end they suggested the $3,000 be turned over to them straightaway as evidence of good faith and high principles.

Reading the handwriting on the wall, the Yangtze Rapid man then telegraphed Shanghai for $10,000 while he admitted to the bandits that he hadn't actually brought the $3,000 with him. Furthermore, in case the bandits were not good at figures, he reminded them that three and five equaled eight.

The hint didn't register; the bandits mentioned that $20,000 was about right, but that $10,000 of it must be turned over first, and the other $10,000 a week later, when Baker would be freed. Baker was still alive and kicking, as indicated by his letter to YR listing his assets as C$4,431 and complaining that he couldn't write more as "my hands are freezing." A much longer letter to Lieutenant Commander Morcott, commanding the *Panay,* reiterated the discomforts of his life, mentioned the imminent prospects of "a horrible death" if the ransom were not forthcoming soon, and added what appeared to be a wholly injudicious postscript: "As they have come down considerable I think they will come down more."

YR's representative next decided to "go for broke," and offered to send ". . . $5,000 by your messenger, after which Captain Baker is to be sent to me and the messenger will receive the balance of $5,000 and everybody will be satisfied." He also pointed out that Baker was getting no younger and that $10,000 would buy a lot of rice. In any case, the bandits had better make up their minds as *Panay* was leaving as soon as *I'Ping* was refloated, after which the negotiations would end, Baker or no Baker.

The bandits replied that $10,000 was only half of what they had demanded, and said they were tired of writing any more letters. "At least you have to raise the sum of $14,000 in order to get Captain Baker released, and we will not deduct one cash from the said amount." They then proposed following the plans of Mr. Case, but asked $7,000 in advance and $7,000 more on delivery of Baker.

"I also am tired of writing letters . . ." replied Case, adding that he was an honest man and did not speak empty words. It should have been clear to the stupidest bandit this side of Tibet that negotiations were rapidly drawing to a close. Case had made this doubly clear by withdrawing the advance payment offer and specifying that the bandits should bring Baker to midriver in a sampan, where he would be met by a small tug with the money. There the swap would take place.

The foregoing exchanges must have overwhelmingly proved that "for ways that are dark and tricks that are vain, the heathen Chinee is peculiar." And what was to follow would certainly not shake that conviction.

The bandits clearly recognized that American patience was wearing thin, but nonetheless they tried one more trick:

We now know that you have generously recognized $10,000 as the amount of ransom for your friend Capt. Baker and have asked us to bring him to the river at Pailochee. In order to show our good will, the same for both sides, we are not going to refuse your latest proposal . . . but we think that your tug boat is of too great horsepower . . . too dangerous for our small sampan. We wish you to send to us therefore $6,000 or $7,000 in advance, the rest of the amount to be cleared when we deliver Capt. Baker to Pailochee.

Case, the YR man, was not about to stand still for this and promptly reiterated his original proposal as final. Whatever happened then, in the bandit camp—the arrival of a big supply of *samshu,* or just plain Chinese perversity—their next letter was so filled with veiled arrogance, intransigeance, and go-jump-in-the-lake independence that it merits extensive quotation:

. . . In our latest letter we have honestly demanded the sum of $14,000 but in your letter you still only agree to pay $10,000 and you repeat that the U.S. gunboat wants to sail immediately and asking us to settle this matter promptly. In addition you stated that the Captain is too old to be of any further use to the company after he is released. By which you try to reduce the amount and fool us. Our idea at present is to get the $14,000 in full and then release the Captain. If our demand is too much for you, please do as you wish and go away as soon as you like. Do not worry. We shall use our own way to fix the Captain. Don't think that our government is in need of this small amount of money. You said in your letter that you will send the silver dollars to us and bring the Captain by using a tug boat. This is plain to us that you are planning to do something to us by using your gunboat force. As we are not idiots and are not foolish children we will never fall into this trick. Regarding the immediate leaving of the gunboat, we note that we will have no more chance to secure the ransom money for the Captain. It is a very slight matter to us. Anyhow, we have the Captain in our hands. We can keep him as long as we wish and fear nothing. Suppose you want to get your friend released you had better send the $14,000 by motor boat.

It was now time for Mr. Case to be conciliatory. He assured the bandits that the gunboat would not shoot, and that in any event the Navy had nothing whatever to do with the ransom racket and such being the case, permission to use the *Panay*'s motor sampan instead of the tug could not be obtained.

The *Panay*'s skipper relaxed his no-interference role to the extent of sending a letter to Baker reiterating Mr. Case's conciliatory position, reaffirming the fact that $10,000 was the limit and that *Panay* would not fire on their soldiers, nor would they allow Nationalist soldiers to do so. But just to be sure the bandits would not themselves take a few potshots at *Panay* in supposition of immunity, Morcott added: "They have their own soldiers every day at Pailochee and we could easily shoot them. But we do not shoot unless we are shot at first."

On 17 February the bandits had either sobered up or had second thoughts: They said Baker was in Pailochee and that $10,000 was just fine. They even agreed to the hitherto unacceptable "high horsepower tugboat" as the means of transport. All was sweetness and light, ". . . hoping you will kindly send the tug boat over at 2 pm without armed soldiers. Please do not worry about anything. We will never try to do anything, but will release the Captain."

The salvage tug which had been sent to Chenglingki for shelter from rough weather was hurriedly recalled. After a half-hour struggle with a balky safe combination the 10,000 Chinese silver cartwheels were counted by the bandit messengers, and messengers, money, and Captain Eisler (the salvage expert from the *I'Ping* operation) were put aboard the tug. Towing the messengers' sampan, the tug then chuffed over to the vicinity of Pailochee.

Then the bandits wanted the money sent ashore before Baker would be sent out.

No dice, said Eisler. How about transferring the money from the tug to the sampan? *Then* Baker would be sent out. So the money was put in the sampan, with the precaution of securing it soundly to the tug with the latter's anchor chain. Now that the money is in the sampan, the bandits proposed in the best spider-and-fly tradition, just let the sampan come ashore and Baker is yours. No Baker, no money, said Eisler.

These negotiations had dragged out until it was nearly dark. The tug was only three hundred yards from shore, where a hundred or more armed men were stalking back and forth in a most unreassuring way. So Eisler, fearing trickery, broke off the palaver and returned to *Panay* with his $10,000 without having seen Baker.

The *Panay*'s mailed fist had been ready, although discreetly out of sight, throughout the afternoon. The executive officer, Lieutenant T. R. Wirth, had been standing by in a motor sampan with several men and machine guns to make a quick dash to the scene. The after 3-inch gun was in effect cocked and primed. Captain Eisler, negotiating from an heroic position in the bow of the tug, had loosely lashed himself to the jackstaff to prevent falling over the side in either the excitement of the argument or anxiety to see Captain Baker. His departure from the forecastle or the firing of a Very's rocket would signal *Panay* that support operations were in order.

By this time the YRS Company boss in Shanghai, Mr. Lansing Hoyt, impatiently radioed Case: "Bandits must accept my terms or get nothing. Company not interested. Surprised that ten thousand required. Send message immediately to bandits that I will withdraw offer unless Baker is released at next meeting. . . ."

This immediately resulted in a sharp exchange of mutually recriminatory letters between Mr. Case and bandits, plus several heart-wrenching notes appealing for speed, action, and $10,000 from Baker and his compradore Chuan Chai-fu, the latter considered to be in the same bundle with Baker for ransom purposes.

Consequently, the tug, Eisler, and the ten thousand much-traveled dollars returned to Pailochee. Captain Baker finally appeared for the first time, in a sampan beached above the bluff, and engaged in heated conversation with a group of Chinese. Next he was moved to another sampan drawn up on shore below the bluff, where further long consultation resulted in Eisler's weakening to the point of allowing the ransom money to be taken ashore.

The atmosphere was tense aboard *Panay,* where the agonizing proceedings were viewed through binoculars, like watching a movie with no sound. "Everyone was rather keyed up," wrote Captain Morcott, in something of an understatement, "either with the prospect of getting Baker or a good fight . . . I'm sure we could have done a good job on that Red stronghold as the crew is eager for excitement. . . ."

Hoping against hope the bandits would not prove to be as perfidious as he rapidly was commencing to suspect, Eisler, still presumably lashed to the jackstaff, watched both Baker and money disappear from the foreshore! The bandits had not only run ninety-nine yards for a touchdown; they had torn down the goalposts and stolen the football. In a note to Eisler still hopefully waiting three hundred yards offshore, they dutifully furnished a receipt for the cash, continuing that, "This is only for the ransom of Captain Baker. As you know already that we have lost plenty while we were capturing the Captain, therefore we will never release the Captain unless you deliver us another fifty thousand dollars to pay the previous loss. Otherwise you will never get the chance to get Captain Baker released."

Understandably, Mr. Case's report to Shanghai closed with something less than exuberance: "I am on the verge of a nervous breakdown myself."

In best Chinese tradition, it was next discovered that a respectable business man of Chenglingki was on brotherly terms with the number-one bandit—in fact, had once been a bandit himself. For a mere $400 he would make himself available as go-between. Meanwhile, it developed through the grapevine that Captain Morcott's preparations had not been ill-founded; if Eisler had not sent the money ashore, the bandits had planned to open fire on the tug and attempt to capture it, Eisler and the SS *I'Ping*'s interpreter, all fresh and presumably negotiable material for ransom. With men in sampans stationed upriver from the bluff, upstream of the tug, and providing *Panay* stood still for it, the game was a fair gamble. But the bandits, like many landsmen everywhere, did not understand that while seapower can be fleets, it can also be a small microcosm such as *Panay;* they apparently had not considered the fact that hidden behind *Panay,* her motorpan had the motor running and eager sailormen were manning four machine guns.

On 31 May 1932, Captain Baker was released through the efforts of one of those remarkable men found in pre-Mao China, Mr. G. Findlay Andrews, missionary, traveler, linguist, and engineer. His activities in connection with dike repair had gained Andrews the appreciation and cooperation of the communists. Without dikes the river overflowed its banks and flooded vast areas of farmland. And in the communist liturgy, fundamental was the thought that, "The people are the water and the guerrillas are the fish, and without the water, the fish will die."

Panay carried Andrews and Baker back to Hankow, and her officers replaced Baker's clothing and supplied him with pocket money. Baker had had enough of China, the Yangtze Rapid Steamship Company, and bandits, and was soon on his way to retirement in Oakland, California. For Captain Morcott the whole affair had been just another incident in the history of YangPat, a ride on the last great surge of the West's gunboat diplomacy in China. It was, in fact, the beginning almost to the very day, of YangPat's final decade.

Up the Yangtze, bandits were still active, but by April 1932, Shanghai had cooled to the point ships once more could come in for recreation. So on 8 April, the old submarine tender *Canopus* shoved off from Manila and wallowed up the China coast, via Hongkong, Amoy, and Shanghai, en route to the summer operating grounds off Tsingtao. This annual trek was described in musical detail by a favorite old Far East ditty which began, "Oh! We'll all go up to China in the springtime!"

The *Canopus* could make eight knots. Her tender old hull, topped by a towering deckhouse, took a permanent port or starboard list, depending on the wind. Chugging along astern came five obsolete "S"-class submarines, the sixth of her brood having been left behind at Cavite for overhaul.

When the *Canopus* reached Hongkong, the British Navy threw a welcoming party aboard the submarine tender *Medway*. During the festivities, dress uniforms suffered in charging manned barricades of furniture, or in skinning the cat on wardroom overhead piping. The Americans played a return engagement ashore in the Hongkong Hotel, known locally as "The Grips." There, $350 worth of round, marble-top tables were surveyed by rolling them down the stairs. Several pedestrians were near-missed by large potted plants dropped from the third-floor balcony, until imperturbable Sikh policemen had cleared the area of strollers. In such tribal rites, the Far East submarine forces of two great navies celebrated the semiannual ceremony of becoming blood brothers.

At Shanghai, the American submariners were on their own. The enlisted men fanned out to the dozens of cabarets on "Blood Alley" and in the rest of Frenchtown, where the luscious White Russians were. Honkew, the site of the old American Settlement and the hangout of sailors before World War I, had been knocked about in the early spring fighting and was out-of-bounds.

The senior officers gravitated to the Shanghai Club, where the 110-foot-long bar was called the longest in the world. In deep leather chairs on the upper floors, one could sleep off a lunch of the horrid stuff the typical Britisher considered to be food, the state of stupor being strongly reinforced by three or four gimlets or pink gins which preceded the meal. If a visitor innocently lowered himself into a chair long the preserve of a charter member, the *gaffe* was soon revealed by an elderly picket, marching up and down, angrily rustling a month-old copy of the *London Times*.

The younger officers might pause for a quick restorative at the Palace or Astor Hotel, a block from the landing jetty—then proceed via taxi, ricksha, or double-deck bus, down teeming, fascinating Nanking Road to Futterer's German Bakery, for gustatory reinforcement. Staggering thence, under a cargo of the best liverwurst, sausages, potato salad, and dark beer outside of Hamburg, a half-hour's walk past rank on rank of shops shook down the burden while bringing the stroller to a favorite rendezvous, *Le Cercle Sportif Français*—the French Club.

Le Cercle's shaky claim to *"sportif"* was lodged officially in several tennis courts and a swimming pool that was decked over in winter for indoor badminton. Otherwise, the term more aptly applied to the spring-loaded dance floor and the adjoining cocktail and dining department.

The clientele of *Le Cercle* was mixed, in keeping with broad French views on such matters. One saw Greeks, Baghdad Jews, and Austrian Jewish refugees, along with the French, British, and Americans. There was even an occasional Japanese, Chinese, or Indian, types as unlikely to be seen at the Shanghai Club or the almost equally stuffy American Club as would be a camel taking the jumps at Aintree. Neither the proper British Shanghai Club nor the American Club admitted Orientals, even as guests. Not aware of this, officers from *Canopus,* being entertained by their Chinese compradore, arranged to meet him at the Shanghai Club. The rendezvous went off as planned, but next day they received a stern admonition from the club secretary, explaining the rules.

One soon learned the clues. The Greeks, Vladivostok Jews, Austrian refugees, Baghdadis, and associated members confined their athletic manifestations to pencilling business details on tablecloths or menus. The athletes were easier to identify. The

Britisher took a cold shower; the American had a hot shower, possibly followed by a symbolic but very brief cold one, accompanied by sounds like a strangling seal; the Frenchman showered not at all, but sat on a bench until the sweat had dried, thus avoiding any of the hazards which accompany superfluous bathing.

There was beauty galore at cocktail time in the French Club—the three gorgeous daughters of the Italian postmaster of Shanghai, two stunning offspring of the Finnish manager of the Palace Hotel—one of them possibly the world's most beautiful of that day—plus a platoon of well-escorted *jeunes filles* who wore exquisite clothes and made extravagant Gallic gestures as they twittered away in liquid French. There were also pretty young Navy wives who had come up on the "Dollar boat" from Manila, for a glorious summer at Tsingtao or Chefoo. And there were sloe-eyed Russian beauties with a trace of Tartar ancestry in cheekbone and epicanthic fold that lent allure and mystique. With ten thousand White Russians in Shanghai, few of them prosperous, one saw at *Le Cercle Sportif* only the pick of the crop; the competition was very keen.

Over the muted rattle of the dice boxes in the club's bar, deciding the signer of the latest drink chit, there was the discussion, among other important things, of dinner. Had the martinis got anyone's courage high enough to go all out and order *escargots bourguignon?* Or something less exotic, like *paté de foie gras aux truffes?* Or pheasant? Or venison? Or a big plate of *bouillabaisse,* for which the club was famous? Or just plain steak? Meanwhile, the martinis continued to come. And saucers of french fries with catsup.

On a table here and there stood a foot-high tripod, like a miniature water tower. Into the little tank on top, a boy poured a measure of absinthe, which dripped, drop by drop, through a lump of sugar held on a wire frame, thence into a glass. While one sipped, a second was dripping full.

And at last, the boy would come to announce dinner. Would the masters please come to dining room table? The next moments were critical. Had the martinis done their work too well? Would the legs take the strain?

Night life in Shanghai was gay, bizarre and cheap, but not kind to the early riser. There was very little stirring in the night clubs and cabarets before ten. Some of the more dedicated *amateurs* hired a hotel room and slept from mid-afternoon until nine or so, to reinforce the tissues for the night's work. By midnight, things were well under way and by 2:00 A.M., festivity was rife. One breakfasted at del Monte's, on the far side of the city, watching the sun rise over open fields.

There was always a last year's alumnus around who knew the ropes:

"If by any chance you get lured away from the crowd by a Russian princess at one of the cabarets, you'd better know how to get back to the ship," he would warn the tyro. "First, go down to the jetty on the Bund and take a look at the tide. If it's ebbing, grab a sampan and tell the coolie, 'Keith Ex.' Stay out of the cabin. It's full of bedbugs. When you've got at least one foot on the lower gangway platform, and not before, hand the coolie twenty cents Mex. But if the tide is *flooding,* for God's sake take the long way around by taxi, or you'll still be sculling away in the same spot three hours later. Tell the taxi driver, 'Felly Loo.' That means Ferry Road. When you see the swayback lines of the USS *Canopaloopus,* you are home. Just holler, 'stoppu!' "

"And what the hell is 'Keith Ex?' " the tyro would ask.

"That means 'Keith Eggs.' Keith Egg Factory. Your nose will tell you when you're close. They don't actually *make* eggs. They process 'em. The average wealthy Chinese farmer qualifies as such by owning a half-acre of ground, a pig and a couple of chickens. The hen hasn't even gotten to the end of her first cackle before Junior's grabbed the warm egg and is hotfooting it for the nearest village to convert it into something less exotic. Like, for instance, cash. None of this foolishness of waiting for a dozen!

"The eggs then percolate on down to Shanghai via wheelbarrow or sampan or coolie back, maybe a little ripe by then. The egg factory blows 'em through a nozzle into a heated vacuum tank and the stuff falls to the bottom as powder. Whole shiploads—thousands of tons—go to the western world."

The tyro stood mute. He wondered if everything in China were this complicated. All he had wanted to know, really, was how to get back to the ship.

Actually, there was much in what the old-timer said. In 1932, China shipped to the United States US$491,235 worth of dried egg yolks, not to mention such other exotic items as dog fur, $2,166,779; weasel skins, $425,935; hog bristles, $1,782,303; rugs, $1,254,641; sausage casings, $471,019; raw silk, $12,624,737; shelled walnuts, $936,518; and lamb skins, $2,185,880. The trade had come a long way since the early Canton days when clipper ships filled their holds with tea, the medicinal-aphrodisiac, and rhubarb, the savior of western nations from mass death by constipation.

Upriver, things were different, to put it mildly. Early in the morning of 1 June 1932, the China Navigation Company's SS *Wan Liu* (British), under Captain J. L. Gamble, went aground at Tai Pan Tze, about forty miles below Chungking. She was bound downriver to Ichang with the usual load of passengers and cargo, plus something far more of interest to the piratically inclined gentry ashore: about M$110,000 in treasure, then roughly equivalent to US$35,000.

British and American Yangtze naval authorities were well integrated. The USS *Oahu,* Lieutenant Commander A. G. Shepard commanding, was closest to the scene, so she was rushed to protect the stranded vessel. *Oahu* and the British merchantman *Kiawo* arrived at about the same time, and the latter took off the passengers and treasure.

The transshipment work was halted at nightfall and the gunboat moored for the night near the stranded vessel. About eight o'clock, a shower of bullets from the beach slammed into the *Kiawo* without causing any casualties. The firing continued for a few minutes, then paused while an order was shouted from the bank for *Kiawo* to send a boat ashore. The *Kiawo* signaled by light to *Oahu,* which came up alongside, at which the overtures from ashore ceased.

Early the next morning, the shooting broke out with increased intensity. Without further delay, *Oahu* opened fire and within a few minutes the gentlemen ashore had bowed out. The transshipment completed, *Kiawo* departed downriver with passengers and treasure intact, while *Oahu* remained standing by *Wan Liu*.

The real wonder of it all was how, under such conditions, any foreign merchant ship could survive. Or if it could, why bother?

In April 1932, at Mopanshih, about two hundred miles above Hankow, the SS *I'Ping* of the YRS Company managed to hit a rocky ledge in midstream and was rammed up on the beach to avoid foundering. The YRS Company manager promptly sent a tug to the rescue, under the direction of a former Navy yeoman who had elected to make China his home, but might better have stayed with his typewriter. The salvage operations were something less than a roaring success. The *I'Ping*'s shallow hull supported a superstructure high above the waterline, wherein were placed the living and cargo spaces. Well up on this deckhouse the tug attached a hawser, headed upriver and commenced to pull. With a downstream current of five or six knots pressing against the underwater body and a brisk wind blowing upstream against the tall deckhouse, the results were easily predictable; *I'Ping* gurgled, groaned, and rolled over on her side.

The left bank, looking downstream, was the "hot" side—bandit country. The *I'Ping* was piled up on the opposite bank, but bandits could be counted on to scent a ship in distress and in short order cross the river in sampans, like so many jackals at a kill. They could also be counted on to loot the cargo, if any, unscrew everything unscrewable and cart it off, and nab any of the ship's company who had foolishly neglected to make themselves scarce. Such unfortunates, if not considered worthy of ransom or useful impressment into Red service, could always be given a cheap ticket to join their ancestors—a shot in the back of the head.

To make things less equal for the Red fraternity, ComYangPat, Rear Admiral Yancey Williams in the *Luzon* at Shanghai, very soon had *Panay* under way from Hankow, escorting the indispensable Captain Eisler and his two small tugs.

With *Panay* standing by for protection, Eisler soon rounded up eighty or more coolies, righted *I'Ping,* set up about twenty windlasses ashore and hauled her out, patched the holes, pulled her clear of the beach, and headed downriver for Shanghai and permanent repairs with a barge lashed on each side of her for support in case the temporary cork popped out.

The whole operation had taken about three weeks, during which the rain slashed down, the wind howled, and the yellow torrent of the river by turns either piled sand deep on top of *Panay*'s anchors or eroded away her holding ground. During the black nights with no lights ashore on which to take bearings, Captain Morcott could only hope that he was not drifting down on the rocks five hundred yards below, where *I'Ping* had come to grief. Alongside his bunk Morcott kept a 3-inch Chinese compass: "It made sleeping easier when I knew the wind was blowing us crosswise to the river current."

When current and wind worked against each other, less strain came on the anchor chains. But with a long upriver reach, the gale would sometimes pile up waves of near-ocean size. In a wide roadstead such as Hankow, the keelless little gunboats could roll a frightening 20 degrees on a side, as they were forced athwart the current and into the trough of the sea by wind and river working in opposite directions. The *Tutuila*, in Hankow roadstead, once recorded rolls of 26.5 degrees to starboard and 25 degrees to port.

An old Yangtze skipper, Lieutenant Commander Albert McQueen Bledsoe, was personally involved in such a situation:

One of the most generous acts I've ever encountered took place several hours after I'd relieved [Commander L. J.] Joe Stecher as commanding officer of Palos. *The ship was moored to a large pontoon off Hankow. A real typhoon came up and it looked like pontoon or* Palos *or both might be blown ashore. Getting under way was out of the question owing to the low freeboard, size of the waves and the wind. With a freezing, driving gale of eighty or so knots, knowing my newness to the ship and area, Stecher came back aboard and offered to relieve me. I thanked him but declined. He stayed aboard, helped get out all the wire and manila we had and we rode it out.*

Bledsoe and Stecher and Morcott and a hundred others . . . of such prime stuff were Yangtze skippers made!

There were times, when Hankow's freezing rain had changed to blistering summer heat, that for days and nights on end the thermometer stuck above the hundred mark. On one such *hot* day, Bledsoe, who carefully underlined "hot" in his account, made an official call, wearing high-collar whites, gloves and sword, on Rear Admiral Hill, Royal Navy, in his flagship *Bee*. "Bledsoe," said Hill, before sitting down to the inevitable pink gins, "let's get into something more comfortable!" Whereupon he peeled off his white jacket, leaving him bare from the waist up and insisted on Bledsoe's doing the same. The latter compromised by taking off all above the "skivvy" shirt. "Who says the British Navy is stuffy?" concluded Bledsoe.

There was a sore point about duty in smaller U.S. ships having only two or three line officers aboard, as was often the case with *Palos* and *Monocacy*. An officer had the duty every other day, or at best stayed aboard one day out of three. No such formalities bound His Majesty's ships, whose two or three line officers could be seen *ensemble* at the tennis club any and all afternoons, with petty officers left aboard to keep the ship. "There was a saying," according to Bledsoe, "that when Greek meets Greek they start a restaurant. When two U.S. naval officers get together they start a watch list, and when two British naval officers meet they start a tennis club."

The American passion for doing things "by the book" carried on into and after World War II, as evidenced by the American destroyer skipper conferring with a British destroyer skipper off the mouth of the Yangtze some time after the Japanese surrender. "Here are my orders," said the American, pointing out three or four telephone-book-sized documents on the table, covering every detail down to estimated consumption of toilet paper. "Do you have your orders with you?" he inquired of the Britisher. "Sorry, old boy," was the reply. "I have nothing written at all. My orders are to be alert and protect the Queen's interests!"

Americans might have been sticklers for documentary formality, but they could also be whimsically imaginative. Bledsoe knew of an enterprising river skipper who hit on a good technique for navigating shoal water; he put a man wading ahead in red shorts. As long as the shorts were not visible he steamed ahead dead slow, but ordered "back full" if they showed.

While he was skipper of the *Palos* in the early thirties, Bledsoe made a call on the CinC, Asiatic Fleet, Admiral Montgomery M. Taylor, in the flagship *Houston* at Nanking. Admiral Taylor was no stranger to gunboats; as a lieutenant he had commanded the *Pampanga* during the Philippine Insurrection. He met Bledsoe at the gangway, with his stubby pipe clenched below his walrus moustache, peered at the *Palos* moored nearby with her hull showing only a couple of feet of freeboard, and inquired of Bledsoe if that was his ship.

"Yes, Admiral. It is indeed!" proudly replied the young skipper.

"Well, I think you should return to her as soon as possible," said the admiral. "It looks to me like she's sinking."

1931-1937

Taels, Cash, and Dollars "Mex"
Report on the "New Six"
Last Cruise of the Palos
Final Port for Palos*—Chungking*

In their "invasion" of Shanghai in 1932 the Japanese accomplished two ends. The first was their objective of taking control of the Yangtze, which began following the 1895 Sino-Japanese War and continued while the Westerners were preoccupied with World War I. The second the Japanese had not counted on at all, and viewed with distaste, if not alarm. That was the solidification of Chiang Kai-shek's Nationalist position as a result of universal Chinese annoyance at the arrogance of the Japanese *wojen* (dwarfs).

The situation stabilized temporarily as each side carefully sized up the other for a good grip in the next round. Meanwhile, at the annual New Year's Eve party at Manila's Army-Navy Club, everybody enthusiastically sang "Oh! We'll all go back to Chinah in the springtime . . ." And they did—Hongkong, then Shanghai, then Tsingtao or Chefoo, with side trips to Peking. Nobody ever called it Peiping, as the Chinese would have liked, now that the capital was Nanking. Everybody saved their money after Christmas so they could ". . . all go back to Chinah on a great big Dollah linah."

Service in China was welcome because it was inexpensive. With duty-free liquor shipped out in barrels and bottled in the Far East, one could hardly afford to be abstemious. Baby-sitters had not yet been discovered, but an amah, the devoted Chinese nurse, faithful unto death, took charge of small fry. It turned out that the amah was not always devoted. When a Chinese chauffeur, on the "outs" with amah, warned "Missy" that "amah no do plopah with baby," Missy was all ears. Baby *had* looked sort of peaked lately. "What fashion amah no do proper?" she inquired. The chauffeur was happy to spike his enemy's guns. "Baby cry, amah put baby head in stove, puttee on gas! Baby finish cry, take head out, turnee off gas!"

They might not have been halcyon days for baby, but for sailors and officers, bachelors and married alike, duty in China was like living in a mild version of the Arabian Nights—something enjoyable, mysterious, and not requiring a great deal of money. And that was well, because in China money could be as mysterious as anything else.

The renowned nineteenth century Jesuit missionary scholar and chronicler Abbé Huc, sipping tea and puffing his pipe in a Chinese inn, once tried unsuccessfully to involve several Chinese in a political discussion. Finally, one elderly Celestial spoke: "Listen, my friend! Why should you trouble your heart and fatigue your head by all these vain surmises? The Mandarins have to attend to affairs of state; they are paid for it. Let them earn their money then. But don't let us torment ourselves about what does not concern us. We should be great fools to want to do political business for nothing."[48]

What, then, *did* Chinese talk about in their apparently incessant chatter? Money! Mostly money—the prime subject in pre-Mao China where eternal vigilance was the price to be paid for immunity from misunderstanding. As Dr. Arthur H. Smith put it:

> *Who is and who is not to receive money, at what times, in what amounts, whether in silver ingots or brass cash, what quality and weight of the former, what number of the latter shall pass as a "string"—these and other like points are those in regard to which it is impossible to have a too definite and fixed understanding.*[49]

To that admirable piece of advice any Old China Hand who ever rode a ricksha, hired a pony, or stepped in a sampan would append a hearty "Amen!"

Early foreign traders in China found the medium of exchange to be silver bullion, generally in molded lumps the Chinese called *sycee,* meaning "fine silk," as the silver when heated was highly ductile and could be drawn out into threads. The Dutch named these lumps *schuyt,* meaning "boat," as they were usually approximately so shaped. Later, the ubiquitous Anglo-Saxons corrupted the Dutch word *schuyt* to "shoe," and that is how these clumsy "coins" came to be generally known to foreigners.

In Imperial days the average lower-class Chinese rarely laid eyes on any *sycee,* conducting his modest affairs in *cash.* These were small bronze coins with a hole in the middle, a theoretical hundred of which made up a standard "string." But very little is ever "standard" in the Chinese scheme of things. Depending on the location, era and state of the economy, a string could have been 99 *cash,* or 98, or 96, or 83—even as low as 33. Commodore Perry, believing cash to be the only medium of exchange in Japan, carried some five tons of these coins on his expedition.

As the Chinese landscape abounded with mythical dragons, so was their monetary maze cluttered with equally mythical units of currency:

10 hao	=	1 cash or li
10 cash	=	1 candareen or fen
10 candareen	=	1 mace or tsien
10 mace	=	1 tael or liang

Of these five coins, only the *cash* actually existed. Land and certain expensive items of commerce were quoted in *taels,* translated into something less ephemeral but more painful when one stepped up to the cashier. Like the British guinea and the American merchandiser's "$99.98," the *tael* served as a psychological boondoggle to help break down the customer's sales resistance.

As late as 1900, the *sycee* "shoes" were in almost exclusive use upriver from Hankow, the "western frontier" of the "Mex" dollar. To make change, one shaved off slivers of the *sycee* in exchange for "strings" of bronze *cash,* the rate being based on a multiplicity of factors guaranteed to drag out the haggling interminably. Items minutely discussed were fineness of the silver, what the current *cash* per "string" rate was, and whose weights were being used on the scales. This last was vital as each merchant had his own scales made up to suit himself, with one set of weights for buying and one for selling. A most important consideration was what the money changer assessed to be the degree of gullibility of the customer.

Travelers and merchants journeying off the beaten path were burdened with a heavy weight of metal money, carried by one or more coolies. Paper money had existed a thousand years earlier in China and had been judged one of the wonders of the medieval world when carried back to Europe. But long before the nineteenth century, Chinese public and governmental moral decay and disruption had destroyed confidence in such a convenience as paper money.

Gold existed in China, but not as a common medium of currency. Between the two World Wars, trading on the gold bar exchange in Shanghai was a genteel form of gambling, something like playing the New York stock exchange. If anyone had insisted, he might actually have taken delivery on some of the finger-sized, dough-soft little bars that lay shining dully in the bank vaults. But such a demand would have struck the staid brokers as no less ridiculous than the Chicago Commodity Exchange would view a request for physical delivery of a trainload of wheat by someone dealing in futures.

By the twentieth century, paper money had made a comeback, although many Chinese preferred the hard, satisfying feel and lovely musical ring of silver coins, minted in China or bought abroad "ready to go" in the form of Mexican or Spanish dollars. The *sycee* "shoe" soon began to be scarce, found only in curio shops. China had come of age monetarily—or *had* she?

For some people, money presents complications even under the best of circumstances.[50] To the newcomer in China during the decade of the thirties it was something more than that; one was faced with a drama of confusion in the pocketbook that eventually was topped only by the roaring inflation—the cigarette and sardine standard—that developed in China, Japan, and Germany following the outbreak of peace in 1945.

A traveler crossing Europe's many frontiers is accustomed to the many changes in money size, color, and texture, and to a reasonable degree is aware of what it should be worth in terms of a U.S. dollar. In China, in 1932, one could equal all these experiences without ever leaving Shanghai.

A galaxy of Chinese central and provincial banks and agencies printed paper money—or more accurately, had it printed—abroad, where too large a proportion would not stick to the machinery. When hot off the presses of the American Banknote Company or its European counterparts, this money was sturdy, colorful official-looking stuff. But like the Chinese navy's ships of that day, paper money was seldom withdrawn from circulation for replacement when overage or worn out; it progressively disintegrated, until the incredibly filthy, pasted and patched fragments presumably simply evaporated, unredeemed and unmourned by the issuing agency.

Metal money was more durable, but it, too, was subject to certain subtle Oriental shenanigans that not infrequently caused the unwary to become the victim of a minor swindle. Silver dollars were sliced in two edgewise and the metal scooped out, to be replaced with baser metal and the two halves then cleverly joined together again—thus forming a "three-piece Mex." "Sweating"—the jingling of a bag of coins together in a stout sack to wear off minute quantities of metal dust—netted the patient operator a modest profit. This was especially true in the case of gold coins (all foreign) which resulted in their having to be collectively weighed in any transaction of more than a few pieces, rather than accepted at face value.

In the latter half of the nineteenth century, much of the silver had been drained out of China. Replenishment was made from Mexico, in the form of Mexican silver dollars. The thrifty Chinese saw no point in redoing a highly satisfactory minting job, so the coins were circulated "as-was." In the majority illiterate, the average Chinaman saw no incongruity in that the rare silver dollar which came his way bore an unknown bird (eagle) perched on a roost foreign to Chinese soil (cactus) surrounded by unfamiliar hieroglyphics. It rang satisfactorily when expertly struck, acted properly when bitten, and had a uniform, recognizable appearance. This was enough.

Only in "officialese" did one hear the term "yuan," the basic unit of modern Chinese currency. In the Far East's pidgin, it had universally become the "Mex" dollar, thanks to those earlier imports.

By 1932, one rarely encountered an original Mexican cartwheel in China, but in Japan and elsewhere in the Orient, one could buy Chinese silver dollars at about a ten percent discount, equivalent to the freight and insurance charges which would have been involved in repatriating them commercially or melting them down for bullion. It was among these bags of dollars that one found many of the original Mexicans, some probably having lain in the local vaults for a generation or more.

Such strikes of unusual coins were not confined to dirty sacks of expatriate Chinese dollars; in 1934 a gunboat officer discovered in a Canton junk shop a ten-pound bag of U.S. pennies, none later than the Spanish-American War and many dating back to the clipper ship days before the Civil War. The owner was delighted to take twice their value as scrap bronze and no doubt considered he had put over a fast one on a more than usually gullible barbarian.

The fluctuating value versus dollars of the Chinese pennies which had replaced *cash* was a minor aberration of the local currency. One hundred coppers did not make a buck. Their value depended on the gyrating price of copper based on silver for that date. One day the tram fare in Shanghai might be thirty coppers; the following day, thirty-two or twenty-eight, approximately half a pound of slugs the size of a U.S. quarter.

It was not often that so wealthy and dignified a foreigner as a lieutenant in the U.S. Navy rode the tram, but when he did, it was even rarer that he paid in coppers. Much more convenient was "big money." Big money was paper money in denominations less than a dollar—cigarette coupon-sized bills of ten, twenty, or fifty cents. One bought them new in packs at the bank and carried them in a breast pocket like salesmen's visiting cards, for easy dispensation as tips and ricksha fare. There was really not any requirement for much other pocket money; with the exception of sampan coolies and ricksha pullers the chit system was universal. One simply signed one's name and military organization or firm and inevitably the chits caught up with the signer—at the end of the month or the end of the cruise. Many of the British settled semiannually. There was not too much cheating; the super efficient Chinese grapevine soon tagged the poor risks.

"Small money" was not so simple. Small money was silver—dimes and twenty-cent pieces. But ten dimes or five twenty-cent pieces did not equal a paper or silver dollar. For a dollar, one received five twenty-cent pieces and a dime, plus perhaps a copper or two—or some equally complicated variation, so that the smaller coins in total equalled the value of the silver content of a dollar. Thus, one could change a dollar into small money, buy a ten-cent newspaper and have 104 cents left. But if one handed his *dollar* directly to the paper boy, this canny operator would hand back ninety cents in change, the latter type transaction resulting in a ten percent loss for the original owner of the dollar.

The paper money circulating in Tsingtao sold at a discount in Shanghai, as did the Shanghai currency taken to Tsingtao, even though both might have been issued by the same bank and look identical except for a local overprint. A trip to Peking on leave meant another swap in paper, and in Hongkong the British issued their own silver and paper based on the pound sterling, which fluctuated daily in relation to the Chinese money just across the border, and the American dollar.

Although the gunboats boasted a doctor in their modest complement, they had no *bona fide* Supply Corps officer aboard to pay the bills and dole out the monthly money to the ship's company. Thus, each gunboat had one harassed junior officer designated as special disbursing agent—SDA—a paymaster in practice who for his term of six or eight months fought a losing battle against the five-foot shelf of instructions furnished by the Bureau of Supplies and Accounts.

The SDA's safe aboard the Chungking stationship in 1939 held Chinese yuan, U.S. dollars, Hongkong dollars, French Indo-Chinese piastres, and a book of U.S. Treasury checks, the various currencies dancing up and down the scale relative to each other like a battery of yo-yos.

Shortly before payday, bids went out to local merchants for offers on a U.S. Treasury check, and soon the SDA's money bag would be bulging with the successful bidder's bundles of sticky, incredibly filthy bills. One trusted to the Good Lord and Chinese honesty that the count was right, the remaining stock of prayers being directed to a safe sampan passage across the half-mile-wide Yangtze without swamping in the roiling muddy waters.

The rate of exchange of the new money was then averaged with the rate of that in the safe and the crew lined up to receive their hatsful of stained, faded paper which would very shortly disappear into the folds of the long gowns of the bill collectors already assembled in happy expectation on the foreshore.

It had not been too long in the past when the gunboats paid in silver. The Old

River Rat allowed his monthly insult to pile up on the books until he had himself a bucketful and the ship had come down to Shanghai or Hankow, where one could manage a lost weekend of genuinely respectable proportions. Such quantities of silver dollars made the paymaster's little safe in the wardroom wholly inadequate for the payroll which came aboard in wooden boxes slung from the carrying poles of half a dozen or more sweating, grunting coolies. The ship's magazine offered the only secure storage for such a bulky treasure, or so it was believed until one enterprising fellow cut his way up from the bilges through the magazine deck for a special money requisition of a type not covered by regulations. The *Luzon* once carried silver dollars in her brig until she ran up on her own anchor, punctured the hull under the brig, and let a stream of coins pour out. It was probably a case unique in Navy annals: running aground on a silver bank.

The silver dollars accumulated by Chinese merchants and banks far upriver could be shipped to Shanghai only at ruinous insurance rates or by taking a chance they would not be lifted by bandits or claimed by the river itself on one or another of its sharp rocks or roaring rapids.

When a gunboat SDA offered a ten- or twenty-thousand yuan check on a Shanghai bank in exchange for locally held silver dollars, Chinese merchants fought for the chance to buy the check at anything up to a 20 percent premium. The check could then safely be mailed to Shanghai in exchange for desperately needed merchandise. "What do I do with the extra dollars premium on my check?" radioed one new and inexperienced SDA to the horrified Patrol Supply Officer at Hankow. Neither knew that Chungking gunboats had long enjoyed this extra bonus in their pay envelopes until the bubble had so innocently been burst.

The Chungking "bonus" was just another of those flukes in the ping-pong exchange of foreign versus Chinese money, profitable for the sometimes forgotten man, the U.S. sailor, legal, but not always spelled out as such by that awesome assemblage of oracular wisdom, the Bureau of Supplies and Accounts Manual. There was a typical example in the last century:

> *Each paymaster of the [East India] squadron, warned doubtless by Perry's predecessor, brought out sacks full of Mexican or Spanish silver dollars. . . . These cost the government less than $1.04 each at home, and the Navy made its personnel a present of the four cents. China merchants were then so hungry for Mexican or Spanish dollars [since the Indian opium exporters would take nothing else except silver ingots which were in short supply], that in Hongkong and Shanghai they were worth $1.50 to $1.60 United States each, and a Mexican dollar, usually 4 shillings 2 pence in sterling, cost the British 5s 8d to 7s 8d. Thus officers and men paid in "Mex" in Hongkong could buy for $100 a Baring Bros. draft on New York worth $150.*[51]

A member of the *Wilmington* crew in 1920–21, Frank Brandenstein recalled that on some paydays they were lucky and were paid in goldbacks "which could mean a few more dollars added to our wealth." Few people now remember those beautiful, large-sized U.S. paper bills, a rich, coppery gold on one side and green on the other, promising to pay the bearer "on demand in gold. . . ."

In the early summer of 1920 Thomas Wheeler Moore was a quartermaster third class in the *Upshur*, at Shanghai, ready to sail for Hankow. "It was payday," he recalled, ". . . not just an ordinary payday, but a real red-letter one, since we were to receive back pay for a raise which had been in effect for several months."

The paymaster came aboard, lugging a heavy black satchel, and we had our payday—in twenty-dollar gold pieces, which was the custom at that time on the China Station.

We got under way immediately. I tucked the four gold pieces which I had received into the pocket of my white jumper and manned my station in the starboard chains. After the first cast of the lead, I leaned over the lifeline to haul in the leadline, and gold pieces started sliding out of my pocket. Before I could clap my hand over my jumper pocket, three of the precious gold pieces had plunked into the murky yellow water of the Whangpoo. I stuck the remaining one in my shoe and continued swinging the lead.

Oh well! Easy come, easy go. And even twenty dollars was a lot of money to a sailor in those days.

The shakings of the money tree had their good seasons and bad for China sailors. The exchange value of local Chinese commodities and later, the relative worth of the Chinese versus the American dollar were correlated with the world price of silver. If there could be said to have been a "normal" rate, it would have been two "Mex" to one U.S. (gold) dollar. This made life slightly cheaper in China than in the United States. But in early 1920, Admiral Gleaves, the Commander in Chief, Asiatic Fleet, cried for help to Washington, requesting some sort of exchange relief—supplementary salary to make up for the drastic dive the U.S. dollar had taken in terms of "Mex." Even the tight-fisted British Admiralty was shelling out to its Far Eastern fleet sailormen. And the American consular representatives up the Yangtze joined in the general recommendation, although taking care to make their views "unofficial," after their traditional cautious manner in policy matters.

From July 1909 to Armistice Day, 1918, the rate had averaged a fairly steady 2.20 "Mex" for US$1. It had been near the "normal" two for one by mid-1916, when the "Mex" had equalled 45.74¢ US. By the end of 1918, the "Mex" cost 60¢ US. But by February 1920, the dollar was quoted at *one for one*. Lean days had come to the Asiatic Fleet.

By 1931 the exchange rate was four to one in U.S. favor. An ensign could afford a private, brass-mounted ricksha, a pony by the month, membership in four or five clubs, confine his interior disinfection exclusively to such superior brands as Johnny Walker Black or Haig Pinch and go dancing with some gorgeous White Russian girl three nights out of four. The fourth was for recuperating with the duty. But by 1932 the United States had followed Britain off the gold standard in wake of the Great Depression. Then Exchange Relief (supplemental pay of about 30 percent to compensate for loss in foreign exchange owing to the U.S. having gone off the gold standard) was dropped. Next came a 15 percent pay cut and a promotion freeze. Young married officers of Shanghai's 4th Marines walked to work and shifted from scotch to UB beer.

Disaster piled on disaster; the American-Oriental Banking Corporation folded when some of its officials looked up at the blue Shanghai sky and saw there no laws nor limits. Most of the Fourth Marines and a good many other modestly financially endowed Shanghailanders banked with the A-O, as it required no minimum deposit. The bank's senior officer got five years, as reported in the *China Weekly Review* for 8 February 1936, but the depositors lost most of their money.

Once again the pendulum swung, so that by 1938, at ten to one, a cook-houseboy was happy to be on call from sunup to midnight, seven days a week, three days off at Chinese New Year, all for the equivalent of US$7 per month. Eighty-five years earlier, Commodore Perry noted that Chinese servants were paid "... from four to six and seven dollars per month, the cooks, however receive from seven to ten."

By 1941 the exchange rate had zoomed up to eighteen to one. The highest paid functionary in the household, the chauffeur, received the equivalent of US$2.50 per month, not counting the inevitable "squeeze"—the judicious liberation of a modest amount of gasoline.

Beyond this, there is no need to stretch credulity nor reawaken the sweet memories of what a bachelor lieutenant's pay check and a state of youthful, ebullient health managed to sustain. It was the swan song era of a duty station that would remain unique in the annals of the U.S. Navy, certainly for over the half century and perhaps for all time.

By 1933, the six new gunboats were well shaken down, and the men who sailed them had reached firm views on their characteristics. They were good enough to cause the CinC Asiatic, Admiral M. M. Taylor, to request the Chief of Naval Operations to provide two more to replace the decrepit *Palos* and *Monocacy*—but with improvements. Everybody agreed that *Guam* and *"Tutu"* were the ideal size, but their draft of six feet, four inches was two feet too much. They would profit from diesel drive with diesel electric generators. A low pressure distilling plant could change two thousand gallons of bacteria-loaded muddy river water into clear, drinkable stuff daily, providing heating for the living quarters into the bargain.

For years, the ships had been subjected to plunging fire from the high banks during the low-water period from 15 October until 1 April, when the river falls steadily. Thus, there should be a machine gun position well up the foremast to reach those devils that popped over the edge of the dike, fired, then pulled back out of sight from deck level.

There were four requirements for the upper river operations which controlled ship design: length, sustained speed, draft, and steering ability. ComYangPat Rear Admiral Yancey Williams felt that the error made in *Luzon*, *Mindanao*, *Oahu*, and *Panay* was that they were too long and too deep. The two small boats, *Guam* and *Tutuila*, could steam all the way up to Pingshan, almost three hundred miles beyond Chungking, during the summer months, and to Chungking at any time except under most unusual and rare river conditions. Luzon could make it up to Chungking only between 1 June and 1 October, departing Ichang with a water level of ten feet or more, the river rising or steady and leaving Chungking at twelve feet or more, with it steady or slowly falling. It was ticklish business. One could not guarantee that bandits or U.S. vessels in distress would cooperate with the requirements of this table.

The proposed new vessels should have one of the two 3-inch, high-angle guns replaced by two 81-mm Stokes mortars, in turrets on the fantail. In April 1933, to test this proposal, *Tutuila* was authorized to have a trial mortar installation, but in the depts of the depression, there was no money to carry it out.

As for the *Guam* and *Tutuila,* the only solid complaint was useless weight which increased draft. The enclosed, armored bridge, extending full across the ship, was judged three times too large. Those big, comfortable mahogany wardroom chairs were grossly overweight, too. The stewards' mates, who had to lug them topside nightly to the movies, would have agreed heartily.

Changsha, perennial trouble spot, was inaccessible to *Luzon, Oahu,* and *Panay* from 15 October to 15 March; to *Guam* and *Tutuila* from 25 October to 1 March; to *Palos* and *Monocacy* from 5 November to 1 February. If a ship was "in," there she stayed. If she was "out," the revolution or what-have-you continued without benefit of U.S. gunboat intervention. Shallow draft was the only answer.

Guam showed how she could kick up her heels in low, low water in February 1934. Inspector Plant, who probably knew more about the upper river than any man alive, had written that at any time below the 2.5 local water mark, all vessels were compelled to resort to heaving at the Hsin T'an. But *Guam* made this wicked rapid at below the two-foot level without heaving, then went on to the passage of the almost equally tough Kuling T'an smartly and without incident.

The same month, the new six received an indirect accolade from someone on another great river, halfway around the world. Perhaps at the suggestion of some Old River Rat returned "stateside," the Commander, Sixth Naval Reserve District, recommended to the Commandant, Ninth Naval District, that for "showing the flag" on the Mississippi and training reserves, a Yangtze gunboat would be just the thing—far superior to the World War I subchasers then in use.

Backing up the argument for the two new gunboats on the Yangtze (the Mississippi project came to naught), the CinC Asiatic wrote to the Chief of Naval Operations on 20 April 1934, groaning over the Navy's inferior position. The British had twelve gunboats, plus a cruiser at Hankow and another at Shanghai, as well as a sloop at Nanking. The Japanese had eleven gunboats, five destroyers on roving riverine patrol, and a cruiser each at Hankow, Nanking, and Shanghai. The Americans, in fact, barely nosed out the Italians and French, whose commercial interests were not a patch on U.S. oil, tobacco, missionary, and other lesser but widespread enterprises.

The U.S. Navy had seven gunboats on the Yangtze in mid-1934. The *Mindanao* had "gone south" in June 1929 and *Guam* had come back. That left five of the new six, plus *Monocacy* and *Palos.* The latter two, for economy and because of their poor condition, had been in reduced commission from 1926 to 1931. Then rising troubles had brought them back. Trim but aging *Isabel,* a hotbox in tropic summer, had been withdrawn from the Yangtze for use by CinC Asiatic as a yacht. The *Luzon* had replaced her as ComYangPat's flagship in February 1929.

For two decades, the *Palos* and *Monocacy* had ranged the Yangtze from Shanghai to the very end of the road, Kiating. Now, to free the new six for brush fires and showing the flag, it had been decided to station the old and rheumatic *Palos* at Chungking. It would be her final port.

The afternoon of 1 October 1934, *Palos* dropped her Shanghai pilot at Woosung and rounded up into the Yangtze.[52] The old gunboat had just been overhauled in the Chinese navy's British-run Kiangnan Dockyard in Shanghai and was replete with retubed boilers, new buff stacks, a green crew, and orders to steam upriver, thirteen hundred miles above Woosung to Chungking, to take over the job of permanent station ship.

The skipper, Lieutenant Commander T. G. W. Settle, was under no optimistic illusions at the prospect of the long pull upriver. On her maiden way thereto in 1914, the *Palos* displaced 180 tons on a two-foot, three-inch draft. As in the case of so many humans, passing years had added heft; now she displaced 340 tons with a four-foot draft. No longer could she safely carry deck loads of coal to eke out her steaming radius. Cruising had become a matter of carefully calculating where the next "filling station"—coal pile ashore or junk load afloat—might be. The previous year, *Monocacy* had run out of coal below Hankow and lay cold and dead until another gunboat came along, towing a junkload of fuel. Any serious delay would lose *Palos* her chance of using favorable seasonal river conditions to fight her way up the rapids between Ichang and Chungking. In November, the water level of the river was neither too high nor too low for a marginally powered ship.

The ship soon began to labor in the trough of long swells coming in from the Yellow Sea. A few men appeared on deck, hanging on for support. "Christ! What a bucket!" they groaned. The ship was rolling twenty degrees to a side, and the Chinese pilot, definitely a smooth water type, had early made his contributions to the fish and was soon followed by men who clung to the lee rail oblivious to the coffee-colored water that sloshed in solid streams over the main deck.

"Now hear this!" bawled the cook, an old-timer with a questionable sense of humor. "All those wot want a wet bread an' raw liver samwidge wit' chawklit sauce, lay aft to tha galley!" The contributors at the lee rail gathered reinforcements. "An' lukewarm coffee wit' a raw aig in it!" cheerfully added their tormentor.

In the engine room the two old up-and-downers clanked and rattled as they ground out four hundred horses each, at high speed and short stroke. They, and all the other machinery, after long years of battling Old Man Yangtze, vibrated, oscillated and danced, each in its own rhythm, and stopped from time to time for a breather to allow river debris to be blown from the water inlet to the main condensers.

Finally, the ship was riding easier; changed to a zigzag course, she was no longer taking the seas abeam, but alternately on the bow and quarter. In a few more hours low-lying river banks appeared and *Palos* was once more in her proper habitat.

Hankow, the last "foreign style" city—the last piped plumbing, the last cabarets with Russian princesses and iced drinks—was reached on 11 October. The six-hundred-mile trip involved a plodding exploitation of backwaters, occasional stops to clean fires, drills at repel-boarders and shift-to-hand-steering. After the usual warship custom on the Yangtze, she moored during darkness. Commercial steamers on the lower river, where time was money, steamed day and night.

The middle Yangtze, Hankow to Ichang, was a superlative example of how a river can meander this way and that over a flood plain. One might see a steamer a thousand or so yards abeam on an opposite heading, apparently gliding over the rice paddies. But both actually would be bound upriver, ten miles apart by water. In flood season the intervening necks of land were submerged and short circuited, so that ships were led into false channels, over the countryside, dodging hillocks, temples, and the half-submerged roofs of drowned villages, until after steaming miles along the wrong side of a dike a passage might be found back out into the real channel.

On a sultry Saturday afternoon, 20 October 1934, sailors aboard the Ichang station ship USS *Oahu* lazed around the ship and the pontoon to which she was moored, digesting their standard inspection-day noon meal of baked beans, ham, and canned pineapple. Then a pillar of smoke, dense and black, was sighted rising high in the still autumn air, distant some ten miles by crow flight and twice that by river. Four hours more and the pillar of smoke, *Palos* at its base, arrived. *Palos* moored alongside the pontoon, her once-beautiful buff stacks scorched a dirty brown.

The veterans and the newcomers milled around on the pontoon and made friends, everybody asking questions at once. The *Palos* men had heard rumors about Ichang, none of them good, in Hankow and from the old-timers aboard.

A dozen or so small Chinese urchins surveyed the scene with interest. Here were some new suckers to work on for *cumshaw*. "No mamma, no papa, no chow-chow! *Poor* little bastard! *Please* give me ten cents!" they would say, then do a cartwheel or two or juggle a handful of sticks as entertainment to justify payment. At liberty time they would offer to lead the lovelorn or thirsty stranger to some temple of heavenly bliss, where the guide would get a cut of whatever the customer could be mulcted of. The boys chattered animatedly among themselves. Those who had been swimming as naked as jay birds in the coffee-colored Yangtze modestly hid behind their hands.

Just outside the Standard Oil compound a small stretch of flat ground tripled as rifle range, drill ground, and baseball diamond. The challenge to a game was soon passed. It would be much better than those between *Oahu*'s deck apes and black gang, where each knew intimately everybody else's tricks.

When the game was over, players and rooters all trooped over to the little building on the far side of the diamond, the enlisted men's canteen.

"Come on, you guys, grab a cue! They's still a couple with tips on. We'll shoot for beers!" Four-inch miniatures of duckpins were set up in a pattern in the center of the table for pinpool. A pin knocked over in the course of a try for a ball in the pocket meant a scratch. A scratch meant the player lost what score he had so far racked up and also stood the rest to a round of drinks. This process was progressive, attested to by the neatly mended slashes in the faded green covering of the pool table, which doubled as the Chinese manager's bunk.

Others pumped the handle of a battered slot machine that swallowed brass slugs, now that Chinese silver twenty-cent pieces had disappeared from the scene.

The results of a series of scratches, jackpots invested in various refreshments, and just plain, over-the-bar commerce could naturally enough be expected to arouse the wanderlust. "Wot's in town?" a newcomer wanted to know. "You gotta hoof it two miles," an old-timer filled him in. "Or ride a ricksha if you can find one. They all pretty well beat up. Use 'em fer haulin' wood, pigs, buffalo crap, anything. Cushions fulla bedbugs. Just tell the guy you wanna go to Cockeye's!"

At Cockeye's Bar and Restaurant* one could partake of dubious culinary and liquid delights while engaging in romantic badinage with such of the local debutantes as found Cockeye's to be a profitable trysting place. Or one could arrange to pick a fight or exchange a treat with some crew member of a French or British gunboat in port, the state of international relations depending upon those nebulous circumstances which sometimes befuddle even the professional diplomat.

"I got me a nice Roosky girl in Shanghai," said a *Palos* man. "I ain't tossin' my money away in any of these here local dumps."

"You wanta be a cowboy for a day, you can rent a pony fer one Mex," a veteran said. A groom, the *mafu,* was part of the deal. "He hightails it along behind so's you don't get lost in the boondocks or steal thuh pony. Like as if anybody would! They ain't fit fer dog food. About four foot high an' they look like they was chewed up by the moths. At the end of the day's run thuh *mafu* ain't half as beat as the pony."

The old-timer had another long draw on his beer. "Wait 'till you get up to Chungking!" he said. "They got real ponies up there. But man, what ponies! They go up an' down those stone steps in the paths like a cat. . . . Those ponies are even littler than these here Ichang crowbaits, but they got real guts. They even race 'em on a sand track on the river bed in low water season."

The Standard Oil manager's residence was a neat little place a hundred yards from the pontoon. Now that the manager had gone without relief, the station ship officers had turned it into a sort of club, rallying around at four or five each evening for a predinner gimlet or two. There would be light conversation, punctuated by a few air pistol shots at rats on the front porch. Or someone might even indulge in a good soak in the managerial bathtub as a variation from the struggle to get the right combination on the shower valves aboard ship. Then all would ease back aboard for dinner and the movies, or a bridge game.

On this Saturday, the night of the monthly covered dish supper at the Ichang club, the officers had the cook prepare a monster pot of baked beans as the Navy's contribution and all but the duty officers helped introduce the new arrivals to local society. The latter included the entire foreign population of the city, perhaps twenty or more souls, less those missionaries who considered tobacco smoke and the aroma of scotch whiskey as issuing from the bottomless pit.

There would be Asiatic Petroleum, British-American Tobacco, Butterfield and

* The establishment featured in the movie *Sand Pebbles* was a dead ringer for Cockeye's. The author of the novel *Sand Pebbles,* Richard McKenna, had been a sailor in the USS *Luzon*.

Swire, and several of their other steamship line colleagues, a squad of Maritime Customs officers, British gunboaters and several of the more liberal of the medical missionaries. The wives skittered around, chivvying the Chinese servants into hurrying with the food. The men played billiards, rolled liar dice for drinks, or sat in the wicker chairs on the veranda and damned the Heathen Chinee.

On Sunday, the athletically inclined could shatter the morning calm with the outboard motor mounted on the pulling sampan, and buzz past His Britannic Majesty's gunboat where the skipper would be conducting divine services. Then seven miles farther upriver, at the very entrance to the first of the gorges, the Sanyutung flowed clear and cool via a deep gully and under a bridge into the Yangtze. In its rocky pools one could bathe in clean water, and unlike the Yangtze, fear no infection. A tug from Ichang would most likely bring up picnickers from the group met the night before, out to restore the damages.

The golf course, if one could be sufficiently charitable to so dignify it, shared the terrain with a Chinese cemetery. Burial mounds provided more than the standard number of bunkers, but in view of the undoubted high density of spirits, benign and otherwise, one at least had a decent alibi over failure to control the ball. The holes, improvised from half-gallon tins, had to be removed after the game and stored in a safe place to prevent their liberation during the night.

At Ichang lay the British steamer *Kiawo,* Captain "Taffy" Hughes, the same ship and skipper which had played a stellar role in the "battle of Wanhsien" almost a decade previously. Hughes was one of the few foreigners personally qualified to pilot an upper river steamer, and no one could better induct a greenhorn gunboat skipper into the mysteries of the rapids, gorges, anchorages, and accompanying demons which he soon would face. So, leaving *Palos* at Ichang where her crew could concentrate on practicing spar mooring, heaving rapids, and inspecting Cockeye's Bar and Restaurant, Lieutenant Commander Settle embarked in *Kiawo* for a trip to Chungking.

Lesson one came early. In the first gorge above Ichang, Hughes blew his siren to warn a deep-laden junk. The blast echoed back and forth across the narrow gorge in a way most startling to a newcomer. "I once made a prolonged blast at a stubborn junk," said Taffy, "and was surprised, upon returning to the spot three days later, to hear it still echoing away in the gorge."

The Chinese Maritime Customs, that solid British-managed enterprise, maintained the Upper River Inspectorate which manned the signal stations at bends and rapids, hoisting shapes at yardarms to warn approaching ships as to what might be coming toward them around the bend so that the two might not slam into each other head on. Cannon fire had been used to signal between stations until a green and trigger-happy armed guard aboard *Kiawo* had assumed one of their shots to be bandit gunfire, and fired on the signal station, scattering its crew. The word had soon gotten around that signal gunning was a high-risk job and that was the end of it.

By the time he returned to Ichang, the skipper, if not an expert, was at least conscious of what lay before him. Meanwhile, the water mark had fallen below optimum level. No time was to be lost. The imperturbable brothers Chow, pilots, together with their native quartermasters, stove, food, opium, and "makee-learnee" boy were rounded up ashore, herded aboard and stowed away in the forward boatswain's hold.

At dawn on 30 October, *Palos* set out, fires clean and bright, coal scoops at the ready and butterflies in the stomachs of those responsible for survival of a venerable ship and green crew in a perilous three hundred and fifty miles that on an annual average claimed the lives of a thousand people.

There had been the usual conference the day before, debating the basic alternatives as to whether to start an hour before daylight and risk an inauspicious if not dangerous envelopment in early morning fog in the first of the gorges, or take departure at noon and make a safe and quiet anchorage at Miaoho, losing thereby a full day en route. Steaming at night anywhere on the upper river would have been sheer disaster.

Captain S. Frandsen, senior River Inspector at Ichang, had impressed on the Chinese pilot the fact that *Palos* was no longer young. This was possibly her last passage of the gorges before the final trip to the boneyard. In the spry young oil-burning *Oahu,* power flowed uniformly and dependably so long as she had fuel oil, but in *Palos* the boilers were coal fired. She would start off below a rapid with safeties popping, fires white hot and clean, but then her strength would ebb and her tail droop as the violence of the effort bled off the steam and her fires turned to dull red. If by then the little ship, shaking and trembling, still hung short of the crest of the rapid, there were only two courses to follow: Drop back—hoping to maintain control, seek a quiet backwater, lie to, clean fires, toss on more coal and an hour or so later, safeties popping once more, try again; or heave—a unique method of upriver travel with which all Yangtze sailors were familiar.

Navy men new to the Yangtze were fascinated by the method of communication between pilot Elder Chow and his Chinese quartermaster. The wheel was spun this way and that, now faster, now slower, for quick and violent reversals of rudder—and not a sound was uttered. Hand resting on the pilothouse windowsill, the long, bony, opium-stained index finger of Pilot Chow moved ever so slightly. To right, to left, a beckoning motion, or no motion at all. In dancing, swirling eddies, roiling boils of water coming straight up, hair-raising slithers across tongues of rapids, the pilot was a man of stone. Only that one finger showed evidence of animation in an otherwise waxwork effigy.

A sort of earlier day human computer, the upper river steamer pilot was a wizard of hydrodynamics and a storehouse of specific intelligence on a multitude of detail that changed in character with each foot of variation in river level. Many had got their start as ordinary crewmen aboard a *ma-yang-tzŭ,* one of the great upriver junks a hundred or more feet long with a twenty-foot beam. Then he was promoted to bow steersman, directing the eight or more men handling the huge, tree-sized spar with a blade on it that guided the bow in and out as the junk was warped upstream by the trackers, or roared downstream through a rapid, or they desperately attempted to keep from going broadside.

With luck at the gambling sticks, incredible parsimony, uncommon good fortune at mere physical survival, and astute investment in piddling bits of cargo carried for his own gain, a man might next become part or even full owner of a junk—or become a junk pilot without this intervening step. As such, he and his fellow pilots would gravely wait at each of the great rapids, along with the trackers for hire, wine salesmen, coolies for transporting cargo around the rapids, and all the other enterpreneurs who were ready to help or prey on the junk folk. It was a rare, foolhardy, or destitute junkmaster who failed to take advantage of the local knowledge of such junk pilots.

From the ranks of these experts came the real VIPs of the river—the steamer pilots themselves. But whether one began as the apprentice of an old veteran or came up from the ranks, the trip would have been long and arduous, a constant struggle with the never-sleeping demons of the wicked river and the sometimes even wickeder fellowmen who fought for bare survival. The steamer pilot would have spent twenty years of an adult life, the total span cut to perhaps half a century by earlier hardship and later opium. When at last he could let fingernails grow long, like gently curving yellow claws, and don the long-sleeved black gown and skull cap of a steamer pilot, one could honestly say he had earned his stripes.

In the 1930s, the monthly going rate for a first class steamer pilot was about C$350, second pilot, C$250, and quartermaster, C$50, all paid whether the ship was constantly on the move or lying doggo the whole time in port. For a man with no starting capital, no chance for official squeeze or plunder, nothing more than iron nerve, an encyclopedic store of river lore and a phenominal rapport with the river gods, this was in Szechwan very good money indeed.

The pilots' curse was opium. They reeked of it. The cubbyhole aboard ship where pilots and quartermasters took their ease absorbed a special sickly stench that persisted for weeks after the smokers had departed for quarters ashore. In Szechwan, the stuff was plentiful and cheap.

The digestive tracts of some of the older pilots had been corroded by the drug to the point they no longer could cope with ordinary food. For these grey-faced, melancholy, sunken-eyed walking bean poles, there was one last resort; where pilots congregated, awaiting commissions from passing junks, wet nurses were to be had for the sustenance of those no longer on solid diet.

By the time *Palos* had ascended the Ichang gorge, all hands had begun to feel a certain confidence. They needed it. In the first rapid, Paitungtzŭ, fortunately mild at that river level, the steering engine failed. Shifting to hand steering in an instant, they climbed up, then got through the Tatung T'an, also mild. But next came the first real test, the Kungling T'an, where the German SS *Sui Hsiang* was wrecked in December 1900, with heavy loss of life, on her maiden voyage.

It was not the water velocity that made the Kungling T'an a killer, although it was fast enough, but the rocks. A house-sized pinnacle towered in midstream. The south channel was impassable at that river level. The north channel, a little downstream of the pinnacle, held several small but wicked rocks—"the Pearls."

Throttles wide open, *Palos* charged straight for the pinnacle, taking advantage of the relatively slack water below it. Then, a nerve-wracking ten yards short of a head-on smash, the rudders were slammed over and *Palos* sheered smartly to starboard. As she straightened out and headed upstream, the Pearls, now dead astern but uncomfortably close under the counter, were left behind as the ship inched past the pinnacle on her port hand. This was not the place to lose steam.

That day their joss had been powerful. Creeping up the last of the fast water into the relative calm of the four- or five-knot current above the rapid, the crowning good luck of the day awaited them—the coal junk.

Arrangements had been made in Ichang for such a rendezvous at each night's anchorage. The state of the fuel bunkers was always of prime concern to all on board, and well it might have been, considering the primitive communications on the river and the odd Chinese conception of time, all compounded by the routine high hazards of junk traffic in general.

Steering engine in top shape, bunkers filled, fires clean, off they started at dawn of a day to remember. The Hsin T'an, "New Rapid," formed only one hundred and fifty years earlier by a vast landslide, served merely as a good warm-up. But next came the Hsieh T'an, running at full strength. Several miles below it, fires were cleaned, wire hawsers broken out on deck and faked down ready for easy running. A fresh gang of stokers was sent below and all hands took up a notch or two in their belts. In the space of a few hundred yards, the ship would have to climb about five feet vertically up a fourteen-knot tongue of rushing, bottle-green water.

There are four basic ways to mount a Yangtze rapid: bull it with superior power; heave it by winding in a wire moored upstream by trackers, provided the forecastle capstan is powerful enough; sheer out into the current at the end of a wire anchored well upstream, then sheer in quickly, desperately heaving in on the slack of the wire, with each sequence gaining ten or twelve feet; or lastly, be towed up at the end of a hawser by the brute strength of a hundred, two hundred, even *four* hundred trackers, flat out on all fours, straining as though their very lives depended on it.

Except for the fact that she had steam engines, the ascent of the Yangtze by the *Palos* was just about what Marco Polo had seen six hundred years earlier, and what Blakiston had described nearly a century earlier:

A very necessary prelude to the ascent of a rapid is for the skipper to go ashore with some strings of cash and beat up all the people he can find to help in hauling the line. Then all hands being ashore, except two or three men to manage the long sweep-oar [which projects from the bow of the boat, and is used to shoot her out into the current, so as to clear rocks, which the steersman unaided would be unable to avoid], and two more to attend to the paying out and hauling in the tracking line and to pole off rocks, and all else being ready, the strongest bamboo line being in use, the word is given and off the trackers start, sometimes eighty or a hundred on the line at once, and the boat stems the troubled waters and steadily ascends. Frequently there is a hitch in the performance, caused by the line getting foul of a rock out in the water, and then one or two of the most daring venture to swim out and clear it, or the junk itself runs on a sunken rock, and then great is the thumping on a small drum by the one-eyed cook, and shouting to stop the trackers' hauling, which, amid the noise of the rapid and the general hubbub, is no easy matter. Then there is a great difficulty to get her off the rock, and the boatmen have often to go into the water in the middle of the rapid. When she starts again, perhaps she goes safely up; or it may be the line breaks, and then away go boat and cargo at the mercy of the current, and probably do not bring up till she is a mile or two below; and not then if she should strike a rock and go down in the middle, as sometime happens.

Captain Settle made a couple of preliminary feints at the foot of the rapid, rank sheers across the stream from the slack water on one side to that on the other—like a matador testing the bull before a serious pass. Next, he eased *Palos* up in the counter current in lee of a projecting cape on the south foreshore. Then, at full throttle, he rounded the cape and charged into the edge of the main flow. Solid water boiled over the forecastle and cascaded down the main deck in a flood. The ship decelerated like a ferry boat hitting the slip. From ten knots over the ground in a backwater she suddenly had plunged into the full strength of the current and was making a momentary twenty-five knots through the water. Momentum was soon dissipated. She crawled up the edge of the tongue into even faster water, until after a couple of ship lengths gained, she stuck and her ground speed was zero.

On the foreshore expectantly waited a hundred or more stark naked "Szechwan bears," villainous-looking trackers who hauled up junks or took ashore and anchored the wires of steamers. They were ready and eager to seize a cable from the *Palos*, secure its end to a boulder a hundred and fifty yards upstream, then throw the bight into the water clear of the rocks. Aboard ship, the hawser was taken to the capstan and hauled taut. Inch by inch the ship forged ahead, the seven-eighths-inch wire vibrating like a fiddle string. Down below, the engines were making all the revolutions the sweating, furiously shoveling stokers could make steam for. A ship's length gained and once again *Palos* stuck. The capstan could heave no more in spite of a full head of steam.

The final maneuver was now in order—sheering. Gently steered out into the stream, the ship forged over and slightly up as the bottom leg of the triangle, of which the wire was the hypotenuse. Then, with rudder thrown the other way, she rapidly came back toward the bank, holding her gain, furiously cranking in the slack in the wire. In and out *Palos* sheered, making a few feet each time, until the wire parted with a zing like a rifle shot. Fighting the current, hanging on once more close to shore, a second wire was passed and on the second sheer, *it* parted at an old nip, a point of weakness.

In the game of sheering, two strikes was out. With fires clinkering, steam pressure rapidly dropping, sweat pouring down the backs of the exhausted stokers, the ship was eased back to the center of the tongue where the current carried her stern first to slower water below.

The retreat had been enlivened by a near miss from a downbound junk, laden to the point of vanishing freeboard. Among the vortexes at the foot of the tongue the clumsy, hundred-foot craft had thrown herself athwart the stream, across the gunboat's bows. It was only by drastic use of rudder and engine that *Palos* cleared the junk and her howling crew "by a dragon's breath," as skipper Settle recorded the event.

With the crew fed, fires once more white hot, the third and last hawser out and ready, *Palos* tried again. Old Chow handled her gingerly; the skipper was certain that each out-sheer would bring the "pop" that announced the limit of the wire's margin of strength. But at last with engines and capstan straining their utmost, the crew sallied forward to bring the bow down over the rapid's lip, by then under her midships, and the ship flattened out and was over. The capstan raced exuberantly as it took in slack wire while *Palos* surged ahead in the glassy water.

Such tussles with the rapids were no mere exercise in highly specialized seamanship. They were fraught with as much personal danger as armed combat with an enemy. The *Siu Hsiang,* wrecked in 1900, was only the first of many lost ships. As a random sample, in 1929, with sixty-seven power vessels operating on the upper Yangtze, there were forty-seven casualties and three total losses. This was not unusual. In March of that year, the *Chi Ping* had eight wires out to four hundred coolie trackers who struggled over four hours to drag her through. Then she tore out her windlass, drifted downstream, and missed destruction by a hair.

Disaster could be near instantaneous. On 25 May 1932, the Yangtze Rapid Company's six-year-old *I'Ling* struck a rock and sank in two minutes. Only a few of her passengers were saved.

Near Kweifu, *Palos* passed the Yangtze Rapid Company's *Chi Chuen,* well up on the foreshore on her beam ends. She had been holed by a rock some months before and beached. The falling water had left her far from the river while the damage was repaired and ways were built for launching her. Then one day, she had broken loose, slid down, and rolled over. With their marvelous ingenuity and ability to improvise, the Chinese workmen would eventually right and float her. Meanwhile, she sat there as fair warning of the tricks the river had in store for the unwary.

Junks fared worse. One in ten stranded on the trip down and had to be laboriously unloaded, cargo spread out to dry, the craft repaired and once more dragged back into the water. One in twenty was a total loss.

The age-old enemy of junks, the river, had been joined by the fire-spewing power vessels. A near incredible but dangerously true phenomenon was the persistence of heavy turbulence in a rapid or gorge after a large steamer had charged through at full throttle. Enduring for as much as an hour, this heavy chop could be fatal to the downbound junk, always loaded to the point of near swamping.

A thousand junkmen a year were lost. Had it not been for the "red boats" of the British-run Chinese Maritime Customs, there would have been many more. This well-organized and efficient rescue service deserves special mention in a country where human life was held as one of the cheapest of all commodities. The service originally had been a function of the Chinese Imperial navy. The large, red-painted rescue sampans, manned by expert boatmen, were stationed below danger points. Crew morale was high through regular payment of salary. The men were the essence of dependability and good performance. When disaster struck, the red boats instantly darted out and did what they could to succor the drowning.

In this respect, the red boats on the upper river were unique. Elsewhere in China, the man overboard would be ignored with utter dispassion by passing river folk unless before disappearing permanently beneath the surface he could gasp out a satisfactory offer of payment for his rescue. Coming to an amicable agreement while one of the parties involved was going down for the third time tended to be difficult, in that it cut drastically short the give-and-take so elemental in Chinese bargaining. Indeed, such lack of enthusiasm to drag out the victim made sense in China; the rescuer not only saved a life, but by tradition he thereby permanently adopted a dependent. Clearly, any deal struck under these circumstances would not be entered into lightly.

The red boats might be compared to modern racing cars: fast, agile, well handled, but uncomfortable. They had only a tiny deckhouse. Travelers who wanted a quick. relatively safe trip upriver before the days of steam hired a red boat for a nominal sum. Ladies and gentlemen accustomed to greater comfort rented a houseboat or large junk, but took the precaution of engaging a red boat to travel in convoy. Then, in the unhappy event the larger craft fell apart or otherwise came to grief, an immediate rescue was guaranteed.

Once safely over the Hsieh T'an, the *Palos* faced the hangover which sometimes follows exhilarating intoxications. On the foreshore were the trackers, howling and gesticulating. There had been no time to bargain before engaging their services and they were out to make the most of it. The head tracker, who had sculled out with the wire hawsers, made an opening offer of 850 cartwheels. No paper money for *these* fellows. This was a fair opening offer, only about four times the proper figure for a gunboat and ten times too much for a merchantman, which paid about C$300 for the Hsin T'an's three stages; C$60 to C$80 for the others.

In proper Chinese style the words and gestures became more violent. Fists were shaken under noses. The ancestry of both parties to the dispute was opened to serious question, with threats by the head trackers to forget the whole thing and shove off with the wires, thus stalling *Palos* at the next stiff rapid.

In half an hour, the drama was played out to the satisfaction of all concerned and a bargain at last concluded. One could be sure that in the verbal fine print, Pilot Chow had not been overlooked. As chief intermediary, he would have subtly seen to that. The wires were handed over in one direction and the bag of silver dollars in the other. Parting company on the best of terms, both sides felt it had been a capital show.

The new crewmen of the *Palos* were now veterans. But the remaining hazards provided their quota of thrills: up Chinchupiao, a mild rapid, but calling for a heave due to last-minute bad "power foresight" and low steam. The ship stuck fast for a time in Wushan gorge. A lone fisherman casting his net from the bank ignored with fine disdain the shouts and requests to take a line cast ashore in order to make a spar moor while cleaning fires and catching a second wind. But in proper unpredictable river form, a soft spot in the millrace developed, a swirl on which *Palos* managed to ease ahead to slower-running water.

Hsiama T'an was heaved and lustily running Paotzu T'an was bulled over, by carefully conserving the steam until the "white of the tongue" was underfoot. But by then, the joss had run out. When they anchored for the night at Kulinto, the river started walking ominously up the marks the carpenter had painted on the boulders alongside. By river custom, the ship's name and date were painted on the nearby rocks. Miaochitsu rapid lay ahead, booming strongly from the rise. Two hundred fathoms of wire, which *Palos* did not have, would be needed to heave it. It could mean a three-day wait, while Skipper Taffy Hughes in SS *Kiawo* brought a wire down from the *Tutuila*.

The men took advantage of the enforced stopover to unwind from the strain. Some hiked along the foreshore, others sharpened their shooting eye on targets painted on the rocks. Farmers brought little baskets of eggs and strings of fish to trade for bits of manila hawser or old clothes. The hawser ends would be unlaid and rewoven into rope. A pair of patched dungaree trousers might soon adorn some coolie woman miles in the interior.

The evening of the second day brought news: at Chungking the water level was falling. "Tomollah mebby can do!" allowed Pilot Chow. Lower water might justify a crack at the rapid without heaving, but in such a marginal situation, why overlook the smallest bit of joss? An observant Skipper Settle had noted that difficulties invariably followed whenever a short release from fireroom coaching allowed the chief engineer to shave off his two- or three-day growth of beard. He had been forbidden to touch a razor since arrival at Kulinto.

During the night, the river showed a "down" tendency. The chief's chin sported a respectable stubble. The *Kiawo* had not yet arrived with *Tutuila*'s wire, but the joys of Kulinto were beginning to pall and the omens looked good. The spars holding the ship off the bank were brought in and loose gear on deck stowed out of the way of the expected flood.

They sneaked up on Miaochitsu at slow speed, as if to catch it unawares. There was no running jump. This was a new tactic. As the frothy tip of the rapid's tongue lapped over the forecastle, throttles were opened wide. Safeties were near popping, fires clean and hot. The fantail danced and rattled as the screws went to top speed. Inch by inch, foot by painful foot, *Palos* climbed up to the top and hung there for an agonizing minute, until with a final wiggle she was clear and over.

Shortly thereafter, *Kiawo* was met coming down with *Tutuila*'s now unneeded wire. From then on, the rapids were kittens; the tigers had all been conquered, and there was no more heaving. There remained only Wanhsien in the way of diversion, the halfway point to Chungking.

Unique on the Yangtze, Wanhsien had ever been a trouble spot. Rival generals fought in and around it. Bandits threatened, floods took their toll of the lower-lying parts, and foreigners clung precariously to their businesses in the face of all hazards and provocations. Mrs. Bird Bishop, the intrepid English explorer, described Wanhsien in 1897 just as it appeared to sailors on the *Palos* thirty years later:

The river narrowed to an insignificant gorge, then came a broad expanse of still water resembling a mountain lake, and then Wan appeared. The burst of its beauty . . . a stately city piled on cliffs and heights . . . a wall of rock on one side crowded with temples, with the broad river disappearing among mountains which were dissolving away in blue mist. It was quite overpowering.[53]

Captain Settle made no mention of the blue mist, but remarked that the morning following their arrival things were foggy. The American tanker *Meiping* had been in port. Captain Gilliberto, known to a generation of Yangtze sailors, had been more than hospitable and the night had been a long one.

Principal considerations beyond Wanhsien were the daily scrounging for coal, and the hope that *Palos* herself would hold together. Pacing the Americans was the French gunboat *Doudart de Lagree,* of equal vintage, but speedier and with longer range because of modernization. The formal warship honors soon relaxed to cheery waves, as they passed and repassed each other several times a day.

At Lochi, the last anchorage before Chungking, a recent freshet had scoured out most of the sand patches that normally provided good holding ground at that river level and uncovered the underlying shingle, on which anchors had no bite. But finally a spot was found where they would hold. Preventer wires were put out to boulders ashore, insurance against yawing, plus instant warning if anchors dragged during darkness.

As the night wore on, the river started an ominous rise: one, three, five, *six* feet! The current increased markedly. An hour before dawn, the usual warm-up of engines commenced. The wires were got in from the foreshore. Then suddenly, the anchors broke out and started dragging over the shingle, the whole ship shaking and trembling. The freshet had scoured away the sand patch.

With one engine warm enough to turn over and steam fed in a hurry to the steering engine, the problem was to use power to minimize dragging while keeping enough strain on the chains to keep the ship's head up river. And in so doing, she must be steered to hold her from twenty to fifty feet off the foreshore in a seven-knot current. In the pitch darkness of a driving rain, with only flashlights to keep the shore in sight, the minutes passed with agonizing slowness. The ship rode ahead to slack chains or dragged astern at uncomfortable speed as slow or fast spots in the swirling current exerted their varying effect. To stay too close to shore was to risk disaster on a sharp rock. To ease too far out was to lose touch and sense of position, then possibly drag off the shoreline shingle bank into deep water. There, the anchors would have no effect at all, leaving the ship blind and helpless to save herself.

Playing this precarious game until they drifted almost the mile downstream to Lochi, there was at last light enough to make out topographical features. With deep relief they hove in the anchor chains and stood up against the evil-intentioned, muddy torrent.

Just below Chungking, on the south bank, an immense gilded Buddha, Tofussu, sits contemplating a Yangtze which at highest flood washes his lap. For the junkmen, this is the end of danger. Here, they chin-chin to Tofussu in thanks for their safe passage, for their victory over their hereditary enemy, the Ta Ho, the Great River and its dragon children. More affluent ones set off strings of firecrackers to emphasize their appreciation. It was not the custom for gunboaters to chin-chin. For one thing, the Chinese would have laughed at such presumption. For another, the missionaries, hearing of it instantly, would have taken even a dimmer view of a group they felt were at best not on too sound terms with Heaven.

But the Chinese were a wise, pragmatic lot. They had survived many vicissitudes. By trial and error over the millennia, they had discovered many truths. Compromise was a virtue they understood and practiced. In any case, why tempt fate? As *Palos* slowed to steerage way with Tofussu abeam, five mex worth of best Wanhsien firecrackers popped and sputtered. Anyone not wholly absorbed in the passing scene might have noted Captain Settle, finger tips together at waist height, make a barely perceptible bow. The pilot joined his hands together in an "A" and placed them under his chin, eyes closed. Then three times he lowered his gaunt upperworks to the horizontal. Few understood or cared what great responsibilities had just been lifted from those frail shoulders. It was high noon, 12 November 1934, and Pilot Chow and his charge were safe home.

The gods thus propitiated, calls made on the top *taipans,* and the club checked out, Settle commenced learning the details of a complex extra duty. In the absence of any regularly assigned American consul, the senior U.S. gunboat skipper was "it." Included in Settle's flock of citizens were hundreds of missionaries, spread all over Szechwan. He could record marriages but not perform them. Business claims and disputes involving Americans were laid at his feet. There was the evacuation "panic plan" to be kept up to date. According to the regulations, if a delegation of seamen from an American ship complained of the food, he must investigate. Was the rice and fish ration aboard the *Mei Lu* and the *Mei Foo* up to standard? The chop sticks sharp enough?

There were other complications, some minor, some worthy of setting down. Such, for example, was the case of the complicated court martial.

There were three officers on board the *Palos*. The skipper, as convening and reviewing authority, would be ineligible to participate. The doctor would serve as recorder. The executive officer would be the deck court officer. But what about legal counsel, to which the accused was entitled?

The doctor, over gimlets at the club, mentioned this *impasse* to his opposite number from HMS *Gannet*. "I will gladly act as counsel!" volunteered the Britisher. Settle could find nothing in regulations against it. Nor did the rules for the government of His Majesty's Navy bear on the subject.

So the trial was held. The Britisher made a noble defense for his client, who was convicted anyway, but felt satisfied that he had had the best possible support. The court recommended clemency in view of the accused's good record and bearing before the court. Convening authority Settle approved and mitigated the sentence, appending a note explaining the unusual circumstances. For his part, ComYangPat, immediate superior in command, found everything in order. But imagine the flurry when the case was reviewed by the Judge Advocate General in Washington. This was something JAG had never envisioned. There was nothing in *Courts and Boards* to cover their embarrassment, thus no exception could be taken. But for future guidance, sternly admonished JAG, foreign officers should *not* be so utilized. Thus ended one more of the many Anglo-American collaborative enterprises at the end of the line.

In August 1936, the river at Chungking was boiling down at near record level, 94 feet above mean. Pushed by a 7½-knot current, floating trees, houses, debris of boats, corpses, and brush fouled the anchor chains, rudders and propellers, greatly increasing the strain on the already fiddlestring-taut mooring. Each new pile of debris would momentarily apply such pressure that the stern pulled down to a mere eight inches of freeboard. A house fouled the bow and before it broke up, water had forced its way up the hawsepipe to the main deck and knocked down men struggling to secure loose gear. There was the added danger of running out of coal before the river had subsided to the point where a fuel junk could be brought alongside.

But life up the river was not all consular bind, court confusion, or flood excitement. There was uncomplex amusement too. In earlier days, before the Russians came to Shanghai and Hankow in their thousands, men found diversion with the Chinese women they affectionately called their "pigs." The glamorous Russians never penetrated as far as Chungking. There were no cabarets or dance halls in the city of "Double Luck." But the "pigs" faithfully trailed their men upriver as the gunboat worked its way, port by port, to the end of the line. Some had gone to mission school and learned a smattering of English, sewing, cooking, and the management of a household on a slim budget. Others were peasant girls to whom the equivalent of US$5 a month was a bounty immeasurably beyond their bucolic early dreams. Some had been selected as sub-teenage orphans by Old River Rats who intended to retire after twenty years and make China their home. Until such girls graduated at seventeen or eighteen years of age, the sponsor faithfully sent his dollar or two a month to the school for their maintenance. The true benedicts were so rare as to be a phenomenon; a seaman could scarcely seriously consider formal matrimony on US$21 a month.

Romantic attachment for Chinese women was not a monopoly of the latter-day gunboaters. In November 1856, a squadron under Commodore James Armstrong assaulted and captured the Barrier Forts, guarding the approaches to Canton. In the squadron was the small, chartered steamer *Cum Fa*—"Flower of Gold"—a name frequently bestowed in China on winsome young ladies. One of her three officers, Charles Henry Baker, commemorated the campaign with an epic timely for all eras:

How oft, since the Harvest Moon silvered the water,
When minstrels of Kwang-Tung were piercing the ear,
I've watched for thy sampan, Oh China's fair daughter
and hoped that kind fates were bringing thee near.

As bright as the light from thine eyes ever beaming
is the flash of thine oar as it dips in the stream,
Like the mantle of night are thy dark tresses streaming
shining like jet in the Queen of Night's beam.

Like persimmons of Shanghai thy full lips are tinted,
as mandarin oranges, golden thy skin.
Thy teeth are like pearls of the rare gems unstinted
that garnish the shrines of the Temple of Chin.

Than the breezes of Fo-Kien, thy breath is more balmy,
than ducks of Ho-ang-se, thy form has more grace.
In the depths of thine eye as it glances so calmly
what wells of sweet love thine adorer may trace.

Alas for the Fan quei, who reft of all reason,
shall hope for thy loving surrender to him,
The flower of that hope, tho' it blooms for a season
shall droop on its stem, and its glory grow dim

And not for these smiles do I long for thy presence,
Altho' thy embrace were perfection of bliss,
Tho' thy tones are so sweet in their clear mellow cadence,
and sweeter than spirit of nitre thy kiss.

Not for kisses and smiles is my weary breast yearning,
for meeting and greeting between me and thee,
But the question to ask which my bosom is burning,
is "When will you bring back my wash clothes to me?"[54]

There was thus not only food for the soul, but sustenance of a more material nature. Though there might be no epicurean establishment ashore in Chungking, a lean sailor could not lay the blame on "cooky."

Those were the days before alarms about cholesterol. The standard *daily* order of eggs, for fifty men, as testified to by the logs, was *five hundred!* To eke out these ten eggs per day per man, a daily supplement from Tong Kee and Company listed one hundred pounds of onions, two hundred pounds of potatoes, one hundred pounds of bread, thirty ducks, one hundred pigeons, thirty pounds of tomatoes—plus string beans, eggplant, bean sprouts, and a gallon of Chinese sauce. That a similar order was received each day, with occasional substitutions of beef (caribao?) for bird, suggests that the parings must have been more than a little thick and the helpings generous.

At Chungking the *Palos* had reached the end of a long road. There, in May 1937, her old bones wracked by twenty-three years of cruising, her hull deteriorated and machinery worn beyond repair, she was decommissioned. She was sold the next month to the Ming Sen Industrial Corporation for US$8,000, and became a hulk for the storage of wood oil. The hulk was still afloat in 1939, a sad reminder of what once had been.

Better had she sunk undirtied at her moorings, or best of all, gone down at sea under the shells of American guns, an honorable target.

If spirits of Old River Rats can return, and benevolent dragons hover over the bones of valiant former antagonists, they must have shed a tear for the inglorious end of a once glamorous lady.

By 1936, the U.S. flag was seldom seen on any merchant vessels up the Great River other than Standard Oil or Texaco craft. Close personal experience with the last American line, the Yangtze Rapid Steamship Company, suggested that the staff work was spotty. An officer arrived in Shanghai in 1934 to discover that the Fourth Marines, to which he was to be attached, no longer furnished volunteer armed guards to the YRSC ships for the six-week round trip to Chungking; the natives were no longer restless en route. Not wishing to miss possibly a last chance to view the wonders of the gorges, he decided to go commercially. A visit to the company on Monday revealed that a cabin was available in the ship sailing Friday. "But check on Thursday for sailing time," they cautioned. On Thursday he checked as directed. Oh! *That* ship sailed *Wednesday!"* they said. "So sorry! There will be another one next month."

The YRSC was not unique in shipping business obfuscation. Some months after the Yangtze fiasco, the same officer conceived the idea of a trip to Taiwan. The clerk in the appropriate Japanese shipping office politely sucked wind through his teeth and produced a book of deck plans for a ship due to sail to Keelung. They examined the merits of various rooms, prices, baggage allowance and many other details, until after a half hour's discussion, a decision was jointly reached on an optimum choice. The prospective passenger, happy that the long negotiations were finally over, pulled out his wallet and started counting out the money. "So very sorry!" said the clerk. "You cannot go to Taiwan."

1937-1942

The Japanese Attack Shanghai
Panay *Incident*
War on the Middle Yangtze
Exit Monocacy
Tutuila *at the End of the Line*
Confrontations with the Japanese
The Navy Leaves the Yangtze
Gunboats at Sea—Shanghai to Manila
War! Capture of Wake
End of an Epoch

With such great interest over the past century in the promotion of American and British commerce in China, there is the possibility of overestimating the importance of that trade in the Far East, which actually was only a small fragment of overall Anglo-American foreign commerce. It was chiefly on moral grounds that Secretary of State Stimson had taken Japan to task in 1931–32, following her first major incursions on the mainland. Stimson was greatly distressed at lack of British support. But the Far East was on the periphery of British interests. British investments there, while substantial, certainly did not warrant a chance of war. And recent Chinese anti-British agitation and boycotts could scarcely be expected to engender British sympathy and warmth toward the Chinese cause.

Perhaps Japan would pull out the foreign chestnuts. Clearly, China trade was vital to Japan. In any case, bluff would not contain her; she would fight. Consequently, with increasing Chinese hegemony under the Nationalists, whose intransigeance had been shared without favoritism by *all* foreigners, and with the Anglo-Americans sitting on their respective hands, a major clash between China and Japan was inevitable.

On 7 July 1937, on the pretext of Chinese interference with a minor Japanese maneuver near the Marco Polo bridge in the Peking area, the Japanese commenced their attack. Undefended Peking fell on 29 July. It soon would be full-scale war, euphemistically labeled by Tokyo "the China Incident." From then on, Americans on the Yangtze and at Shanghai and Canton watched with mixed fascination and consternation a train of events that would not end until Japan surrendered to the Allies on the USS *Missouri* in 1945.

On 11 August, Japanese marines landed at Shanghai, the advance force of a full-fledged army. Next day, "Bloody Saturday," Chinese bombers attempted to hit the Japanese flagship *Idzumo,* moored off the Honkew bund. They missed by as much as a mile, some of the bombs landing in a group of amusement buildings and others between the Palace and Cathay hotels, near the bund, well inside the foreign settlements. Nine hundred and fifty people were killed, and hundreds more injured.

On that same afternoon, the USS *Augusta,* flagship of the Commander in Chief, Asiatic Fleet, Admiral Harry Yarnell, moored off the Shanghai Club.

Human nature being what it is, old times, old memories, and old things frequently seem more attractive than the new times, new things, and the newcomer. Thus, when the USS *Houston,* a spanking new 10,000-ton "treaty" cruiser arrived in the Far East to relieve the old armored cruiser USS *Pittsburgh* as flagship, the Chinese deprecated the replacement. "New 'Merican ship have got only two stacks! Old ship have three! Old ship more strong!" commented Chinese on the bund. There might have been something in what the Chinese said. For in-fighting against Chinese ashore, *Pittsburgh* had a 6-inch armor belt, turrets and casemates, whereas *Houston* was a "tin plater," with 3-inch belt and 1½-inch gunhouses. But *Houston* had nine 8-inch guns to *Pittsburgh*'s four.

Nevertheless, *Houston* won her way into the hearts of the Chinese compradores and tailors, and above all, the White Russian girls on Shanghai's Blood Alley and in the Tsingtao and Chefoo cabarets. So, when the *Augusta,* a sister ship, relieved the *Houston* in November 1933, she was for a time derisively called *"Capusta"*—cabbage—by the bereaved girls. Then they warmed to her too, and when the going got rough, whether it be on the dance floor of the French Club or more serious business, *Augusta,* no longer *"Capusta,"* was not found wanting.

Prior to the extension of hostilities to Shanghai in mid-August, *Augusta* had made, as the Russians say, "le tour de grand duc." After a month of watching spring buds pop out on trees along the Shanghai bund, she underwent two weeks of overwhelming hospitality in Vladivostok as the first American man-of-war to visit that great Siberian port since 1923. Then she sailed to that combination Palm Beach and Cannes of the Far East, Tsingtao, for the height of the summer holiday season and a reunion with the wives.

But the events of early August had taken such a serious turn that the Tsingtao reverie was cut short and the ship commenced a speed run back to Shanghai through heavy seas that left her somewhat battered and missing a whaleboat. She arrived at 4:40 P.M. on "Bloody Saturday," and was greeted near her moorings by two bombs twenty yards off the port bow. They had been dropped with better than usual Chinese accuracy by a pilot who later confessed he mistook the American cruiser for a Japanese. Fragments showered the ship but no one was injured.

From then until her departure for Manila on 6 January 1938, *Augusta*'s log read like a condensed history of war. A day's resume picked at random suggests that the wildest shoot-'em-up at the topside movies or the bloodiest encounter in the cabarets ashore would be tame stuff compared to "A Day in the Life of USS *Augusta* at Shanghai." Any day from August to January, guns boomed, bombs burst, and rifle fire cracked within easy spitting distance of the ship's 64 officers and 760 enlisted men. The *Augusta*'s first battle casualty, first in what could well be termed the Great Pacific War, came on 20 August 1937, at 6:38 P.M., as men gathered topside to enjoy the fresh air, watch the river traffic, and find places for the movies. A one-pounder

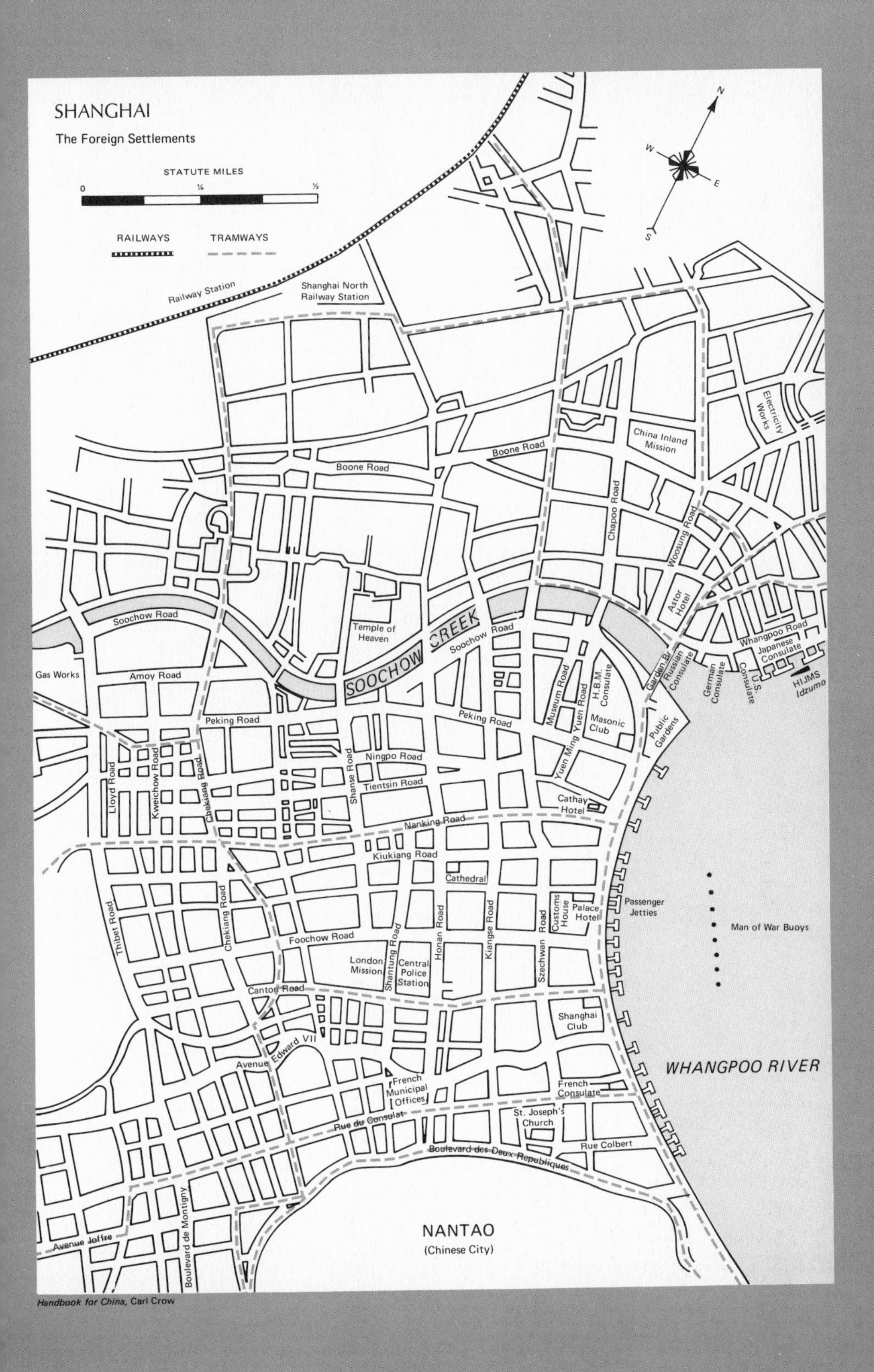

Handbook for China, Carl Crow

shrapnel shell, source unknown, burst on the well deck and killed seaman first class F. J. Falgout. Seventeen men were wounded.

In a little publication prepared aboard ship titled "USS *Augusta* Under Fire," printer first class C. E. Polley and yeoman first class M. O. Hill tabulated the course of events, day by day. They were all very much the same, so only one need be quoted here:

0009 Drone of plane motor heard over *Augusta.*

0009 Five planes dropped white flares in vicinity of *Idzumo* upon which Japanese men-of-war opened fire with antiaircraft batteries, and red tracer machine gun fire.

0010 Three bomb explosions in same vicinity, Japanese tried to locate planes with four searchlights, gunfire of ships deafening.

0012 Another bomb explosion downriver from *Idzumo.*

0014 Japanese men-of-war ceased firing and turned off searchlights. At start of gunfire shells burst to starboard of *Augusta* over Pootung Point. Just before finish of gunfire, shells burst in vicinity of broadway Mansions. Bombing and machine-gunning continued until dawn [very little sleep].

0339 Japanese antiaircraft fire bursting high over *Augusta.*

0756 Bombing in Kiangwan area resumed by Japanese planes.

2255 Three tracers seen over Pootung on starboard beam of *Augusta,* evidently fired from plane. [At night the heat was terrific and most of the men slept on topside, having to seek cover during heavy gunfire.]

During the period from 13 August 1937 until 6 January 1938, the *Augusta* logged the comings and going of an impressive list of warships:

AMERICAN

Augusta	cruiser
Marblehead	cruiser
Alden	destroyer
J. D. Edwards	destroyer
Whipple	destroyer
Barker	destroyer
Parrott	destroyer
Bulmer	destroyer
Edsall	destroyer
Stewart	destroyer
Pope	destroyer
Pillsbury	destroyer
Peary	destroyer
J. D. Ford	destroyer
Paul Jones	destroyer
S-36	submarine
S-41	submarine
Bridge	storeship
Chaumont	transport
Henderson	transport
Goldstar	transport
Ramapo	tanker
Blackhawk	destroyer tender
Canopus	submarine tender
Tulsa	gunboat
Sacramento	gunboat
Oahu	gunboat
Isabel	yacht
Finch	minesweeper

BRITISH

Cumberland	cruiser
Danae	cruiser
Folkestone	sloop
Falmouth	sloop
Dainty	destroyer
Daring	destroyer
Delight	destroyer
Duchess	destroyer
Bee	gunboat
Ladybird	gunboat
St. Breock	tug

FRENCH	
Lamotte Picquet	cruiser
Primauguet	cruiser
Tahure	sloop
Savorgnan De Brazza	sloop
Dumont d'Urville	sloop
Doudart de Lagree	gunboat

ITALIAN	
Montecuccoli	cruiser
Lepanto	minelayer
Ermanno Carlotto	gunboat

One of YangPat's outstanding but nearly hidden assets was "NavPur," the U.S. Naval Purchasing Office, Shanghai. Six or seven flights up in the "skyscraper" Robert Dollar building on Canton Road, NavPur did everything but baby-sit for the transient wives. It was during such periods as the 1932 "war" that NavPur shone. Although the Shanghai Municipal Council and the French settlement were ready and had erected barricades around the perimeters of their territory, the city was caught completely unaware; there was only about a three-day supply of food for the entire population, including the various armed forces. All of the compradores had vanished in panic and foodstuffs were cut off.

Meanwhile, fleet reinforcements were arriving in numbers. All told, NavPur was faced with logistical support for some five thousand men. This was solved by having the Dollar Steamship lines bring in balanced rations for five thousand men every fortnight. After the Chinese were convinced the Japanese had no designs on the city itself, the compradores gradually screwed up their courage and came back. Cattle were brought in from the hinterland and slaughtered locally. Local refrigerator plants and storage godowns for dry provisions were requisitioned. The storekeeper that began it all in 1852 would have stood wide-eyed indeed could he have seen how his venture had grown. There were two officers, some dozen enlisted men, mostly chief or first class petty officers, and a large crew of Chinese.

NavPur made all contracts for food and fuel for the fleet in China, exchanged the fleet's money, arranged for commercial travel, packed and shipped household effects, operated a postoffice, secured pilots, negotiated contracts for ship repairs and carried registered publications. More familiar to the rank and file, they handed out duty-free liquor permits, mail from home, and club cards for the many organizations which gave service people honorary membership.

There were from time to time interesting *denouements* not covered by the Bureau of Supplies and Accounts: rounding up errant wives thought possibly to have been kidnapped, equipping warehouse guards with baseball bats for protection when guns were forbidden, fighting off blatant attempts at "squeeze" and bribery, and overlooking the pecadillos of some of the younger officers caught ashore under unusual circumstances.

By 20 August 1937, 8,700 foregn fighting men were ashore in Shanghai; 4,000 British, 1,500 American (including the 230-man landing force from *Augusta* and *Sacramento*), 1,200 French, and 2,000 mobilized Volunteer Corps. More were on the way. It began to look like 1927 all over again. Of the 4,000 American civilians in Shanghai, some 2,000 were preparing to evacuate.

The Chinese aviators were still poor on accuracy and identification. On 15 August, they dropped bombs a hundred yards ahead of the tanker USS *Ramapo* and twenty yards from *Sacramento,* shaking them both to their keels.

On 27 August, the submarine *S–37* was near-missed. Three days later they came close enough to the Dollar Lines *President Hoover,* anchored in the Yangtze, to wound ten of her people.

On 2 September, a CinC request for reinforcement by heavy cruisers *San Francisco, Tuscaloosa, Quincy,* and *Vincennes* was vetoed by the State Department as too provocative and placing them in too much danger.

On 19 September, the Sixth Marines, thirteen hundred strong, arrived aboard the venerable USS *Chaumont* to join the Fourth in manning the American sector, the two regiments forming a brigade under the command of beetle-browed, no-nonsense Brigadier General John C. Beaumont. No stranger to Shanghai, he had once commanded the Fourth Marines there. Two days before his arrival, it had been amicably agreed with the British that to save their senior officer, Brigadier Telfer-Smollet, from being outranked by Beaumont, he would be made a temporary major general.

Also on 19 September, the Japanese proclaimed that any third party not wishing to undergo a thoroughly good bombing had better make himself scarce in Nanking on 20 September. Taking them at their word, American Ambassador Nelson T. Johnson moved aboard *Luzon,* anchored in mid-Yangtze. He missed the bombing but not the after effect, reporting that, "My withdrawal from the Embassy premises has aroused the scarcely veiled resentment of the Chinese and the open disapproval of some of our citizens." But let it be said in his favor that his being used as an umbrella to shield the Chinese capital neither made good sense nor did it coincide with American official policy, which was one of strict impartiality. Indeed, until the *Panay* affair, Americans were enjoined by Washington not to become Chinese mercenaries in any sense of the word.

On 22 September, Admiral Yarnell's wings were lightly clipped by the State Department, which was "concerned" over the admiral's statement of policy. Yarnell had stated that he intended to station a naval vessel in ports wherever Americans were concentrated, there to remain until it was no longer practicable or until the Americans were evacuated. He added that most American citizens in China were engaged in business which was their only means of livelihood, and he meant to see that they got protection. Henceforth, said the State Department, clear this sort of thing with us first. The following day, Japan contracted for 500,000 tons of American oil. Three days later, by which time pacifists in America were flooding the mails with demands to withdraw *all* U.S. forces in China, the Japanese pasted Nanking properly with a hundred dive bombers.

On 12 October, Admiral Yarnell reported that the Japanese had landed from seventy to one hundred thousand men so far. He was strongly impressed by the formidable-looking heavy Japanese destroyers, mounting six 5-inch guns. They drilled day and night, wrote Yarnell, instead of polishing brightwork. No serious accidents or groundings had occurred, in spite of the large numbers of ships involved; they were crack ship handlers. As for his colleagues, British Admiral Little and the Frenchman, Admiral le Bigot, they were fine chaps. But the Italians were "hand in glove with the Japs."

On 14 October, Yarnell was on the bridge with chief of staff Captain McConnell when a shrapnel fragment wounded a radioman standing between them. Perhaps smarting under this indignity, Yarnell let the press know his sentiments; if fired upon, he would return the fire. That sat very badly with the State Department, which protested to the Navy Department that "this is a great embarrassment," which had brought an avalanche of protests from peace societies and isolationists.

If Ambassador Johnson politely declined to be a lightning rod at Nanking, the State Department forbade Yarnell to sound off independently on policy, and Americans were enjoined not to become partisans or mercenaries, at least *somebody* was standing by the beleaguered Chinese; in the autumn of 1937, the Soviet Union sent four fighter and two bomber squadrons, completely staffed and equipped. By the end of November about fifty of these planes were on Chinese airfields. General Chennault reported their pilots as tough and determined, with enormous vitality, but keeping strictly to themselves in their own hostels, with their own guards, vodka, and prostitutes.[55] On 23 November, four of them were shot down defending Nanking. Not impressed too favorably either by the air defense or the Nationalists' chances of holding Nanking much longer, Johnson boarded *Luzon* and moved to Hankow, some four hundred miles upriver. By 4 December, the Russians had got the hang of it; their hot, speedy little I-15 and I-16 fighters knocked down three Japanese without loss to themselves.

On 8 December, the Generalissimo and Madame Chiang quit Nanking, along with such of the government as had not already pulled out.

On 12 December, it began to look like it might be *anybody's* war. At 9:00 A.M., *HMS Ladybird* was hit by four Japanese artillery shells, which killed a seaman and wounded several others. *HMS Bee,* nearby, was shot at but unhit. Late that afternoon an air attack missed *HMS Cricket* and *Scarab,* with a convoy twelve miles above Nanking. But the event that was to make huge black headlines worldwide, even though already history *before* the *Cricket–Scarab* affair, would not be known until the following morning.

Sparked by the 12 December attacks on the British and aided by what must have been finely tuned extrasensory perception, Ambassador Johnson at Hankow composed a telegram to the Secretary of State, the embassy office at Peiping, and the American consul at Shanghai. Dispatched by *Luzon* at 10:15 P.M., it in all probability had been in preparation about the time of the very disaster it uncannily predicted. Johnson's message gave details of the attack on the British convoy, pointing out that the merchant ships were loaded with civilian refugees, including some Americans. He urged the Department to press Tokyo to issue instructions stopping this sort of thing. Then he added these appropriate words:

> JAPANESE INFORMED BRITISH AT WUHU TODAY THAT JAPANESE MILITARY FORCES HAD ORDERS TO FIRE ON ALL BRITISH SHIPS ON YANGTZE STOP UNLESS JAPANESE CAN BE MADE TO REALISE THAT THESE SHIPS ARE FRIENDLY AND ARE ONLY REFUGE AVAILABLE TO AMERICANS AND OTHER FOREIGNERS A TERRIBLE DISASTER IS LIKELY TO HAPPEN

The object of Johnson's remarkable prediction, the *Panay,* had not been forgotten. On the forenoon of 12 December, Admiral Yarnell sent her a message, rather ambiguously giving permission to depart Nanking at discretion:

> PRESS REPORTS INDICATE PANAY BEING SUBJECTED TO CONSIDERABLE DANGER FROM SHELL FIRE PERIOD HAVE ALL US NATIONALS ASHORE BEEN GIVEN FULL OPPORTUNITY TO BOARD PANAY AND HAVE THOSE STILL ON SHORE REFUSED TO AVAIL THEMSELVES OF THIS OPPORTUNITY IF SO IT IS NOT INTENDED THAT YOU SHOULD REMAIN IN AREAS WHERE CASUALTIES TO SHIPBOARD CREW ARE LIKELY TO RESULT

Actually, ComYangPat had, the day before, given *Panay* permission to proceed at discretion.

Shells had been landing with monotonous regularity uncomfortably close to *Panay* early Sunday morning, 12 December, where she lay fifteen miles above Nanking. Consequently, to get out of the backstop end of the shooting gallery, Lieutenant Commander J. J. Hughes, her skipper, anticipated the message from the CinC by several hours, and got under way at 8:25 A.M. for upriver. Following in convoy were the Socony-Vacuum tankers *Mei Ping, Mei Hsia,* and *Mei An.*

Panay's complement of five officers and fifty-four men was swollen by a mixed crowd of embassy personnel and American and foreign civilian refugees. Most fortuitously, there were also aboard several topflight newsreel cameramen and correspondents who had been covering the fighting at Nanking. As if guided by an omniscient fate, they had just completed a short documentary on the *Panay,* to show the folks back home how life was lived in a U.S. warship in the danger zone. They had filmed a perfect prologue for the high drama soon to unfold—the slashing attack itself, the actual sinking of the ship, and the equally stirring three-day struggle of the survivors to bring their dead and wounded out of the swamps to friendly hands.

As the convoy steamed upriver about 9:40 A.M., they spotted groups of Japanese soldiers on the north bank waving flags. The *Panay* hove to, while a boatload of some twenty armed Japanese came out to the ship. Lieutenant Sesyo Murakami climbed aboard, followed by his sword bearer and two armed soldiers. Hughes met him, noted the presence on deck of the armed men, but made no issue of it in view of his instructions from ComYangPat to overlook certain departures from accepted form when dealing with the Japanese army because of their ignorance of seagoing etiquette.

Murakami was interested in the movements of *Panay,* the character of her convoy, and the activities of any Chinese troops sighted en route. The former matters were clarified fully. Concerning troops, Hughes explained that the United States, as a neutral, was not at liberty to disclose the military situation of one opponent to the other. Murakami then left the ship, inviting Hughes to return the call. The latter, under the circumstances, begged off.

Proceeding upriver about five miles, the convoy anchored in a wide space in the river, chosen as least likely to become the scene of another artillery duel or arouse the suspicions of either side.

The weather was calm and the sky cloudless. The *Panay* was painted normal

Yangtze gunboat white and buff. Two large U.S. flags were painted horizontally on her upper deck awnings. The biggest U.S. ensign in her flagbag hung limply at the gaff. The spot seemed so remote from trouble that after lunch eight men were allowed to sampan over to *Mei Ping,* where the U.S. Navy's rules against strong drink did not apply.

About 1:37 P.M., the lookout reported two aircraft in sight at about 4,000 feet. Hughes and chief quartermaster John Lang went to the bridge for a better look. Captain Frank Roberts, USA, assistant military attache, had just come back from *Mei Ping,* where he and two others had gone to hear the one o'clock Shanghai news broadcast. Roberts, naturally enough, was interested in what type planes they were. Several of the correspondents, scenting a possible story, had come out on deck, too. Thus, a good many knowledgeable, credible witnesses were on hand.

Early that same morning, Lieutenant Commander T. Aoki, naval liaison officer on the staff of Lieutenant General Prince Asaka, who commanded the forces attacking Nanking, learned that the army had spotted ten loaded troop ships fleeing Nanking.

The Japanese army had only a few aircraft, mostly reconnaissance types, around Nanking, but the navy had in the area twenty-four fighters, twenty-four level and eighteen dive bombers.

Heretofore, standing orders had forbidden any attack on ships in the river, for fear of hitting neutrals. But here, there was no room for doubt. It was the first time eager young navy pilots could get into their proper element and attack *ships,* after all these months of going after land targets. Chinese air opposition had practically disappeared from the skies.

An attack force was formed up: nine fighters and six dive bombers from the 12th Air Group, and from the 13th, three type 96 Mitsubishi level bombers under Lieutenant Shigeharu Murata and six dive bombers under Lieutenant Masatake Okumiya. The level bombers carried six 130-pound bombs each, and two, including Okumiya's own plane, carried a single 550-pounder. In their enthusiasm and haste, no strike plan was set up. Each flight leader wanted his unit to cover that hundred miles first and be in a premier position for a unit commendation. On takeoff, several of the planes rocked dangerously in the prop wash of those ahead, so great was their haste to get airborne. With no oxygen masks, the dive bomber pilots flew at about 12,000 feet. Murata's three level bombers flew 1,500 feet below them, in V formation, and so they got there first.

In view of the series of attacks on British ships, the Anglo-Americans were naturally enough in an apprehensive state of mind, and any break in routine communications was cause for concern. Thus, at 4:30 P.M., the CinC, Asiatic, queried ComYangPat:

> *PANAY* UNHEARD SINCE 1342 WHAT IS NATURE OF CASUALTY ARE YOU IN CONTACT WITH *PANAY* VIA BRITISH

Thirty minutes later ComYangPat replied that he had had no communication with *Panay* since 1:35 P.M., when a message she was sending broke off abruptly. The British were trying to find out what was going on, but none of their ships was in the immediate vicinity. Three hours later, an 8:00 P.M. message from ComYangPat to the CinC gave *Panay*'s last known position as mileage 221 above Woosung.

Aboard *Panay,* all hands had spent a peaceful Sunday afternoon. Some of the men drank beer aboard *Mei Ping,* others sacked out for a holiday snooze, digesting the whopping noon meal. All were hoping that Nanking would soon fall, so they could return, unload the fifteen bored, cramped passengers, and resume a normal existence.

As Hughes looked out the pilot-house door to see what the planes were up to, he was astonished to see them rapidly losing altitude, heading his way. Seconds later, an explosion flung Hughes across the pilot house and he momentarily lost consciousness. When he came to, he found the bridge a shambles and himself with a broken thigh. The forward 3-inch gun was wrecked, foremast down, and radio room smashed. These bombs had come from Murata's high level attack. Then very shortly after their fall, those from Okumiya's dive bombers started pelting down, the near misses splattering the area with whistling shrapnel.

Panay's response was immediate, but of small effect. The machine guns chattered away, directed by Chief Boatswain's Mate Ernest Mahlmann, who was truly stripped for action. Displaced from his CPO bunk by the "refugees," he had been snoozing in a storeroom below decks. The first bomb rudely interrupted his reverie and opened cracks through which water started to pour. There was no time to dally. Sartorial perfection was less important than getting out before the next bomb set off the magazine. He managed to get on a shirt, but through the rest of the action Chief Mahlmann fought pantsless. In this he had company. The first explosion had caught Ensign Denis Biwerse on the forecastle, where the blast very nearly stripped him of every vestige of clothing.

The machine guns, eight 30-caliber World War I vintage Lewis, were in armored shields on swivels. They could be elevated to near vertical, but stopped short in train about 15 degrees of full ahead or full astern. And the planes were diving from dead ahead, right out of the sun.

The main battery, two 3-inch, 51-caliber, high-angle guns with telescopic sights, were not bad against low speed aircraft, but incredibly there was no ready ammunition for them on deck. It was simply beyond the realm of imagination that *any* aircraft would ever deliberately mount a sustained attack on these gunboats. To bring up ammunition in the middle of a fight would have meant opening hatches below decks which would have hazarded the ship's watertight integrity. In any case, it would have been useless. The forward 3-inch was wrecked by the first bomb. The after gun was caged in awning stanchions and could not have been fired.

After Ensign Biwerse's defoliation by the first bomb, he had picked himself up, injured mainly in dignity, and started up the ladder to the upper deck, where the machine guns were. But about this time a second bomb hit the radio shack, completing its devastation as well as knocking the remains of the foremast over the side. It also knocked Biwerse back down the ladder.

The executive officer, Lieutenant A. F. ("Tex") Anders, hit in the throat and unable to speak, wrote instructions on bulkheads or on a navigation chart in pencil, sometimes blotted out by blood—his fingers had been hit by shrapnel—as he fitted a pan of ammunition on a Lewis gun.

Lieutenant (junior grade) J. W. Geist was badly wounded in the leg, but kept limping around the decks, doing his job. A former all-American soccer player, "Gunner" Geist was not that easily stopped.

By the greatest good luck, the medical officer, Lieutenant C. G. Grazier, was unwounded. Throughout the whole ordeal, afloat and ashore, "Doc" Grazier took care of the wounded in magnificent style.

Propped up in the galley doorway, his face black with soot from the smashed "Charley Noble," Hughes lay helpless and in heavy pain. There was no necessity for him to give orders. Discipline and training had taken care of that.

The newsreel cameramen and correspondents, now thrust into the middle of an explosive situation they had previously viewed only as spectators, reacted in true professional style, grinding away with cameras and making notes.

From his dive bomber, Okumiya looked down at the four ships and noted that apparently only two bombs from the level bombers had hit. A large plume of steam from *Panay* suggested a ruptured boiler. Actually, the engineer on watch had tripped the safety valves, to prevent an explosion in case a boiler was hit. But an oil line had been cut, so no steam could be raised and the ship could neither be beached or use pumps to cope with the rapidly rising water.

Okumiya was boring in on *Mei Ping* with his dive bombers even as bombs from the level bombers landed. He released his bomb at about 1,500 feet. Looking back, he was chagrined to see that not only he, but all of his remaining five planes had missed.

Next, the 12th Air Group's six dive bombers and nine fighters bombed and strafed, with most attention to the tankers. *Mei Ping,* anchored about a hundred yards on *Panay*'s starboard bow, as well as the *Mei Hsia* and *Mei An,* got under way early in the attack. The *Mei Hsia* made a valiant effort to come alongside *Panay* to take off survivors, but they frantically waved her away. Life expectancy aboard the gunboat was low enough already without having a highly flammable and explosive tanker alongside. Then *Mei Hsia* and *Mei Ping* secured to a pontoon at Kaiyuan, on the south shore, while *Mei An* beached on the north bank.

The eight *Panay* sailormen who had been enjoying cold beer aboard *Mei Ping* instantly turned to fighting fires and backing up the panicky Chinese crew in getting the ship under way.

Japanese soldiers at Kaiyuan had entered the confused situation by desperately waving small Japanese flags to avert final attacks on the two tankers, and several of them were killed or wounded for their fruitless pains. And in the unpredictable and inscrutable way of the samurai, they had first driven the Americans back aboard the doomed ships, then let them come ashore, then had hightailed into the hinterland when there was no doubt the planes meant business.

In view of such behavior, the Americans struck out inland. One more refugee had been added; *Mei Hsia*'s stern had swept so close to *Panay*'s bow in the rescue attempt that newsman James Marshall had leaped across to the tanker.

After the attack, Okumiya felt the whole show was highly unsatisfactory. There certainly would be no unit citation. As far as he could determine, all the ships were still afloat and not even seriously damaged. He flew back to base at Changchow, near Nanking, to recommend another strike.

Rearmed, the six dive bombers returned to what Okumiya thought was the scene and found nothing. The game had escaped. Then he tried Wuhu, upriver. Again drawing a blank, he turned back downriver toward Nanking. There they were—four ships! Peeling off from 12,000 feet, the dive bombers attacked. But in the pullout, Okumiya was again horrified—not by misses, but by a British Union Jack on one of the ships. They were, of course, HMS *Scarab,* HMS *Cricket,* a Jardine hulk, and the SS *Whangpoo,* the latter two loaded with refugees.

Okumiya, still with no qualms about any possibility of mistaken identity in the first attack, breathed a prayer of thanks on having missed the second group. The planes dumped their bombs on any old target near Nanking and went home.

Aboard the *Panay,* less than half an hour after the first bomb, it became apparent that she was doomed. The forward starboard main deck was awash. Water was six feet deep in some compartments. The transverse bulkheads could let go any minute. The two boats, a motor sampan that doubled as the captain's gig and the officers' motor boat, and a pulling sampan with a temperamental outboard motor, would have to make five or more trips each to carry all hands ashore.

The first trips were machine-gunned by low-flying planes, punching holes in the boats' bottoms and inflicting more wounds on the already wounded first evacuees. At 3:05 P.M., the last man, Ensign Biwerse, crawled under the lifeline and left the ship. As a final gesture of courage, Chief Mahlmann (still minus his trousers) and machinist's mate first class Gerald Weimers went back for stores and medical supplies. While they were returning shoreward, two Japanese boats loaded with armed men approached *Panay,* briefly machine-gunned her, boarded, then quickly left. Shortly thereafter, at 3:45 P.M., the ship rolled over to starboard and slipped slowly beneath the surface, bow first. In the eighty-three years of the Yangtze Patrol, she was the first U.S. ship lost to enemy action.

A further grueling test of *Panay*'s beset people was yet to come—the three-day limbo in reedy swamps and tiny Chinese villages, ignorant of any knowledge of Japanese intent or whereabouts. They were not clothed for the near-freezing nights, had only a few dollars, meager medical supplies, and in the beginning, no food.

The rest of Sunday was a nightmare of buzzing Japanese planes and snooping Japanese boats. In the eight-foot-high reeds and ankle-deep mud, they lay like hunted animals.

Hughes, Anders, and Geist were incapacitated by wounds and Biwerse by shock. "Doc" Grazier was fully occupied with the sixteen wounded. So Hughes and George Atcheson, the embassy's second secretary, agreed that Captain Roberts take command. His general good sense, knowledge of the Chinese language and of tactics ashore made him not just a logical choice, but as it developed, almost an indispensable factor in securing their safety.

Consul Hall Paxton, his knee disabled in the action, was sent off on a horse to somehow notify the Navy of the attack. In his wake, as he pressed on to Hohsien, eight miles distant, were radioman first class Andrew Wisler and wardroom messboy Wong. Wisler, hobbling on a twisted ankle, would be useful if some radio facility was found. Wong understood the local dialect, with which Paxton's Mandarin might not be able to cope.

After a message shouted through the telephone to "somewhere," Paxton set out by ricksha for Hankow. Wisler, mounted on the horse, returned to Hohsien to tell the rest of the party what had been found. Hohsien was a poor little village with scant resources, but the magistrate had been friendly and willing to receive them, though desperately fearful of Japanese retaliation.

After much palaver and disagreement, some Chinese coolies had been recruited as litter bearers. The little column straggled out along the muddy path, traveling after darkness to escape discovery by what they still believed to be the enemy, the Japanese.

Japanese planes flew low over Hohsien the following day—search-and-rescue, so it was learned later. But the survivors only had their fears confirmed: the Japanese were hoping to exterminate the evidence, they angrily told each other as they gratefully ate rice and drank bitter tea.

The Japanese "reconnaissance" and Chinese fears persuaded them to push on. Accompanied by much Chinese confusion, a further move was made that night by canal junk twenty miles to Hanshan. There, the morning brought welcome sunshine to thaw the frost out of their bones, and good news—a telephone call from Rear Admiral Holt, Royal Navy, at Hohsien, where HMS *Ladybird* and *Bee* were present. USS *Oahu* was boiling down from Wuhu to join them. The Japanese were back-pedaling like mad, he told them, sending a naval escort and seaplanes with doctors. Aboard *Bee* were most of the beer drinkers from the *Mei Ping,* whose hulk still smouldered across the river. Two others were safe in Japanese hands.

That last painful day, with fever mounting in the sick and feet turning sore on the unwounded, had been spent on narrow, devious footpaths and tortuous canals, returning to Hohsien, which was more accessible to the river. Near midnight, the bedraggled survivors, including the walking wounded and those in litters, were aboard the little Anglo-American "fleet," wolfing sandwiches and drinking proper coffee. Strong "Limey" hands had helped them aboard and strong Limey grog cemented bonds that had been forged up and down the river in many a barroom brawl.

There ended the physical trials of *Panay*'s River Rats, but not the story. Nor has it ended yet. What is the truth? How did it happen? Who or what was the villain?

The Japanese army and navy were crack military outfits, but poor when it came to mutual cooperation. Hence, at Nanking, confusion and cross purpose might have been expected.

Some army units had swept around Nanking to upriver Wuhu and had gotten some boats. Unknown to the navy—or apparently even to their own troops farther downriver—they had started probing the area where *Panay* was in trouble. The Japanese navy planes not only attacked the *Panay* convoy, but also bombed the

Standard Oil tankers at Kaiyuan while Japanese soldiers nearby frantically waved flags trying to stop them. The soldiers had suffered some casualties. But the stoic—or perhaps stupid—troops in the boats didn't report the incident, with serious later complications.

Meanwhile, the troops in the boats, seeing planes attacking the ships and knowing there were no Chinese planes in the air, naturally assumed the ships to be enemy and machine-gunned *Panay*. Then, and following her sinking, the navy pilots mistook both the Japanese army boats and those which had been shuttling back and forth from ship to shore, evacuating the crew, to be Chinese. Thus, after all were ashore, Japanese planes continued to swoop low over the riverbank, looking for the "enemy," giving the *Panay* survivors the uncomfortable feeling they were being hunted to the death.

Concerning the *Panay*, well-established facts were presented thirty years later and are summarized here: The Japanese people and government were aghast and deeply contrite. The American people and government were angered, but not to the point of wanting war; pacifists and isolationists blamed the United States for putting the ships in a perilous position. The Japanese navy, on the command level, was innocent of any complicity or fault. Colonel Kingoro Hashimoto, the artilleryman who gave the order to "fire on anything that moves on the river" was a dedicated, ascetic, supernationalist member of a secret organization that was above government—The Black Dragon Society. Hashimoto was proved by events to be untouchable, both by General Matsui, the CinC in China, and by the War Minister. Despite their denials, it is hard to conceive that the Japanese pilots would have failed to distinguish the American flags painted on *Panay*'s awnings. Norman Alley's magnificent motion pictures show the planes less than several hundred feet distant. This was so obvious a conclusion that President Roosevelt asked Alley to delete these sequences prior to public release, to avoid overly inflaming popular opinion. The undeveloped negative of this highly incriminating documentary evidence, insured for $350,000, was rushed to Manila by destroyer, thence to the United States via Pan American clipper, and guarded like the gold at Fort Knox.[56]

Immediate and profuse Japanese apologies and explanations went far toward cooling American tempers. But a new development less than two weeks after the original incident brought the affair to a boil once more. This was the issue of the machine-gunning of *Panay* by the two army boats. In the original explanation, the Japanese navy and government denied that any such thing had happened. Clearly, any Japanese suggestion that a boarding party would not have recognized *Panay*'s nationality would have been farcical. Actually, the army had failed to report the incident and denied it when approached by the Japanese navy with a request for clarification. But the truth came out, once more arousing American tempers and suspicions. The public showing of Alley's motion picture films and still photographs in the newspapers at the end of December kept the wound open. But the State Department proclaimed the incident closed, and closed it was. On 22 April 1938, the Japanese government paid the United States $2,214,007.36 as settlement for the *Panay*, three Standard Oil tankers, personal losses, and personnel casualties. If the U.S. wanted another gunboat to replace *Panay*, the Japanese said they would be glad to get the contract. And also, might they salvage the gunboat and three tankers? After all, they had paid for them. To these pragmatic but ill-timed requests the answer was no.

The Japanese may have bought off a war, but the episode in the long run cost far more than the indemnity they paid. From then on, American sentiment commenced to harden in favor of China. The build-up of the U.S. armed forces proceeded at a faster pace, with the thought that *Panay* might indeed be just a prelude to something bigger. Perhaps she and her little band of River Rats had served better than they thought.

Following Nanking's fall to the Japanese in early December 1937, Hankow became the provisional capital of "Free China," as Chiang's territory had come to be called. By the following spring, the new capital was a boom town. Restaurants and cabarets and hotels were packed with noisy, apparently well-off Chinese refugees from downriver. Competing for space were hundreds of diplomats and other foreigners, following the government. German General Baron von Falkenhausen and Soviet General Asanov were incongruous but near neighbors in the ex-German concession, both advisers to the Nationalists. Retired U.S. Army Captain Claire Chennault, a major general and adviser in Chiang's air force, and American pilots from the China National Aviation Corporation played poker for table stakes and watched with glee when the "Szechwan boys" (Soviet Russian "volunteers") knocked down intruding Japanese planes. Each Friday, a pot of Navy baked beans went from *Luzon* to Mme. Chiang Kai-shek, as a reminder of her American college days and an indication of the warming American attitude toward China after the *Panay* episode.

In the hospital, wounded Soviet airmen were happy to receive copies of *Life* magazine but had trouble distinguishing the advertisements from the news. They were having identification trouble with American gunboaters as well. U.S. officers and enlisted men wore identical uniforms—white shorts and shirts with no distinguishing marks of rank or rating. Officers wore white shoes and plain white topees. Enlisted men wore black shoes and topees with the ship's name ribbon. With hats off and feet under the table, which was the admiral and which the ship's cook?

The Soviets had no such identification problems with the Japanese. On 29 April 1938, they carried out one of the classic booby trappings of all time. Having ostentatiously stripped Hankow of fighter planes on 28 April, they sneaked them back in quietly at low altitude that same afternoon at dusk. Next day at breakfast time, the air raid sirens wailed. Fifteen Japanese bombers were on the way from Nanking and a large fighter formation en route from Wuhu, all no doubt lulled into a sense of security by reports from their Hankow spies.

At the Hankow fields, twenty Chinese-piloted Soviet planes were scrambled to patrol south of Hankow, the standard Japanese approach route. The main force of forty Soviet fighters loafed around to the east, awaiting developments of the plot.

For the Japanese fighters, the round trip from Wuhu stretched their fuel capacity to the limit. Much of their meager surplus was used in tangling with the Chinese fighters. As the Japanese, fuel now critical, turned their noses homeward, the forty

Soviets swooped in. Of thirty-nine Japanese, thirty-six were shot down. The Chinese lost four pilots and nine planes; the Soviets two planes and no pilots. Hankow crowds danced and shouted in the streets as the magnificent show went on. The indignity dealt the enemy had been double. Not only was it a tremendous air victory at a time when Chinese spirits were low—it was also something radically different from what the Japanese had planned for their emperor as a present on his birthday.

The big Hankow Race and Recreation Club was still going full out, tennis courts crowded, and "swimming bawth" heavily patronized by the girl watchers. Bright lights shone on "Dumpenstrassen." Faithful White Russian hostesses still hung on at "Ma" Jones' International Cabaret. Johnnie's Moderne Restaurant catered to the solid trenchermen. At Rosie's Tea Room, the only air-conditioned place in town, one could pick out through the smog at one time or another all the notables in town, including the staff of ComYangPat, restoring their tissues after a hot afternoon on the golf course.

In short, Hankow in mid-1938 was Free China's final fling before the retreat to Chungking and the dark night of total war.

The Japanese amphibious juggernaut continued its inexorable crawl up the Yangtze. The *Oahu* lay at Shanghai. At Kiukiang, the *Monocacy,* in a sort of riverine no man's land, had her crew on deck with bamboo poles, frantically fending off tenderly fused Chinese floating mines. Her provision locker and coal bunkers were close to empty. She had been scheduled to go out of commission 31 March 1938, but stuck as she was in Kiukiang, this was impossible. The Japanese refused permission for the *Oahu* to go upriver on a relief mission, or for the *Monocacy* to go down: "Consent to *Monocacy*'s passage must be withheld for the time being because of undisclosable tactical considerations." They offered to supply food, fuel, and transportation for personnel, but this was refused as weakening the American case for her withdrawal.

Chungking, uncovered since *Palos* was decommissioned, was due a visit by *Tutuila* in November, but *Tutu* hit a rock in the Paotzu T'an, again demonstrating the capriciousness of rapids and the temerity of man to oppose a determined dragon. The hole in the hull was patched with cement, but not before all the ship's records stowed there for years were reduced to pulp. The dividend in this otherwise lamentable incident was that henceforth, vexing inquiries from on high could happily be answered by: "The records unfortunately have been destroyed by flooding. No action, therefore, is feasible."

If *Tutuila*'s ghost should return in some future century, she might be in a position to laugh at Paotzu T'an's frustrated dragon. For many years, it has been China's great hope to build a 400-foot-high dam at Sanhsia, near Patung, in the lower end of Wushan Gorge. The lake would reach nearly to Chungking, submerging the worst rapids and drowning some of the most magnificent scenery on earth. Other dams tentatively are planned for Chungking and Suifu, to wholly tame what for millenia has been periodically one of China's great scourges when floods make central China a vast lake. If China's antlike millions could almost with bare hands build a Great Wall, dig the Grand Canal, dike the Yellow River, and hack out the Burma Road, it would be foolish to predict China could not or would not dam the Yangtze.

In mid-1938, *Tutuila* lay at Ichang, the port probably deader than at any time since *Monocacy* first showed the U.S. flag there in 1877. No ships moved. With the Japanese cork in the river, there could be no movement of goods or people. The sole remaining link with the outside was the German airline, Eurasia, from Hankow to Hongkong, and the tenuous Hankow-Canton railway. Japanese planes swooped down on the line continually, forcing passengers to leap for their lives out of car windows long since bomb-blasted free of glass. One of the air liners had been shot down.

Tutuila lay moored to the Standard Oil pontoon, several miles below Ichang. The departed resident manager had turned his house over to the gunboat. This provided a club for the officers, so that the highly polished "cocktail sampan" which used to ride astern of the station ship to provide predinner gimlets could be dispensed with. It also kept thieves out and the rat population down through the marksmanship of the officers with the wardroom air pistol. The men played baseball, threw dummy hand grenades, shot on the little rifle range, and drank a bottle or two of rationed beer in the evening.

At the end of June 1938, Lieutenant Commander E. H. Gilmer, *Tutuila*'s commanding officer, suddenly and unexpectedly died, which at last jarred the ship loose from her immolation, to drop down to Hankow with his body. There was no way to ship the skipper's personal effects home, so they were auctioned off. The crews of *Luzon, Guam,* and *Tutuila* crowded on the mooring pontoon. The auctioneer, a young officer in an unaccustomed role, stood on the upper deck. The ricksha coolies had of course straggled down to gawk.

"What am I bid? Dollar, dollar dollar! Dollar and a half, half half! Two, two! Going, going, going, GONE!" With increasing amusement, the coolies took it all in. For weeks afterward the erstwhile autioneer had only to appear for the coolies to go into a hilarious burlesque of the auction, holding up one or another's sorry rags for bids from the others. "Dollah, dollah! Watchee give? Ding hao stoffee! Ai yah! No give littee mo? Allight! SOLE!" All broken up, the coolies would shriek with laughter and come close to rolling on the ground. Considering the conditions under which those wretches existed, their general good humor was a source of never-ending wonder.

At Hankow, the roiling, muddy river's gently shelving banks and 25-foot seasonal rise-and-fall ruled out the use of docks. Out from the river's edge to the pontoons where the ships lay, 150-foot catwalks swayed on slender piles driven into the bottomless Yangtze ooze. Over the catwalks, steel muscles standing out in taut cords, human beasts of burden manhandled the drums of oil, pieces of machinery, bales and boxes going upriver ahead of the inevitable fall of Hankow.

The gunboat sailors watched the ever-changing scene with interest. "*Yei*-ho! *Yei*-ho!" went the sing-song cadence of the coolies, their sweat-shiny, sun-blackened skins stretched over gaunt frames, each miserable creature a small grain in the giant powder keg that one day would threaten India, then Russia, then the world.

During the almost daily Japanese air raids, a different sort of freight swarmed over the catwalks; thousands of Chinese crowded down to the pontoons to be near the magic totem of a foreign flag, supposedly safe from Japanese bombs. The wail of the air raid siren was a signal for the little warships to lay out steam and fire hoses, man repel-boarders stations and stand by to prevent being rushed by panicked natives.

Gunboat sailors soon developed an indifference to man's inhumanity to man—the coolies' beastlike burdens, the fat rice merchant amiably fanning himself under his sunshade, bulbous belly exposed, while a beggar lay dying of starvation in the gutter before him. There were small children, their legs twisted into grotesque shapes by their parents so that their cries for alms might be more appealing as these cripples crawled the streets. And there was the old coolie, fished out of the river by a gunboat's motor sampan. A Chinese sampan had first approached him, while its occupant palavered, then sculled placidly away. "Why didn't that sampan pick you up?" a gunboat officer inquired through the interpretation of a steward's mate. The old coolie gasped and sputtered.

"He say he old man!" translated the steward's mate. "He say sampan man want five dollah pick he up. He no want pay five dollah. He say he no worth five dollah. He too old man!"

As the summer wore on, the lights on "Dumpenstrassen" went out more and more frequently, while Japanese planes droned overhead. Slugging their way up the river, Japanese would soon be in Hankow on the land as well as in the air. Consequently, much of the Chinese government had pulled out of Hankow for the end of the line, Chungking, beyond the near militarily impenetrable gorges and rapids of the Upper Yangtze.

Following the government to its new provisional capital, U.S. Ambassador Nelson T. Johnson chose not just the only *reliable* but the *only* means of transportation—a U.S. gunboat. On 3 August 1938, he embarked in the flagship USS *Luzon* with his small suite for the 750-mile, 7-day trip upriver.

Tagging along astern, *"Tutu"* had her flat fantail loaded with cargo nearly as precious as the person of the ambassador himself; forty cases of UB beer, ballast for the expected long immolation in Chungking.

Gunboats sometimes carried more than beer, as one officer found to his dismay:

Our mail clerk was transferred to another ship the other day, and being communication officer made me ex-officio acting postmaster. The post office is so full of whiskey I can hardly get in the place, and caused someone to remark that things had come to a pretty pass when an officer of my rank and experience could be seen sitting on a case of John Haig in the post office of a man-of-war, counting out two-cent stamps. The decks are so jammed with cases of beer that there is no other safe place for the whiskey, the magazines being stuffed full of gin. The note continued with the hope that, *the Japs don't elect to pull a* Panay *on us as it would be embarrassing to explain to the newspapers we couldn't get at our powder supply due to the confusion between the shell boxes and the gin cases.*

His *aide memoire* concluded with the observation that "things go on in YangPat that would scarcely bear repeating outside this fleet."

The trip upriver to Ichang was uneventful, with a day's stop there to inspect the Standard Oil installation and the enlisted men's club, now devoid of liquid cheer or smiling young Chinese girls bent on instant conquest.

Beyond Ichang, from ambassador on down, first tripper or veteran of many passages, all watched in awe at the stupendous sights of the gorges and rapids. "Chinese pilots direct the course," wrote the communication officer, evidently out of the post office at last, "and take us shooting from one side of the river to the other, and when it seems certain we're about to run directly into some enormous rock, they kick our triple rudder over and we dash across the rapid like a chip. We were making our best speed of 14.7 knots and at times in some of the worst rapids, seemed almost to be standing still. The water repeatedly boiled in a solid yellow mass right up over the forecastle. On sharp turns and in whirlpools, the ship heeled over until the water was rushing along the low side of the main deck like a millrace."

ComYangPat, Rear Admiral David M. LeBreton, returned downriver in *Luzon* after a few days of observing the amenities in the new capital and unloading the embassy's freight. "It was a good thing the ambassador and four or five of his spare pumphandles chose to ride *Luzon,* or we would be hanging from the overhead by hooks trying to find a place to stow them. As it is, the beer, furniture, toilet paper, etc. necessary to last six months is practically putting us below the surface."

On 25 October 1938, Hankow was abandoned by the Chinese. A vacuum existed until the Japanese arrival four days later. During this hiatus, the *Guam* landed two officers and twenty-two men at the oil installations—Texaco, Asiatic Petroleum, and Standard-Vacuum—to keep out looters.

Life for the *Guam* and *Luzon* soon began again under new shore management. The *Monocacy* at last escaped from Kiukiang and was decommissioned on 31 January 1939.

On 10 February, a quarter of a century after she slid down the ways, she was sent to the bottom of the China Sea. Admiral Yarnell, CinC Asiatic Fleet, ordered the hulk sunk and a young Marine lieutenant, Victor Krulak, who was ready to do anything in those days, volunteered his services as a demolitions expert to do the job. A Norwegian tug hauled the *Monocacy* out to the Saddle Islands where Krulak, with TNT and a ten-minute length of fuze, descended to the engine room and happily set about the job everyone has fancied at one time or another—blowing up the ship. With the fuze lit, he boarded the tug and they scurried off to a safe distance to watch the fun.

But fifteen minutes passed and nothing happened. Uncertain as to whether the charge had misfired or whether Krulak had underestimated the situation, they knew only that *Monocacy* was still floating like a duck. Very cautiously the tug crept back to the ship and even more cautiously Krulak ventured down to the engine room, because nothing is more embarrassing to a demolitions expert than to have a charge go off while he is trying to find out why it didn't. However, the TNT *had* gone off, and merely blew a bathtub-sized hole in the bottom. Krulak's calculations had apparently been out by a decimal point and he needed ten times as much explosive as he had used—and he had no more. At the rate water was coming in, she wouldn't sink for a week. It would be a lot longer than that before Krulak felt like telling the admiral he had muffed the job and needed ten more cases of TNT.

What to do? Finally he talked the tug skipper into backing off and ramming the poor old gunboat right in her rusty midriff. With that crowning insult, the *Monocacy* rolled over, died, and went down in ten minutes. Lieutenant Krulak then returned to the flagship and reported to the admiral, "Mission accomplished!" There was no point in bothering that busy gentleman with details he didn't need to know.

Generations of Yangtze sailors regarded Chungking as never-never land, the Shangri-la that many aspired to, but relatively few ever reached. And of those who did, some soon began to wonder why they ever had wanted to come.

It was at Chungking in the early days of World War II that the *Tutuila,* the last U.S. gunboat on the Yangtze, hauled down her flag, although she was not the last of her breed to fly the United States colors; three others fought on for a few weeks more at Corregidor. Thus, *"Tutu"* became *de facto* the *last* of her line at the *end* of the line when on 19 January 1942, she was turned over to the Nationalist navy at Chungking.

For those who never sailed to Chungking, and now, of course, never can, as well as to revive the nostalgia of those who have had that unique experience, it is worthwhile to flash back in memory for thirty years and spend a day aboard a gunboat at Chungking.[57]

The *Tutuila* at Chungking in 1939 presented in some respects an anachronism reminiscent of the early nineteenth century—a small warship wholly on its own, a symbol of naval presence more than a force in its own right, but nonetheless an outpost of the national territory with honor and uniqueness deserving recognition and remembrance. Without the *Tutuilas* and the *Monocacys* and the other tough, resourceful little ships that sailed the Yangtze, the nation would have been scarcely less powerful, but it would have lost some of the subtle difference between a mere collection of vessels and a Navy.

The ship swung comfortably at her winter moorings in Lungmenhao Lagoon, on the south bank of the Yangtze. One anchor chain led to a submerged rock ahead and a heavy wire hawser stretched across to a bolt cemented in a river-edge cliff. The flood waters from last winter's Asian snows had subsided, so that the river now flowed placidly thirty feet below its spring crest. A spur of rock, like a rooster's comb, rose from the river bed fifty yards to "seaward." The huge timber rafts and unwieldy junks that often threatened to come crashing down across the gunboat's bow in time of flood water must now pass outside the little ship's snug, almost landlocked basin. Across the yellow-orange Yangtze, Chungking sprawled in giant panorama up a series of ever higher hills. The city was now the overcrowded, bomb-torn provisional capital of Chiang's harried republic.

The sailor on deck watch knocked on the wardroom screen door and respectfully cleared his throat, awaiting recognition from the three officers within.

"Sir, Toesy's a-bringin' a foreigner out to the ship!" "Toesy," whose ducklike feet each boasted six toes, was cheerfully available from dawn to midnight to scull people back and forth to shore in his sampan, for thirty cents a day.

"Ain't never see the guy before," the deck watch added. "He's got on a elephant hunter hat an' hoss pants an' he come down to the foreshore in a sedan chair with four bearers. All the dogs is a-barkin' at 'im. Must be a stranger!"

Strangers were rare in those parts. The officers in the wardroom mentally noted the position of the counters on the acey-deucey board and all trooped out on deck to watch as Toesy eased the sampan out to the gunboat from the riverbank.

The stranger saluted the quarterdeck in the self-conscious manner of one not accustomed to it and announced himself as the assistant military attache, a U.S. Army captain, just arrived in Chungking and looking for a place to work on some code messages and to clean himself up. He smelled strongly of sweat, campfire smoke, fried rice, and rank Chinese cigars.

The attache was a cavalryman, with a good store of horse stable directness and vocabulary. He was in a black mood, having recently been sent packing by a Chinese general who obviously had no sense of humor. The officers led the captain into the wardroom to unburden himself and have his first cup of real coffee in three weeks.

He lit up a foul-smelling Szechwan cheroot and commenced his story. "This fat bastard of a general had a whole damned corps surrounding an understrength Jap division commanded by General Doihara—*you* know—that famous Jap general they call 'the Lawrence of Manchuria!' " The attache continued his story while further befogging the atmosphere by burning the worksheets of his code messages in the ash tray. "We were sitting around the campfire one night, swilling their interminable tea, after I'd waited a week for the promised kill. I was awash with tea and it gave me a brilliant idea. 'General!' I said, 'Why don't you have all your troops simultaneously empty their bladders and *drown* all those Japs down there in the valley?' Early next morning my transportation was waiting. The general wasn't even there to say goodby. Now I find that when Doihara got good and ready, he tied his girls and his cooking pots on his pack horses and simply marched out!"

Doc Lebon, the ship's medico, dropped in from sick bay, the look in his eye suggesting he might have been checking the quality of the latest batch of medical alcohol to save a trip to the club for a pre-lunch gimlet. He noticed the khaki topee on the table and, without waiting for the formality of an introduction to the owner, burst into a snatch of song from a recent Marx Brothers movie: "Hooray for Captain Spaulding, the African explorer—who goes into the jungle to gain undying fame!" In the best traditions of Madison Avenue, the name stuck. From that time on, few people in Chungking ever used the attache's real name and some never even knew it. He was thenceforth and to all, "Captain Spaulding, the African explorer."

Besides the ignominy of the bum's rush from the field, "Spaulding" was having a bad time getting his message through in Mandarin to his Szechwanese chair coolies. The two branches of the language were close enough together to give a Pekingese confidence but far enough apart in some vital respects to get gloriously mixed up. Thus, when Spaulding was ready to leave the ship, his chair coolies had disappeared and were probably miles away, relaxing in some noodle joint, having misunderstood his instructions to wait.

"I've got the duty today," said the first lieutenant. "You are welcome to borrow my pony." Each of the officers and many of the enlisted men hired one of the wiry little Szechwanese ponies by the month, both for transportation ashore and for exercise. Everybody was required by YangPat regulations to walk or ride an hour a day, to toughen the feet or backside in case a speedy overland exit via Indo-China became necessary.

Spaulding eyed the four or five little beasts lined up on the foreshore awaiting the afternoon recreation party's coming ashore. "I'm a cavalryman by trade," he said, "but I'll be damned if I ever saw a riding horse with a crupper under its tail!"

"That's to keep the saddle from sliding over its head when he goes down steps," said the first lieutenant. "Up steps! Down steps! Just like a dog. I rode him up to the third floor of the embassy the other day—the embassy is in the Standard Oil Building, just over the officers' club. Your general's story reminded me of it. I parked the pony in the secretary's room and stepped out to find the fellow. The pony meanwhile got restless and upset a basin of water, probably looking for a drink. The secretary came trooping in and concluded the pony had urinated all over his floor. Just shivered with emotion. Couldn't utter a word. Never saw a man, not even a State Department man, so wrought up!

"I must say, not all diplomats are so highly strung," he continued. "The other day, the exec and I were shooting at a map of Japan on the club wall. I let one go by mistake right into the ceiling—one of those trick throw shots. Next day the minister-councillor called me in—pointed out a splintered hole through the floor right alongside his desk and kept on working. Never said a word. A real gentleman!"

"A real gentleman is right!" said the exec. "Did you hear about his little whirl in Hankow? It seems the old boy was sitting around by himself swilling a beer in the outdoor restaurant in the French concession when he noticed a couple of neat but threadbare Chinese ladies close aboard, having some ice cream. He figured they probably were refugees from Nanking down on their luck, so he motioned the boy over and told him to bring him the chit for the ice cream and present his compliments to the ladies, who appeared to be mother and daughter. The chit appeared, and the old gent was astonished to find it was for five Mex. He felt that was a shocking price to pay for a couple of dishes of ice cream but forked over anyway, bowed and made his way out. The boy caught up with him outside and wanted to know what was the matter with the girl; he had paid to take her home with him and now here he was beating it without her. She would lose a yard of face over it, the boy said."

When the laughter died down, the exec got back to the original subject. "We've got to lay off this shooting business in the club before somebody gets hurt!" he said. "Last night the first lieutenant let go at that big brass gong with his 22 Woodsman instead of hammering it to call the boy. The damn' bullets went right through it. There was an immediate thunder of hoofs next door in the enlisted men's club and we charged over there and found the Chinese manager had been renting out sleeping space to his friends after enlisted curfew. The Chinamen were abandoning ship ten abreast. Those bullets went right through the wall after they rang the gong. It was lucky nobody got hit!"

"What with these crazy chair coolies and a bunch of trigger-happy sailors, it looks like I'd be a damned sight safer chasing Doihara!" said Spaulding. "I'm staying with one of the embassy secretaries on the third range and I haven't the foggiest idea how to get there."

"You can't get lost!" the chief engineer assured Spaulding, who clearly viewed much of what he had seen and heard with suspicion. "The *mafu* trots along ahead of you on foot to clear the way with a stick in daytime and a lantern at night, shouting 'Ma lei loh! Yang Kweidzah!' That means 'A horse is coming! Foreign devil!' They don't really mean anything bad by it. The kids in the mission school greet the priest at breakfast with, 'Good morning, Father Foreign Devil!' "

The chief had one final word of advice. "The pony's name is Handlebars, because his ears stick straight out sideways. He's a single-minded stallion. If you spot a mare en route, you'd better by a damn' sight hop off pronto and grab his bridle or you are in for an unusual experience!"

The chief then made his farewells and went below. He was busy writing stories for pulp magazines and learning to play the Chinese flute. None of the stories had ever been accepted. As for the flute, its screech was so ear-splitting that even the chief himself couldn't stand it; he poked the business end of the instrument into the voice-tube that led from his stateroom to the engine spaces. It was all harmless fun and helped keep his mind off his advancing baldness.

Aside from providing some tricky transportation, Handlebars had furnished quite a start to one of the elderly Britishers at the Chungking Club, the gathering place for all the foreigners thereabouts. The first lieutenant had come back from his daily canter, parked Handlebars in the club's back yard and, to dry the sweat out of his gloves, put them on Handlebars' ears. Then he went to join the crowd for a predinner martini or two and a round of snooker. The Britisher, who by this juncture had polished off several stiff gins, happened to look out the window in the deepening twilight. "Christ, chaps! Come quickly!" he gasped. "There's a bloody moose in the compound!"

A spiral of black smoke rose from the funnel of a small Chinese tugboat that lay alongside *Tutuila*. The gunboat had just enough fuel in her tanks to drive her a couple of hundred miles in case further withdrawal upriver became necessary. The tug supplied steam for *Tutuila*'s evaporators, generator and galley bean kettle. Fuel presented a constant problem for the chief engineer. He roamed the river banks daily, trying to buy a basketful here, a sampan load there, from the tiny mines that dotted the Szechwan landscape and produced some black looking stuff, largely rock, that went locally as coal. A wag in the foreign colony had suggested that what he needed was a coal sniffer, a coal hound, a breed unique to Szechwan. They could turn up coal like a beaglehound can a rabbit, he said. The chief wasted some precious coal-hunting hours checking this canard out, led on by the basic reluctance of the Oriental ever to say "no."

Adding to the chief's woes, the Chinese firemen had a habit of throwing quantities of coal on the fire, then proceeding to take a snooze, whereupon the supply of steam gradually dwindled to zero and the lights went out as the generator ground to a halt. In their more ambitious moments, they would toss coal on the fire, then before it had really well caught, shake it down into the ashbox, where they could sift the unburned coal out of the ashes and recover it for their own purposes.

The coal problem also spilled over on the first lieutenant who, as Special Disbursing Agent carrying out the functions of supply officer and paymaster, was having a feud with the General Accounting Office back in Washington, over why an oil-burner was sending in all those piddling vouchers for coal. Besides, said the GAO, they were signed in what appeared to be Chinese, which was illegal. It took six months for an exchange of letters with the Navy Department—time enough so that complaints of the Great White Father in Washington became secondary to such pressing problems as to whether or not to throw a Thanksgiving party for the foreign colony.

The latest such international whingding had been on the Fourth of July. All the officers had helped mix the punch, starting with half a barrel of local tea, then adding several bucketsful of delicious Szechwan orange juice, a case of port wine and bourbon whiskey and a large cake of ice.

For an event of such national significance, even the missionaries, who seldom appeared, came in numbers. The basic tactical error of the guests in mistaking the punch for iced tea, abetted by the blistering, cloudless heat of the day made the party literally a howling success. One could hear full-throated hymns as well as more familiar ditties of the day floating down from the hills as the guests were carried homeward in their sedan chairs.

There was nothing available now, said the skipper at lunch, but Szechwan gin, beer and cointreau, all concocted from local oranges by Pop Schwer, a retired chief machinist's mate who ran the local ice plant. Although the sailors managed to get it down, the orange beer tasted strongly of low-grade shellac. The gin and the cointreau went down more easily. But gin at Thanksgiving seemed incongruous, the skipper said. Anyway, the Chinese might take offense. There was not much for them to be thankful about at the moment and their intelligence boys knew every move made aboard ship. What about their knowing in advance and protesting when the *"Tutu"* planned to shoot off skyrockets when the birth of Doc Lebon's baby was announced, thus pinpointing Chungking for the Japanese bombers? Well, at least the answer to *that* one had been found out; the former officers' club bar boy had been seen strutting around Chungking in a major's uniform.

Also, with cool weather approaching, it was time to start thinking about a regrowth of beards which had been whacked off for the summer's heat. "You know," said the first lieutenant, "after I cut off my beard last spring people passed me without recognizing me. They even stopped playing the *Marseillaise* on the phonograph whenever I went in the Hazlewood Cafe. And the dogs started to bark again like I was a stranger."

The exec expressed scepticism. "More likely it's that fish oil soap you've been using. I feel like barking at you myself after a sniff of it."

That reminded the exec to sniff the butter. "This butter is not so bad," he announced. "The mess boys whipped it up with chopsticks from carabao milk."

"Matter of fact, the carabao's offspring hasn't been making such bad veal, either," said the SDA, "but I'm having the usual hassle with the GAO over it. They inform me that nothing in the manual provides for the purchase of live animals. Only sides of beef are allowed, they say. They kicked like hell over the charge for skinning, too. The next time I'll just call the whole thing a ton of coal and consolidate the suspensions in my account."

The skipper lighted up a cigarette made from local cheroots, soaked in water three or four days to remove some of the sting, then dried, shredded and rolled in rice paper. "Let's get cracking on the afternoon's program!" he said. "We want to get back aboard before the next raid. The last time I got caught ashore the gendarmes herded me along with a bunch of coolies into one of those stinking caves until the raid was over three hours later. Couldn't even talk. 'The Japanese pilots will hear you!' they whispered. They seriously believe the Japs have special gear to do this."

"Think I'll take up the general's invitation to play tennis," said the chief engineer. The general was a retired Chinese who ten years before had been leading violent attacks on the foreign colony of Chungking, under colors which were a sort of combination of the red of communism and the black of anarchy. Now, his claim to validity supported by numerous scars and a heavy limp, he presided over a sumptuous villa and a harem, carefully selected in Hongkong on a combination basis of pulchritude and ability to play tennis, a sport the general delighted to watch. But like a great many Chinese, he felt the sensible thing was to pay or invite someone else who was foolish enough to work up a sweat by getting out there on the court in person.

On the question of violent physical exercise, the skipper's views coincided closely with those of the Chinese general. He had decided on an afternoon's program of poker, bridge and snooker at the Chungking Club, which could be counted on to net him a generous financial dividend. Losses and gains were totted up on chits and reflected in additions or subtractions from the monthly club bill, with the club steward deducting a modest ten percent for wear and tear on the equipment.

The skipper would find no lack of convivial company at the club. Foreign business had ground to a near standstill because of the blockade across the Yangtze valley just below the militarily impenetrable gorges. The foreign *taipans* now crossed the river to their Chungking offices three times a week, and not even then if the day were clear and thus good bombing weather. There was plenty of time for relaxation.

In winter a cheery fire in the main clubroom was ringed by a sort of little United Nations—British from Imperial Chemicals, Hongkong-Shanghai Bank, and various export-import firms; Americans from Standard Oil and Texaco; Russians from British-American Tobacco or feather and pig bristle businesses, a Norwegian or two from the Chinese Maritime Customs; plus an occasional Swede, Italian or Swiss exporting sausage casings, or in some dark enterprise about which the less said the better.

In summer tennis buffs sweated on the courts while others sat watching and drinking shandygaff—half-and-half beer and ginger ale—while awaiting their turns. At the clubhouse, a small Chinese boy lay half-asleep on the porch rail, with a rope tied to one foot, swinging the *punkah* that fanned a gentle breeze over the billiard table.

"Things might be worse," ruminated the SDA as he sat before the cheery open fire at the club, restoring his sagging spirits with some of Pop Schwer's orange gin. The aroma from his "Szechwan cigarette," if one could be sufficiently charitable to so term it, masked the odor drifting over from the nearby graveyard, where victims of recent bombing raids had been hastily buried in makeshift coffins.

The club members had been in unusually good form when the SDA had arrived from his afternoon horseback ride. The British military attache's glass eye was already in its place of safekeeping for the evening—a tumbler of water at the end of the bar, a precautionary move which generally never was made until at least several rounds of drinks had been accounted for. Their excitement concerned one of the British vice consular types who had been knocked flat by a dismembered leg hurtling over the embassy compound wall. Aside from the indignity of it all, there was the sticky question of what to do with the several hundred Chinese dollars found tucked inside the stocking.

"If the rapeseed contract goes through," said the SDA to his heavily bearded British colleague, the Number One of HMS *Falcon,* "we can dispense with that dirty little tug for awhile, and I will be spared this damnable business of scouring the waterfront every day for a few baskets of decent coal."

The SDA had encountered the usual difficulties with the General Accounting Office over the proposed new makeshift fuel: "The referenced contract for rapeseed oil is not clear," they had written. "What *is* rapeseed oil and for what purpose is such a large quantity intended?" After this sally, the suspension in his account of a mere $7,000, unaccounted for crew's pay slips, was something of an anticlimax.

The SDA had finally found a way to get a night's sleep uninterrupted by the shouts of late-homing Chinese mah-jongg addicts, bellowing for a sampan to take them across the river. One simply whanged a pebble from a slingshot toward the beach, while the deck watch simultaneously fired a blank round from his Springfield. There would be a moment's silence while the alerted Celestial tied splash and bang together, then a clatter of stones under flying feet as he departed at flank speed. The word would soon get around the all-night noodle parlors that the Yankee gunboat was not one to trifle with after dark.

Relations between the British *Falcon* and American *Tutuila* at Chungking were excellent. There was much fraternization amongst the crews. The *Tutuila*'s wardroom officers were regular guests at the *Falcon*'s Saturday noon curry tiffins, a treatment guaranteed to lay the strongest man out until time for a predinner restorative. There was also much mutually beneficial collusion in establishing a common front against the local Chinese produce vendors, on whom the gunboats were wholly dependent for food.

The *Tutuila* had arrived at Chungking on 10 August 1938, and the only supplies of any kind received since then were small items air-lifted from Hongkong. But help was on the way. "The newest addition to our family," a *Tutu* officer wrote home on 21 May 1939, "the Navy truck which is supposed to shuttle between Chungking and Haiphong to keep us from starving to death, our radio clacking and the grog barrel replenished, started its maiden voyage a week ago. To our horror, instead of a light, easily handled affair of several tons, it turned out that the Cavite Navy Yard had constructed us a cross between a tank and an armored car, the whole body made of bullet-proof steel, the boxcar-shaped body fitted with indirect lights, bunks, green leather seats and sliding panels with gun slits like a Brink's money wagon. Needless to say, the thing had to be practically torn to pieces and rebuilt at Haiphong in order to squeeze through the various city gates en route, traverse the appalling 'roads' and cross the rickety bridges and ferries."

To everyone's astonishment, the face-lifted truck, promptly christened "Bessie" from her opulent shape, arrived at Chungking after two harrowing weeks, having used half her capacity to accommodate her own fuel, but carrying treasure beyond local price: cigarettes, spare parts, coffee, medical supplies, and several cases of whiskey.

The French gunboat, an ancient little spitkit named *Balny*, based two miles downriver from *Tutuila* abreast a commodious establishment officially named *Caserne Commandant Odent*, but universally known as "The Bastille." On those not too frequent social get-togethers with the Anglo-Saxon River Rats, language difficulties generally tended to channel activity less to conversation and more to that universal means of communication, the bottoms-up toast.

One visit to *Balny* by *"Tutu"* officers had left the state of Franco-Chinese relations in the Bastille area at a low ebb. There had been a luncheon on the veranda deck, and red wine had flowed freely. An enormous flock of walkee-walkee ducks had chosen this inauspicious moment to swim by. Possibly the Americans instigated the massacre by some reference to the ducks' military appearance reminding them of Nazis. In any case, the Frenchmen had opened up with a machine gun and had slain ducks by the dozen, which they then had to pay for. Being Frenchmen they were not about to throw them away. Eating all those ducks before they spoiled in the summer heat must have been monotonous. For a long time after that one had only to quack in the presence of a French gunboater to insure a highly satisfactory Gallic outburst.

The gunboats at Chungking were looking out a window on war, with some attendant danger. *Tutuila* was near missed once—the blast holed her waterline and smashed the officers' private outboard motor skimmer. The British gunboat had likewise been so favored. The rescue party from *Tutuila* got through the pall of smoke to find the skipper dazed but coherent, his cabin door blown out and debris scattered about. "Those bloody bahstahds!" he roared. "They've bloody well broken my lahst bottle of decent gin!"

It would be remiss to describe Yangtze gunboat life without mentioning the Chinese steward's mates—messboys. As the climate demanded they wore blue or white Chinese style button-up-at-the-neck jackets, U.S. sailor white hats, and—as well they might—expressions of supreme satisfaction. After all, they were paid U.S. sailormen's wages, not much below those of a Chinese colonel.

Quite a crowd of Chinese had collected one forenoon on the pontoon to which

Tutuila was then moored. "We must have missed the air raid siren," someone said. "Oh, no!" said the mess treasurer, "I just put an ad in the local paper that we needed to recruit a messboy. Those are the applicants."

Treatment at the Race Club that afternoon was cool. "You blighters!" said a British friend. "None of us got any tiffin today. All our cooks were down to your bloody gunboat trying to get a job!"

The Chinese messboys never missed a cue, whether it be the freshly pressed uniform neatly laid out, paste squeezed on the toothbrush, razor ready over the steaming bowl of water—all prepared in feather-light silence before bringing in the cup of hot coffee as a reveille eye-opener. Or it might be the light banter at breakfast:

"Ching! Bring two fried eggs; one fried on one side, one fried on the other! Can do?"

"Mastah, I aska cook!"

Ching then would repair to the galley and engage the cook in lengthy consultation, not all of which was in Mandarin Chinese.

"Chao ching erh *goddam!* Wong jao jui *hell* toh jer! Shu shoh *sonnabitch* ah sur shao!" (Or words to that effect.)

Finally Ching would return to report: "Cook say can do, Mastah!"

Ah, yes! That, like many others in Chungking, was a very good day.

Tutuila's withdrawal to Chungking had taken her well beyond operational range of Japanese aircraft, and put her under the constant Szechwan winter cloud blanket. But by spring of 1939, the Japanese had moved upriver to Yochow, and with the coming of clear weather, one got the feeling that the provisional capital's number would soon be up. In the first week of May it happened. A gunboat officer who had seen over two hundred Japanese air raids on Canton in 1937–38 described the curtain raiser that would come to be a common occurrence from then on out until the final Japanese collapse.

I knew that if I stuck around long enough my patience would be rewarded, and what happened last week was far beyond expectations. The weather finally cleared enough to let the Japs through and they grabbed the opportunity to stage the greatest mass slaughter I've been privileged to witness so far in the war. Canton was nothing! *We had been sitting around for half an hour or so, with the crew on deck, expecting the air alarm to be the usual false variety, when there was a roar like a million bees approaching and when we looked up, there seemed to be bombing planes in one continuous line from one horizon to the other. Forty-five of them, wingtip to wingtip. The explosions were beyond description, and the whole thing done before our eyes from start to finish. The first bombs landed on the same side of the river as* Tutu's *anchorage, the line of explosions crossing the river and walking up the side of the hills on which Chungking sits.*

We didn't miss a thing, except possibly the few bombs that dropped while we were diving for cover down off the top of the bridge into the armored pilot house.

It was all so unexpected we didn't have the shutters down, tin hats on or anything else ready. We just gave vent to a wow! of dismay and amazement, grabbed our hats and down we went. The concussions made the ship toss like a cabbage leaf, even though the bombs landed half a mile away. The whole show was over in five minutes and the Jap planes were disappearing down the Yangtze. There were three big trails of smoke through the sky where burning planes were dropping and two parachutes swaying from side to side like kids' balloons at the fair. They looked so close we even had the motor sampan ready in case they should land near us in the river. Actually, both landed a mile or more downstream. One drowned and the other, half burned to death already had the job completed by coolies, even though he was a Chinese.

Enormous fires literally leaped up as soon as the smoke and litter from the bomb explosions had died down and the flames seemed to reach ten times as high as the houses that were burning. The pall of smoke was beyond belief, and made the bright blue day almost black, hiding the sun for hours.

The raid on the following day was even more spectacular and beautiful if possible, (if one can say such a thing about mass death) than the one the day before. They came in almost without warning, after several false alarms during the day had got the people less alert. Twenty-seven of them came over, just at sunset, brilliantly silhouetted against the bright evening sky. There was a veritable hell for about five minutes—the colossal explosions of the bombs, one after another, the crash of the antiaircraft guns, and the bursts of the shells in the air. They were shooting tracer shells and in the deepening twilight each of them looked like a rocket, flying out of a ball of flame at the gun's muzzle, then gradually slowing down and curving over like a flower stem, until the bloom, the bursting shell, the ball of black smoke appeared at the end. The same fires this time, but infinitely more terrible at night, with the whole city on fire and outlined in flames like a map lying draped across the stairstep hills. The electric lights were out, of course, and the only illumination the red glow from the blazing city.

The fires burned for two days. They are still collecting bodies and carrying them across here for burial, seven days later, so you can imagine what the air is like. The Chinese go around with cotton stuffed in their noses. The Skipper went over the day after the second raid and was upset for three days by what he saw. There were hundreds of dismembered bodies on the foreshore, desperately anguished relatives trying to sort out legs and arms and heads to fit torsos so the dead would have all their proper parts in the hereafter.*

When the American ambassador, after a year in the narrow confines of Lungmenhao, the foreign colony on the south bank, flew out to Hongkong for a breather, a stateside parent noted the fact in the newspaper and wrote his son in *Tutuila* that with the ambassador no longer there, perhaps she would "get out of there and go someplace."

* Lieutenant Commander Bradford Bartlett, who made excellent motion pictures of the mass atrocity, which were rushed via PAA Clipper to Washington, where a newsreel company prepared them for wide distribution to awaken the U.S. to the true nature of the Japanese threat. But the Navy Department refused release; it would be too offensive to the Japanese, and might cause international repercussions at a delicate stage of relations.

The reply, two months en route via muleback from Chungking to Kunming, thence by truck or rail to Haiphong and slow boat to the U.S.A., made clear why *"Tutu"* was not about to go sightseeing.

"And *where* would we go?" wrote the young man in Chungking, to which the Japanese bombers were now commencing to find their way with uncomfortable regularity.

A few hundred miles below us is No Man's Land, with Chinese barriers and minefields making impossible for ships to pass. Take a look at the map of China! The Japs are as far as Yochow, at the mouth of Tungting Lake, well above Hankow. The Chinese barriers are fifty miles above that. The only other city on the river we can reach is Ichang, now nothing but a smoldering ruin. Chungking is as high up the river as we can go. We are stuck here, blockaded, if you will, unable to move in any direction even if we had the fuel to do it, which we haven't. The ship is here until the war is over. The officers and men are here until they can make arrangements to ferry us out a few at a time by air or truck to Indo-China, almost a thousand miles of near roadless land scarcely out of the middle ages.

In any case, although the ambassador is not here, the embassy is functioning without him. There is a staff of six carrying on and keeping in touch with the Chinese foreign office. We are the guardian angels of the embassy, with radio to the outside world for their voluminous despatches, with emergency rations, motor boats to ferry them across the Yangtze, and a refuge in case of a breakup in China, should the Chiang regime someday blow up. That is what gunboats are for.

In spite of the gloomy connotations of the young gunboater's letter home Chungking, for anyone who liked his pleasures simple, was not too bad a place that fall and winter of 1939. The Chungking Club's New Year's Eve party, in fancy dress after the custom of the predominant British, included some seventy-five assorted souls. With the exception of the Soviet Russians, who kept strictly to themselves, almost every major and some minor nationalities were represented. The "sitzkrieg" already simmering in Europe did not prevent the local nationals of the warring powers from drinking at the same waterhole, but like cocks in a barnyard, the belligerents kept a dignified distance from each other, buffered by neutral common friends.

There was a saying in Szechwan that if in winter the sun shone, dogs would bark at the unaccustomed sight. The constant cloud blanket hugging the hills and the near freezing rain turned the paths into sticky muck and made local travel uncomfortable, but it also kept the Japanese bombers away. One could sampan across the river, at its quarter-mile-wide winter low, climb the several hundred stone steps slimy with the spillings of water coolies, and have a leisurely tour of the city without the grisly prospect of being shoved into some stinking cave to sit out an air raid.

Even as the provisional capital of Free China, Chungking city proper saw so few foreigners that a small crowd would collect outside the door whenever a gunboater entered a store to browse among the brass water pipes, garishly embroidered brocade, cleverly made items of bamboo, human embryos in jars of alcohol, or uncomfortable-looking caricatures of chairs and beds fondly believed by their makers to be "foreign style."

In China there was always bamboo, as an observant visitor, William Edgar Geil, noted:

A man can sit in a bamboo house under a bamboo roof, on a bamboo chair at a bamboo table, with a bamboo hat on his head and bamboo sandals on his feet. He can at the same time hold in one hand a bamboo bowl, in the other hand bamboo chopsticks and eat bamboo sprouts. When through with his meal, which has been cooked over a bamboo fire, the table may be washed with a bamboo cloth, and he can fan himself with a bamboo fan, take a siesta on a bamboo bed, lying on a bamboo mat with his head resting on a bamboo pillow. His child might be lying in a bamboo cradle, playing with a bamboo toy. On rising he would smoke a bamboo pipe and taking a bamboo pen, write a letter on bamboo paper, or carry his articles in bamboo baskets suspended from a bamboo pole, with a bamboo umbrella over his head. He might then take a walk over a bamboo suspension bridge, drink water from a bamboo ladle, and scrape himself with a bamboo scraper [handkerchief].[58]

On shopping expeditions, a gunboater became the focal point of the intently concentrated gaze of a dozen or more pairs of beady black, shoe-button eyes set in immobile, expressionless faces. Eventually he had the uncomfortable feeling of being skewered on the point of a pin like some rare bug. One could simply move on for a repeat performance next door, or take advantage of the Chinese horror of being made a fool of. The beleagured shopper could walk back to the knot of silent Chinese, pick out a random victim and begin a spirited discussion about his buttons, while carefully inspecting one, then perhaps gently lift his hat, look inside and have a mild chuckle at something he clearly found to be ridiculous. And so on. As an ice cream cone dropped on a sunny pavement soon begins to lose shape and substance, so would the resolution and curiosity of the Chinese object of *fanquei* attention. In very short order he would elbow his way back through the crowd and disappear. And then a second victim of scrutiny would lose his nerve, and then a third and soon each would have decided that the lightning might strike him next and the little crowd would be gone.

The U.S. military attache in Chungking at that time described a citizen's reaction to what must have appeared to him to have been just another of those unpredictable, unfathomable and slightly wacky *fanquei*. The attache and a "*Tutu*" officer, Ensign William Lederer, were strolling down a Chungking street, when they met a Chinese gentleman in a well-cut western style suit. Lederer seized the Chinese by the lapels. "Where did you get that suit?" he said, in English in a rather penetrating voice. Whether or not the Chinese gentleman understood the question, he left as though the devil himself were after him, and to this day no doubt is regaling friends, descendants, and the spirits of his ancestors with the story of the *fanquei* madman.

The six-foot-four medical officer was a stellar attraction, as any Szechwanese who topped five and a half feet could have looked over the heads of his fellows. In addition to the natural and more or less universal human proclivity to gawk at something out of the ordinary, the Chinese had a total lack of appreciation of privacy in general. With the exception of the wealthy few, they were born, grew up, procreated, bathed, dressed, lived and died in a small space shared with half a dozen or more others of both sexes and all ages. The commercially-fired and repression-fed American obsession with sex was wholly absent in China, where crimes of passion were so rare as to be phenomena. A lady wishing to create a sensation by walking topless across the Chinese countryside would have been in for a sad disappointment.

Those Chinese who had contact with westerners were at a loss to comprehend this peculiar *fanquei* modesty on the one hand and prurient fascinations on the other, as typically demonstrated by a Shanghai Palace Hotel room boy. In the customary informal manner, he had entered the bedroom of a newly arrived foreign lady. "Boy, you had better knock before coming in. Maybe I no have on clothes!" "Oh, maskee,* Missy!" he assured her. "My allatime lookee through keyhole before come in!"

With *"Tutu"* staked out immobile at Chungking on the west side of the fence, there remained on the other side *Luzon, Oahu,* and *Guam*. The fuss kicked up over *Panay* had been probably the biggest for so small a vessel since Cleopatra's barge had made a rendezvous with Marc Antony's. As a direct result, relations between the ever cockier Japanese and the westerners were cooling by the day.

Admiral Yarnell was trying to hold the line. On 18 December 1937, he radioed Washington a recommendation to try their best to have Vice Admiral Kiyoshi Hasegawa retained as the Japanese CinC, even though tainted with the *Panay* affair. He once had been naval attache at Washington and was favorably disposed toward the United States. Admiral Yarnell felt that any replacement might increase the difficulties of communication.

One day earlier, Yarnell's flagship, the *Augusta,* had seen direct evidence of the Pacific war, even though Shanghai was now largely "pacified." The *Oahu* had come alongside, bearing the dead, wounded, and most of the survivors from *Panay*. With the *Oahu* came HMS *Ladybird*. In the highest traditions of the Royal Navy, *Ladybird* and *Bee* had moved in to rescue survivors without pausing to check Japanese tempers. After the events of the preceding weeks, one was at a loss to say whether the Japanese on the Yangtze had at the beginning of December declared *de facto* war on the Anglo-Americans, or simply gone locally berserk.

At 9:30 on 19 December, Admiral Hasegawa boarded *Augusta* for a half-hour call on unsmiling Admiral Yarnell. General Matsui, supreme army commander, calling five days later, lingered somewhat longer—almost an hour.

Whatever comments Admiral Yarnell might have had to offer General Matsui must have been delivered during the foregoing visit; when Admiral Yarnell returned the call the following day, Christmas, it must easily have been the shortest official call on record. Leaving *Augusta* in midstream at 10:52 A.M., for the barge trip ashore, auto ride to Matsui's headquarters, and return, Yarnell and five staff members were back on board in five minutes less than half an hour.

* Never mind.

Yarnell's flag lieutenant, Lieutenant John Sylvester, remembered that the call was most formal and that they felt General Matsui had been ordered by higher authority to initiate the exchange.

My chief recollection is that when Admiral Yarnell returned Matsui's call, the Japanese army had troops on each side of the road from Shanghai to his headquarters, apparently to show that he wanted to insure Admiral Yarnell's safety, as they were all facing away from the road.

Sylvester felt that Hasegawa was deeply concerned over *Panay*.

His relations with Admiral Yarnell were in general excellent. I know Admiral Yarnell thought him a fine officer and gentleman. Hasegawa was most apologetic, and I know that as time went on, Admiral Yarnell was positive that he had nothing to do with the attack on Panay.

By January 1938, the Japanese position in and around Shanghai had stabilized. The Chinese were down but not out, although as early as mid-November, Chinese bombing and exchange of gunfire with the Japanese was no longer a part of the daily routine as seen by *Augusta*. The entries for 6 January in *Augusta*'s log reflected the calm which had settled not only climatically but militarily on so recently superheated Shanghai:

0928 Seven Japanese bombers flying over Chapei.
1105 Commander in Chief, Asiatic Fleet, shifted flag to *Isabel*.
1327 *Augusta* underway and standing out, for Manila.

The ship hurried to the Philippines in effect as a dispatch boat, to meet the transpacific clipper plane at Manila and send a full transcript of the *Panay* testimony to Washington. She set a record: Shanghai to Manila in 39½ hours averaging over 29 knots for 1,170 miles.

Calm had returned to the International and French settlements. Jai-alai and the dog races were back to normal. Sampans and junks once more plied the Whangpoo and Soochow Creek. From the U.S. naval buoys off the bund where the station ship lay, it was only a five-minute boat ride to the cheery open fire of the Shanghai Club. There, British *taipans* stood at the "longest bar in the world" buying each other drinks. As they discussed "those bloody Jap bahstahds" who were stifling trade on the Yangtze, the *taipans* were unaware of the fact that their world and their "Empah" were at that very moment in the prelude to the final crackup.

Friendly or no, Vice Admiral Hasegawa soon began throwing his weight about. On 21 December, nine days after the *Panay* attack, Admiral Yarnell received a letter from the Japanese CinC. It covered in detail Japanese views on movement of shipping on the Yangtze, and concluded ". . . in view of the fact that minesweeping operations as well as mopping up of scattered Chinese troops are still going on along the river, *it is the desire of the Japanese Navy* that foreign vessels *including warships* will refrain from navigating the Yangtze *except when clear understanding is reached with us*." [Italics supplied.]

Admiral Yarnell, in no mood to temporize, replied that "We cannot . . . accept the restriction *suggested* [italics supplied] by your letter that foreign men-of-war cannot move freely on the river without prior arrangement with the Japanese and we must reserve the right to move these ships whenever necessary without notification." Yarnell's letter was cosigned by the senior officers present of the British, French, and Italian naval forces.

Rear Admiral William A. Glassford, who had relieved D. M. LeBreton as Commander, Yangtze Patrol, took the pragmatic view that the high-riding Japanese could be neither bluffed nor bullied, and that an effort must be made to fall in with their desires as far as reasonably possible.

Although an anglophile to the point of wearing knee-length white hose with his uniform shorts, and sometimes even a Royal Navy type black patent leather chinstrap, Glassford had some bitter arguments with his British counterpart, Rear Admiral Holt, who claimed that the Americans had let the British down in not demanding their "rights." "But," as Glassford set down in his notes, "we persisted in a cooperative attitude with the Japanese and the British reluctantly fell in line, even though they felt it was humiliating."

Holt may have had a point, but when it came to accurately calling the card, Admiral Yarnell wasted no blandishments in a long, farewell analysis of the Far Eastern situation. Dated 20 July 1939, one paragraph expressed what a good many top Britishers from Queen Victoria's able foreign minister Lord Palmerston to Winston Churchill have frankly admitted about England:

> *Her foreign policy in the Far East has been dictated by her imperial and economic interests to a marked degree. She has been willing to support the United States when it was to her advantage to do so, and to support other nations at the expense of the United States, regardless of the ethics of the case, when she felt it the better economic procedure.*

Cordell Hull acknowledged its receipt with thanks, reporting that "it has been read with interest."

To bolster his position *vis à vis* the Japanese, Glassford had a picture of President Roosevelt hung in a dominating position in his cabin, pointing out to Japanese dignitaries that the top shogun in Washington would stand no nonsense. The Japanese immediately countered with a far larger picture of the Emperor, not even troubling to point out what God would or would not stand for. Relations none the less remained excellent, with friendly social entertainment back and forth.

On Glassford's first visit to a newly promoted full admiral, the banter following the formal congratulations reflected the good feeling. Did Glassford intend to remain on the Yangtze, the new admiral wanted to know, at the time making a reference to "die Wacht am Rhein." "Yes, I will continue my watch on the Yangtze," replied Glassford. "But especially I will watch you." This was greeted with much Japanese merriment, as they sat in HIJMS *Idzumo*'s flag cabin, drinking premium scotch.

Unfortunately, the Japanese army, controlling Shanghai, displayed considerably less humor, if indeed, any at all. In July 1940, a Japanese general decided to drop in unannounced on that part of the International Settlement patrolled by the Fourth Marines. Colonel DeWitt Peck, commanding the regiment, got wind not only of the projected visit, but of a plot to assassinate the visitor. The ramifications of the plot would have baffled Machiavelli, but basically the Japanese planned to do in their own general. Alerted Marines picked up sixteen sputtering Japanese gendarmes in "plain clothes," as blatantly apparent to the trained eye as a Soviet NKVD agent. Some of the heavily armed gendarmes were Formosans, which fitted in well with the assassination theory. Had the plot succeeded, it would have provided an excellent excuse for the firebrand younger Japanese officers to press for a takeover of the entire settlement.

One would suppose that secretly, the general might have been grateful that his skin was saved. Publicly however, it was an enormous loss of Japanese military face to have their gendarmes of *any* stripe humiliated before hundreds of grinning Chinese by being nabbed and disarmed by U.S. Marines.

After this incident, American-Japanese relations cooled perceptibly. Admiral Hart directed Glassford to henceforth be unyielding to the Japanese.

Honkew, for a couple of years in Shanghai's infancy the "American concession," was a Japanese preserve. A Japanese sentry stood on the Garden Bridge, over odoriferous Soochow Creek, which separated Honkew from the rest of the International Settlement. Foreigners were expected, on pain of a possible slap in the face, to bow gently from the waist when passing the sentry. Chinese coolies grunted, groaned and yei-hoed, pushing heavily loaded carts up the bridge's steep approaches. An occasional bayonet thrust into a bale or a prick in some tender part of a coolie's anatomy reminded everyone who was boss. Although Honkew was a part of the International Settlement, the Settlement taxis and rickshas were not allowed there. One had to hire a ramshackle vehicle especially licensed—or walk across the bridge, bowing en route, and pick up a conveyance in Japanese "territory."

In October 1940, Admiral Hart, who had relieved Yarnell on 25 July 1939, brought his flagship *Houston* into Shanghai and told Glassford he would not return north again. Hart was close-mouthed about war plans, natural enough in view of Glassford's exposed position. Glassford requested that unity of command be established over the China elements—YangPat, the Fourth Marines, and the naval components of the Marine units at Tientsin and Peking. The latter two were unique in American military annals in that they were under the direct military command of a civilian, the U.S. ambassador.

Hart demurred in the request for unity of command, but felt the Yangtze gunboats should make ready for the open sea. Glassford commiserated with himself in his diary over Hart's refusal to give the unified command: "I could only recommend; I could not direct." Finally, in October 1941, after repeated recommendations by Glassford, Hart authorized him to "assume direction of the 'operation' of all forces and activities in China concerned, 'administration' to remain as before." In his notes, Glassford added that in his insistence for unified command "it is quite possible that we were over-zealous in our effort to formulate joint plans and that the High Command had us ever in mind. If this was the case, we on the spot were not made aware of it."

Shanghai had acquired a new vitality following Hitler's expansion in Europe, in the form of thousands of Jewish refugees, principally from Austria. Most had professional talent of some kind, a few had money. Small, intimate *bistros* sprang up, where the wiener schnitzel and string combo were reminiscent of the best of old Vienna. In exchange for lunch, a distinguished but hungry professor was happy to give an hour's daily lesson in German. The women peddled beautiful handmade gloves from door to door, or gave singing lessons. In some respects, it was a repeat of the White Russian "invasion" of the early twenties, but without the Russian "live, laugh and love, tomorrow we may die!" syndrome. These Jewish refugees, mostly middle-aged, were sad, bitter, and filled with a sense of hopelessness. Five hundred years before, the Jewish colony on the Yellow River was well on its way to assimilation and loss of identity, the first and last such case. This was China. Given time, it could digest the indigestible.

By the end of 1940, distress flags were in the air. On 16 December, *Time* lugubriously reported:

The first sting of winter hung over a dying city. Its tide of fleeing foreigners has reached flood last month with the evacuation of U.S. citizens; its foreign colony has shrunk to a scattering of bitter enders: U.S. taipans *unwilling to leave, White Russians and anti-Nazi refugees unable to leave, British nationals with no place to go.*

The roulette tables at Joe Farren's, the Park Hotel's Sky Terrace, Sir Ellis Victor Sassoon's Tower Night Club had none of their old sparkle. Industrial Shanghai was sinking fast. The tea and silk trades are at a standstill. The cotton mill output is down to thirty percent.

This, too, was China.

During 1941, the United States began preliminary but very definite moves to prepare for war in the Far East in the not too distant future. In late 1940, all Asiatic Fleet dependents, some two thousand women and children, were sent home. As a result, it no longer was possible to find sufficient volunteers for a station where formerly there had been an enthusiastic over-sufficiency.

Thus, in August 1941, a former gunboater, China-bound again as executive officer of the USS *Wake,* found his companions aboard SS *President Harrison* to be some fifty replacement officers for the Asiatic Fleet who had never served a Far Eastern tour and were now "draftees." Ironically, those who had never wanted to go even in the halcyon days (bad for promotion!) were sent at the critical time, and half or more of them subsequently were killed or captured by the Japanese.

The former pleasant stops at Yokohama and Kobe were no longer on the itinerary. From Honolulu, the ship steamed straight to Shanghai. There, at dinner in the Palace Hotel, the *Wake*'s incumbent executive officer, Lieutenant David W. Todd, turned his job over to the new XO, while white-gloved "boys" wheeled in the hors d'oeuvres tray, the salad cart and the roast beef wagon. In the background, a Jewish refugee orchestra rendered Strauss waltzes to an almost empty room.

At dessert and coffee, the briefing continued with an added member, the strikingly beautiful, doll-like wife of the Japanese consular liaison officer at Hankow. She and her husband were *nisei,* American-born Japanese, pressed into service while on a visit to Japan and not allowed to return home. Hankow was dull, said Miura-san. She had come down for a taste of life, and found Shanghai disappointing. But Americans would find the Japanese navy cooperative, she said. They must play the Japanese army and navy against each other; they were very jealous partners, and each would try to outdo the other if cleverly handled.

The gloom and decay of Shanghai described by *Time* had deepened. The rate of exchange had tripled in a year, up to eighteen to one. During the days of waiting for transportation upriver, only the companionship of an old and very dear Shanghai friend helped the *Wake*'s new XO forget the depressing influence of constant rain, grim *fanquei* faces, and empty tables in the once gay spots of a once fabulous city.

Miura-san's prediction of Japanese navy cooperation soon took tangible form. In a few days, transportation was made available to the new XO from Shanghai to Hankow aboard a river steamer carrying Japanese troops. He was the only non-Oriental on board. In the large dining saloon, no sound arose other than the strangling noises which accompany oriental tea drinking and the click of chopsticks as thirty-five or forty officers crouched over bowls of rice, pickles, vegetables, and a small scrap of meat or fish. These were reserves, called from farm and village, innocent of any knowledge of pleasurable dinnertime small talk. When they stared at the single Occidental in their midst, it was not out of hostility, but simple curiosity over how he handled his chopsticks, or in clumsy foreign fashion, fractured the etiquette of eating. One doubted that they were any more enthusiastic over their final destination than were their opposite numbers, the dragooned U.S. Navy replacements aboard *Harrison,* over theirs.

The five-day trip allowed time to read in detail the two or three months' old *North China Daily News* discovered between mattress and spring. There was no other English reading material on board. At each stop, mild diversion was provided by Chinese arriving for steerage passage. The Chinese had been assured by their propaganda organs that the Japanese were out to sterilize them and decimate the race. The injection method was a common one, they had been told. Consequently, when the new arrivals were immediately collared by a couple of medical orderlies in dirty white smocks and face masks for a cholera injection, the result was some highly spirited action on the part of those who felt their manhood in jeopardy.

The drab, flat lower river banks dragged interminably by, reducing one to such desperate expedients as working the British-oriented crossword puzzles in the dog-eared Shanghai paper. The stoic Japanese seemed relieved at nearing their destination, and on the fifth day they produced large bottles of good Asahi beer in celebration. It was the occasion for the first conversational contact with their solitary western shipmate. "I export—import bissniss," volunteered a shaven-pated warrior who appropriately was named Endo. He offered a swallow out of the bottle he had been carrying around the deck by its neck. Several of his friends produced more bottles. Mr. Endo's voice got louder and his conversation less intelligible. Rocking back and forth on the balls of his zori-shod feet, he took a last swig, fecklessly tossed the bottle overboard, and in a voice that could have been heard on the fantail, shouted, "I ruv ribberty!" This was the end of the soliloquoy. He fell back insensible into a wicker deck chair. Whatever happened thereafter to poor Mr. Endo, dragged off from "bissniss" to war in a flyblown, filthy, hostile land, probably only the heavenly custodians of the Yasukuni shrine could say.

The USS *Wake,* which had been the *Guam* until 5 April 1941, when she had to relinquish the name to a new cruiser, waited in Hankow. Hankow of September 1941 was a far cry from the Hankow of 1930. Or even of 1938, when the XO had last seen it. "Dumpenstrassen" was cold and dark; the cheerful, friendly Russian girls had flown. Rosie's air-conditioned tearoom was no more. In the vast emptiness of the Race and Recreation Club, a half dozen or so disconsolate members rattled around where in the good old days five hundred people at a time had filled the place with sound and gayety. On the lawn, several twosomes drank shandygaff or played tennis. There was no waiting time for courts any more. The several young bachelors from *Wake* were lucky if they found more than one curvesome lady in the "swimming bath" at teatime. Of the gay set, only the aging wife of the manager of the Hongkong and Shanghai Bank still held court, on the cool roof, six stories up, where a foot of earth supported a luxuriant garden and lawn. Several young, unattached misses, Eurasian or of indeterminate national origin, worked as secretaries for the few foreign firms whose doors still opened out of pride or nostalgia; there was no business.

Also evident was a social leveling generally found only in times of major disaster. In palmier days, the secretaries would have been considered socially acceptable by proper *taipans* only in the general cameraderie of the New Year's Eve masked ball. Now they were avidly sought after by the sad little squad of civilian bachelors too blind, feeble or deaf to be in the European trenches. Even the two beauteous teenage Eurasian daughters of Hankow's postmaster had left for home: England.

"Godown Hankow," the supply facility set up to sustain a half dozen or more gunboats, now served only one, but was still amply stocked with food, spares, clothing, coffins, and odds and ends such as one finds accumulated in the attic of any long-occupied family dwelling. What was not available there, the Japanese obligingly transported by ship or plane. Unlike *Tutuila* at Chungking, oil was no problem. Following Miura-san's good advice, the word would be let slip at the next meeting with the navy's Captain Fukuda that the army's efficiency was really impressive. Why, the last shipment of mail they handled would put the U.S. post office to shame! It was a pity, though, that there was some hang-up on fuel oil. "What! Who say trubber on oiru? When you want oiru?" Tomorrow there would be oiru!

Just what were the Japanese navy and Captain Fukuda doing six hundred miles up the Yangtze in solid army territory? If Captain Fukuda could not explain in 1941, he did much better in 1968. "My function in Hankow," wrote Rear Admiral Teizaburo Fukuda of the former Imperial Japanese Navy, "was management of policy, economy and culture in the occupied area around Hankow. The administrative heads of the army, navy and the consul general in Hankow, so-called the three heads of government, consulted once a week on the above items. We knew the tense situation between America and Japan in those days but were never informed about the war soon to come. I, myself, was confident there would be a compromise with America through Admiral Nomura, then in Washington. He was a peacemaker, we thought. And I personally took care to make no trouble with the American gunboat because there had been some trouble in Chungking by bombing. We hoped there would be no war, and had no plans to capture the American gunboat in Hankow during the period I was there."

Admiral Fukuda reiterated the views of many other senior Japanese naval officers interviewed in recent years: that the Hull note of 26 November 1941 was to the Japanese an insufferably blunt ultimatum to withdraw their troops from China "where three million Japanese civilians would then have found themselves and their property without protection."

Admiral Fukuda suggested the desperately tight Japanese security on the Pearl Harbor plan in adding that the Japanese CinC at Shanghai received orders to be prepared to seize the U.S. gunboat there (USS *Wake*) only one day prior to the attack.

Yet discerning Japanese in many places must have been able to see the inevitable shape of things to come. A typical example is a 1963 letter from a former captain in the Imperial Japanese Navy:

> *I was required to send receipts for a vast amount of invasion money [bank notes] to be spent in the Philippines, Indonesia, and French Indo-China, of which I knew nothing but which it was claimed had been packed like potatoes and transported together with the potatoes. After checking in the clammy store room, I was surprised to find their claim was correct, and was convinced that the war must have been planned and prepared long before.*

This was several weeks before Pearl Harbor, and suggests that although the Japanese might have withdrawn the Pearl strike force had a last-minute compromise been reached, war with the U.S. was considered by the Japanese to be a matter of not "if" but "when."

Captain Fukuda had been cooperative and friendly, so on his promotion to rear admiral, the *Wake*'s officers tendered him a party. The Race and Recreation Club's big dining room looked cavernous and deserted. Perhaps two or three tables were occupied, other than that one seating the four Americans and six or seven Japanese. Most of the lights were out to conserve electricity. Little place cards carried crossed flags, the Japanese navy's rising sun and rays and the American Stars and Stripes.

The dinner was rather sombre, the Americans thought. Perhaps the Japanese were ill at ease with western food and confused by the shiny array of surgical instruments flanking the plates. No doubt the wine was strange to their palates; saké would have been more to their taste. Conversation funneled through bilingual Miura was slow and tedious. Nobody drank very much.

"The Admiral would like to take you to a Japanese place," said Miura, when dinner was over and the hosts were wondering what to do next. The heretofore bland

Japanese faces broke into smiles. This would be more like it. They could climb down from those damned chairs, peel off their jackets and shoes and sit comfortably on the floor.

There is no need to describe the events of the next few hours to anyone who has enjoyed the beauty and atmosphere of a superior Japanese tea house. Those not so fortunate might think of the contrast in a trip from the seamier side of Somerset Maugham's "east of Suez" to the joys of Lafcadio Hearn. What the evening must have cost did not reflect the color of the contents of a Japanese navy pay envelope. If garrison duty at Hankow failed to pall on Japan's naval contingent, one could scarcely have been surprised.

As far back as 10 August 1937, the Secretary of State had pronounced basic policy:

The Armed Forces in China have no mission of offensive action against authorized armed forces of China or any other country, nor is it one of coercion of foreign governments. The primary function is protection of American nationals; secondarily, American property. [They] are not expected to hold positions at all hazards, nor against a responsibly directed armed force of another country operating on express higher authority.

On 31 August 1939, Admiral Glassford translated this policy into specific instructions for his gunboats:

. . . to inflict greatest possible damage on enemy property. Ordinarily you should not be diverted by seeking out an enemy man of war for the sole purpose of engaging in action. You should engage enemy forces of equal or inferior strength which oppose such operations as you may be able to conduct for the destruction of enemy property. You are not expected to engage an enemy of superior strength except that needless to say, you will defend yourself if so attacked.

By 31 October 1941, the Japanese spectre had become so ominous that Glassford had to review and revise instructions:

In case of war our major effort *is to preserve our personnel and ships* as *much as possible for subsequent action . . . by retiring . . . to the Philippine Islands . . . secondarily . . . inflict greatest possible damage . . . and if necessary to protect against great odds*—*use* fire *for maximum destructive effect.*

In case one has never read a gunboat organization bill, "fire" means burn up or blow up the ship.

Glassford had already gotten Hart's permission on 3 October to take the Naval Purchasing Office, Shanghai, and the Fourth Marines under his command. He and Colonel Howard immediately put their heads together to plan how best to save the military wagon in China with two such disparate horses, plus a "general" in a silk hat—the U.S. ambassador—commanding the Marines in North China. There were two basic plans: defend themselves at Shanghai, and forced retirement into Free China. Both were about as hopeful as lighting a candle in a typhoon.

On 7 November, part of Glassford's load was lifted; the President, "on advice Navy and State" approved the withdrawal of the gunboats after consideration of the function each performed in connection with the nearest diplomatic agent. The following day, the rest of the burden came off when the President "approved the withdrawal of all Marines—except those necessary to perform communications for State. . . ."

Admiral Hart was heading for a photo finish. *His* staff could unscramble certain of the Japanese high level code messages, with the same "purple" coding machine with which Washington was reading negotiator Nomura's instructions, while the Commander in Chief, U.S. Fleet, Admiral Kimmel at Pearl Harbor, was not even aware of the existence of such messages or machines. Accurately reading the purple-tinged tea leaves, Admiral Hart recommended on 18 November the immediate withdrawal of the gunboats.

Wake, deepest in China, at Hankow, feverishly commenced liquidation of the Hankow godown stocks by sale to local American citizens. The *Luzon* and *Oahu,* at Shanghai, had been told on 11 November to ". . . make quietly all preparations *within* the ship for a cruise at sea. . . . Stowage in storerooms, magazines, prepare to house topmasts, etc., etc." The word "quietly" brought a smile. In Shanghai, it was rarely possible to pierce the official secrecy veil over ship movements. But the XO of the *Wake* had a friend who could always get detailed and accurate information from her hairdresser.

Local moves about the harbor at Hankow had heretofore been announced by the *Wake* and agreed to by the Japanese in advance. But *this* time, there would be no notification of departure, although only blind men could have failed to note the preparations. By evening of 23 November, all was nearly ready. The schedule was desperately tight to meet the planned departure date of the two bigger gunboats. The Japanese would certainly slam the gate shut, and soon. There was no time to lose.

Officers enjoyed a last round of drinks in the little bar on the pontoon, then divided the few remaining bottles between them. The executive officer dropped his share at the apartment of one of the Eurasian secretaries, then continued on to the faded opulence of the Shanghai and Hongkong Bank as a guest at the last candlelight-and-wine dinner for many a month. He walked back alone along the deserted bund for a last nostalgic look. The damp chill and dark silence pressed in on him. Spiritually and physically exhausted by the events of the last week, he gratefully found his bunk. Tomorrow was the day of departure. For most of the close-knit, isolated little ship's company, it would be the first step of a devious, delayed passage to prison or death. Only the exec and one other, by the most remarkable circumstance, would escape both. As for the *Wake,* ex-*Guam,* she was fast approaching her first change of flag and second change of name.

Shortly before 2:00 P.M., *Wake* started singling up lines. All but one was in and the brow from ship to pontoon was aboard when down the catwalk to the pontoon at full run clattered a Japanese commander, sword a-jangle, cap askew.

"Stoppu! Stoppu! You cannot sair!" Without pretense of ceremony he leaped the yard of open water from pontoon to the main deck. Puffing and red-faced from the exertion, the highly agitated commander addressed himself to the officer and little knot of sailors on the quarterdeck. "You mus' have escort!" he said. "It can arrange in one week."

The skipper, Lieutenant Commander A. E. Harris, had not come down to greet the unexpected guest. He appeared at the pilot house door and shouted, "Unless you want a free ride to Shanghai, I advise you to go ashore *now!*"

"Wait! Wait! Give me boat to my ship! I escort you myself at once!" pleaded the Japanese.

During the forenoon, a large Japanese sloop had gotten under way and ominously anchored half a mile downriver. The Japanese commander's own ship, a gunboat the size of *Wake,* had been sending up smoke from both stacks since breakfast. In twenty minutes, all three ships were under way, *Wake* bracketed between the two Japanese, and all three at battle stations.

The five-day trip downriver was tense. The rounded objects passing close aboard could be lightly fused Chinese floating mines—or the upturned buttocks of drowned coolies. The two escorting ships exercised at battle stations far more often than appeared justified by doubtful claims of Chinese military activity on the banks. The Japanese custom of using *Wake* as a moving target for gunpointing drill in no way reassured a small American crew which expected the next radio message to be a declaration of war.

It was to ease this tension somewhat that skipper Harris arranged what must certainly be the last social event between the U.S. and Japanese navies. Shortly after the little convoy anchored for the night, Harris took the motor sampan over to the senior escort. On his approach, her few lights blinked out. The unmistakable sounds of "battle stations" rang out across the chilly expanse of dark water. With considerable anxiety, his shipmates awaited his return—with news that nine Japanese officers soon would be over for movies and light refreshment.

Wake reached Shanghai the afternoon of 28 November 1941, to find *Luzon* and *Oahu* had their main deck doors and windows covered by watertight steel shutters. The fireroom blower intakes, on the main deck a few feet above the waterline, were protected (it was hoped) by cofferdams. Would the condensers leak and salt up the boiler water? No one knew. The ships had never before ventured into salt water.

The departure of the Fourth Marines for Manila in late November, left a large hole in Shanghai's defense perimeter. Nobody had troubled to let the Japanese know what, if any, measures were being taken to fill this gap, even though Japanese members now sat in heavy percentage on the Municipal Council. To remedy this omission, Admiral Glassford, as senior military commander, called without prior notification at the Japanese naval headquarters the day before his planned departure. It was no farewell call; none of this "Wacht am Rhein" business or jollity over a glass of Johnny Walker black.

Glassford, in his car, drew up at the headquarters and announced the purpose of his visit, but was not invited to enter. In about ten minutes, the Japanese admiral appeared outside. Glassford remained seated while the Japanese remained standing on the sidewalk, probably momentarily too stunned by this studied affront to move. Glassford told him the Shanghai Volunteer Corps would expand its coverage to include the former U.S. Marine territory. Then he told of his own imminent departure from Shanghai, and forthwith drove off. If any "sayonaras" were involved, Glassford failed to record them.

Between the time *Wake* arrived in Shanghai, and the time the bigger ships departed—at midnight—frantic preparations went on: transfer of all *Wake* personnel and belongings to the others, the optimistic stocking of *Wake* with stores to support a long period of independent duty, removal of all but small arms ammunition, and anguished farewells ashore. Lieutenant Commander Columbus D. Smith, USNR, and a reduced crew, largely of Shanghai U.S. Naval Reserves, took over responsibility for the ship, which would remain there as communication ship for the consul general.

Shortly after midnight on 29 November, two small white-and-buff warships slipped down the Whangpoo, past the ancient, sleeping Japanese flagship *Idzumo,* the fat Italian liner *Conte Verde* taking sanctuary from enemy warships at sea, and the ubiquitous squadrons of fishing junks running for shelter from the approaching typhoon.

In the darkness and without a pilot, the *Oahu,* following *Luzon,* momentarily ran aground and her people spent some agonizing moments wondering about the future. The myriad red lights of anchored junks had been taken to be port running lights of craft coming upstream, and a turn to starboard to avoid them had set *Oahu* into shallow water.

A murky dawn saw the two ships safely past the Yangtze fairway buoys, southward bound at 10.5 knots. Or better, *perhaps* 10.5 knots. The *Wake*'s former commanding officer, "Squire" Harris, had become "guest" navigator of the *Luzon,* and *Wake*'s executive officer was the "guest" navigator of the *Oahu,* and the only officer embarked who previously had experienced the vagaries of a flat-bottomed, keelless Yangtze gunboat when caught broadside by a strong wind in the open sea. He had sailed in *Mindanao* from Hongkong to Macao years before and found she could make 15 degrees of leeway with ease. To magnify the problem, ship speed had never been accurately calibrated in the capricious currents of the river.

In fact, some years before, one gunboat skipper had come to grief in an attempt to remedy this defect. Assigned from the outside fleet, he had been preceded by rumors of eccentricity which bore watching. In the speed-checking project, he and the executive officer had equipped themselves with some cut-up swab handles and stop watches and taken themselves to the forecastle, where the idea was to throw over a piece of swab handle, punch the stop watch as it hit the water, then trot aft as it progressed down the side and stop the watch as it passed the stern. Simple calculations involving time and length of ship would give knots through the water.

The swab handle went overboard, the watch was started, and captain and exec took off at a fast trot to keep abreast of it. The exec went ahead, running interference and shouting "Gangway! All hands stand clear! Watch out for the Captain!" as he charged down the deck. A couple of paces behind him came the skipper, gripping a spare piece of swab handle. In the galley doorway halfway down the main deck, the ship's cook heard the commotion, spotted the chase, and sized things up instantly: "Kee-ryst! This is it! The Old Man is off his rocker and after the exec with a belayin' pin!" He reached back into the galley, picked up a large iron skillet, and when the skipper drew abeam, laid him out cold.

No such spectacular hindrance to navigation marred the efforts of the little convoy that first boisterous day at sea. Everybody, including Cooky, had he been in the mood to swing a skillet, was too busy holding on. But Squire Harris, navigating *Luzon,* was chagrined to discover his position as checked by evening landfall to be fifteen miles off, which would have located him well inland on a passing island. His ex-exec, navigating the *Oahu,* was dead on, having cranked into his solution a fantastic but essentially correct drift to leeward.

In all probability for the first time in the Patrol's long history, its commander was taking it to sea. Under the circumstances, he scarcely could be blamed for pessimism, soon to be doubly confirmed. On departure, he had radioed Admiral Hart:

> SHALL MAKE EFFORT TO RUN STRAIGHT MANILA INSIDE PESCADORES ARRIVING IF ALL GOES WELL ABOUT 4 OR 5 DECEMBER. SHOULD WAR CONDITIONS RENDER NECESSARY SHALL FOLLOW COAST AWAIT OPPORTUNITY FINAL LEG MANILAWARD. IF UNABLE MAKE MANILA PROPOSE MAKE FOR HONGKONG IN HOSTILE EMERGENCY

The apprehension over the ability of the gunboats to make the trip was such that two minesweepers left Manila to rendezvous off Formosa and if necessary tow the gunboats or take off their crews. The typhoon making up through the Formosa straits had already put the two minesweepers in deep trouble. The *Pigeon* had a rudder casualty and lost one anchor. The *Finch* had lost both anchors and, unable to anchor, was towing *Pigeon,* which was unable to steer. They were limping into the lee of Formosa to make repairs.

Mounting seas and winds pounded the two gunboats unmercifully. The *Oahu*'s inclinometer recorded an almost incredible roll of 56 degrees to starboard and 50 degrees to port. In Admiral Glassford's words:

> *The 2nd and 3rd of December will never be forgotten in all their grim details. Not only were we harassed by sweeps of Japanese aircraft overhead and by insolent men-of-war ordering us to do all manner of things that we could and would not do, but as we approached the Straits of Formosa and later got well into it, we experienced a heapy, choppy sea which was almost the undoing of the personnel if not the little craft themselves. We were tossed about as by a juggler, now up like a shot to a crest from which we would fall like a stone. The ships were rolling 28 to 30 degrees on a side, with a three second period. They were taking green seas over the forecastle and even more dangerous, surging seas over the stern.*
>
> *Speed had been reduced to little more than steerage way, but even so the engines raced violently, the ships shaking and trembling.*

What disturbed us most was whether or not the human being on board would be worthy of these incredibly stout little ships. For nearly 48 hours there was experienced the hardest beating of our lives at sea. There was no sleep, no hot food; one could scarcely even sit down without being tossed about by the relentless rapidity of the lunging jerks. The very worst of all the trip was after clearing Formosa, with a quartering sea. I recall just before dawn on the 4th December, while clinging to the weather rail of the bridge deck, that our situation could not possibly be worse and wondering just how much longer we could stand it. Not the ships, which had proven their worth, but ourselves.

To that account, the "guest" navigator of the smaller *Oahu* appended a heartfelt "Amen!"

On the dawn of 5 December, one could scarcely believe the ship was in the same ocean. A light sapphire, cloudless sky arched over a table-flat, turquoise sea. *Luzon* charged ahead at sixteen knots, leaving *Oahu* behind. The "rescue" ships *Pigeon* and *Finch* were still farther astern—the walking wounded limping home.

On that same day, the two-star flag of ComYangPat came down. *"YANGPAT DISSOLVED"* was the short message epitaph. Just twenty-two years after its formal investiture and eighty-seven years after the first American warship had felt its uncertain way up the Great River, YangPat had become only a page in naval history.

Manila was already a city at war. At night, few lights showed. Filipino vigilantes with light trigger fingers patrolled the darkened streets. The harbor was rapidly emptying as shipping headed south.

Luzon and *Oahu* were assigned the Inshore Patrol, under Captain Kenneth M. Hoeffel, and in the shallow waters of Manila Bay the old river boats felt very much at home.

It must have roundly irked those Japanese who knew the war would commence in a matter of days, to see a rear admiral and two gunboats slip through their fingers off Formosa. But bluster, zooming, low-altitude overflights, rings of warships at general quarters, and futile orders to Admiral Glassford to enter a Formosan port was their limit. To have snapped shut the dragon's jaws would have killed the surprise planned for Pearl Harbor.

One can be positive that this bellicose Japanese reaction to the passing gunboats was not lost on President Roosevelt. Nor could he have forgotten the clamor for action that followed the 1937 *Panay* incident. It was not Roosevelt's purpose to take an unprepared, pacifist-riddled nation to war over *Panay*. After a preview of Norman Alley's film, he had asked him to snip out a thirty-foot sequence of Japanese planes swooping less than a hundred yards from *Panay,* whence the pilots could have counted the stars in the flag, as too provocative for the American public.

But in 1941, things were different. By December, Roosevelt thought it imperative that the U.S. enter the war before potential allies were defeated in detail. Thus, on 2 December, Roosevelt personally directed that ". . . as soon as possible . . . charter three small vessels to form a defensive information patrol. Minimum requirements to establish identity as U.S. man-of-war are commanded by a naval officer and to mount a small gun. . . ." He added that Filipinos might be employed ". . . with minimum number naval ratings." This "defensive" information patrol was to be sent to spots specified by Roosevelt, one at the entrance of Japan's base of major fleet concentration, Camranh Bay, the others along the Indo-Chinese coast.

Roosevelt designated the old Yangtze gunboat *Isabel* as one of the ships to be used. The others were hastily chartered inter-island schooners, but they were commanded by Yangtze veterans, officers from the *Wake*. The two officers eventually made their way to Australia, one taking his schooner, *Lanikai,* all the way to Fremantle after the war broke out, and the other riding a submarine. They were the only members of the *Wake* crew not captured by the Japanese when Corregidor fell.

Weighty circumstantial evidence, deduction by experts, and the opinion of one naval officer contemporarily near the top, in a position to know, clearly indicate that in late 1941 there was a move to create a *Panay* type incident.[59] As Secretary of War Stimson put it, the President at a cabinet meeting in late November had said that, "The problem was how we should maneuver them into the position of firing the first shot."

YangPat's displaced veterans were not involved in a *casus belli* of major magnitude thanks to the accommodating Japanese who, at Pearl Harbor, provided the President with his "first shot" in a setting infinitely more spectacular than would have been the case with the "three small ships."

At 2:30 A.M., 8 December 1941, *"AIR RAID ON PEARL HARBOR . . ."* crackled over the radio circuit at Navy Headquarters in Manila. The receiving operator recognized in the distinctive keying of the hand-sent message the "fist" of an old friend at Pearl Harbor. It was a solid guarantee that the rest of the message, *". . . THIS IS NO DRILL,"* meant that the long-expected war was at last under way.

A little after 4:00 A.M. on the same day, Lieutenant Commander Columbus D. Smith, USNR, commanding the USS *Wake,* was brought bounding out of his apartment bunk ashore by a telephone call from his highly excited quartermaster, telling him of the raid on Pearl Harbor.

Smith, a merchant mariner by trade, had skippered a U.S. Navy subchaser in World War I. For six years he had been one of forty river pilots bringing oceangoing ships to Shanghai, the last tricky fourteen miles being up the narrow, crowded Whangpoo. Before that, he had put in a lucky five years and 122 round trips as skipper of an upper river merchantman between Ichang and Chungking, without hitting a single rock.

He had been called to active service in March 1941. On 28 November, a few hours prior to the departure of *Luzon, Oahu,* and Admiral Glassford for Manila, Smith took command of the *Wake,* which had arrived that afternoon from Hankow. The ship's normal crew of fifty-five had been reduced to fourteen. Eight of these were radiomen, to maintain the consulate's communications and to operate short wave equipment in Smith's office ashore that tied in with eight other similar stations scattered over all China.

Smith knew Shanghai better than most people know their own home town. He spoke passable Mandarin, and could get by in Japanese. For some time past he had had a feeling that his Japanese friends had shown a certain reticence—a change in attitude. Not unfriendly; simply withdrawn, but withal a trifle apologetic. Perhaps it was only imagination, he told himself. But when a Japanese naval officer friend phoned him on 7 December (6 December Hawaiian time), offering him a couple of prize turkeys, Smith was puzzled. The officer wanted to know where Smith would be at 11:00 A.M. the next day, so he could make delivery. "Aboard *Wake,* most probably," he was told.

Like any professional seagoing man, Smith was fully awake in seconds after receiving his quartermaster's call, and in minutes was dressed. Taxiing down to the waterfront in blue naval uniform, he was turned back four or five times before finally managing to get through to the pier. "You cannot pass!" he had been told by groups of Japanese soldiers with fixed bayonets. At the pier there was another detachment of Japanese troops, but like the others, they made no effort to detain him. No, they were very sorry, but he could not be permitted to go out to *Wake.*

As Smith stood there in a sort of surrealistic vacuum, the river darkness was pierced by cannon flashes. Japanese field guns on the French bund and on Pootung Point, across the Whangpoo, were firing on HMS *Peterel.* A destroyer alongside the French bund and a gunboat at the Customs pier soon joined in.

Peterel, about *Wake*'s size, was also manned by about fourteen sailors, and for the same purpose. Her skipper, sixty-three-year-old Lieutenant Polkinghorn, was her only officer.

About 4:00 A.M., Polkinghorn had been alerted by British intelligence that Pearl Harbor had been attacked, so he had ordered general quarters and brought a boat alongside in readiness for any eventuality. About 4:20, concurrent with Smith's rude awakening ashore, a Japanese navy captain had boarded *Peterel* and formally demanded her surrender. On Polkinghorn's curt refusal, the Japanese hauled off some hundred or so yards and fired a Very's star. That touched off the cannonade. Soon *Peterel* was in flames. Smith watched in anger and disbelief as sailors left the sinking ship in their small boat, then swam after the boat was wrecked by gunfire. Six of them, including Polkinghorn, survived. They were all wounded, some by rifle fire while swimming.

Meanwhile, Smith could see the *Wake* lying cold and silent. A Japanese boat had come alongside with a boarding party, an officer had poked a gun in the ribs of the deck watch, and that ended another chapter in the life of the *Wake.*

She was now HIJMS *Tatara,* third name, second flag. It was well timed. While the short, short drama was being played out on the quarterdeck, the duty radioman had charged out of his shack with news of Pearl. A few crew members managed to jump over the side and swim to a short freedom. But the demolition charges that would have blown out *Wake*'s bottom and wrecked her machinery lay undetonated. The large bottle of saké, "presento" of the Japanese gunboat on the downriver trip in return for the sandwich and movie hospitality, presumably still lay intact in a drawer in the executive officer's cabin, where he had abandoned it.

The bizarre performance continued. Smith, having seen the end of *Peterel,* had other things to do. It was 5:30 A.M., breaking daylight. Walking through the cordon of Japanese soldiers, he jumped in his waiting taxi and rode to his office in the American consulate. From there, he phoned the four radiomen quartered ashore. By 7:00 A.M., they all were busy destroying records and codes. Nanking, Hankow, Tientsin, Peking, Tsingtao, and Chefoo had been contacted by short wave and given the chilling but not unexpected news, details of which had been filled in by a message from Cavite.

About 11:30 A.M., the long-awaited Japanese party appeared on the street below and commenced a systematic search of the building. On the fifth floor Smith awaited them and ordered men with sledge hammers to destroy the radio equipment. The Japanese arrived after the equipment was smashed and Smith was marched out in the custody of a Japanese naval officer. He wanted Smith to wear a sword, so people would know the Japanese navy treated its prisoners with honor. He felt it a pity that Smith's sword was aboard the *Wake*. The four enlisted radiomen meanwhile had picked up some civilian overcoats in the consulate, put them on over their uniforms and walked, unconcerned and unchallenged, through the cordon of Japanese troops below.

No army or navy, even if staffed with morons (and this, the Japanese were not), could have handled the Shanghai sequence just described in such an awkward manner unless they had been caught utterly unprepared psychologically and physically. Clearly, the Pearl Harbor secret was infinitely closely held, as suggested earlier by Admiral Fukuda. Even psychologically, the common soldier must not yet have been indoctrinated to hate or despise the white man; it would take the debacle of Singapore to clinch that.

An American consul, M. B. Hall, had seized a U.S. flag from two Japanese soldiers who were hauling it in from the staff outside Smith's window. With some twenty officers and men looking on, Hall grabbed the flag, carefully folded it, put it in the safe and twirled the combination. Then he casually walked out of the room while the Japanese looked on in stupefied amazement. The long-held gospel, reinforced by indoctrination after the *Panay* "error" that white skin was more or less inviolate, still held. But to the discomfiture of Smith and others, this aberration soon passed.

For some time after 8 December, Americans and British continued to run their banks, newspapers and other businesses, under Japanese supervision, although Japanese army officers now stood at the longest bar in the world or sat around the big circular table in the American Club. Then, these western civilians were rounded up and imprisoned like the military. The diplomats, loosely confined, were repatriated, in exchange for Japanese.

Smith's mild early captivity under the Japanese navy was soon exchanged for the brutal treatment of the Japanese army in a camp containing American prisoners from Wake Island, the Peking Legation Guard, and the Tientsin Marine garrison. Smith was finally incarcerated in the huge old municipal prison on Ward Road, along with thousands of murderers, thieves, and various common criminals. But along with Commander John B. Woolley, Royal Navy, and a U.S. Marine, he made a fantastic escape—700 miles on foot, mostly through Japanese-held territory—to freedom.[60]

As for the epitaph of the *Wake,* nothing is known of her wartime career under the name of HIJMS *Tatara.* But after the Japanese surrender, she was recovered, weatherbeaten and neglected, and once more wore the U.S. flag. In 1946, she was turned over to the Nationalist Chinese, as RCS *Tai Yuan.* In the debacle of the Nationalists' withdrawal from the mainland, her blue, white, and red Kuomintang ensign was soon replaced by the solid red of the communists. According to *Jane's Fighting Ships,* ex-*Tai Yuan* is still afloat, although her new name is not available. Obviously, such an important military secret would be closely held. But it is reasonable to expect that along with her *fifth* change of flag, ex-*Guam,* ex-*Wake,* ex-*Tatara,* ex-*Tai Yuan,* is now bearing her *fifth* name. The *Wake* had no claim to glory nor to infamy. In durability—almost indestructibility—and in the affection of those Americans who served in her, she is at least worthy of mention. But in her propensity for change of name and flag, it is fairly safe to say she has no peer in the world's naval history.

Tutuila rounded out her last days under the U.S. flag more in isolation from her own kind than in security. In a word, minor disasters multiplied. On 3 October 1939, her people were just about ready to tackle their breakfast eggs when it seemed the grandfather of all river dragons swished his tail under them.

In the quaint words of the vernacular press, as rendered into "Enlish" by the wardroom's Chinese teacher,

There was a thick fog in the morning, so the public ferries have been unable to start running at the usual time. But the passengers have insisted to get going. As soon as the pilot brought the ferry to the Lungmenhao side almost blindly into the stationary American gunboat, which made a hole as big as a human body. The crew of the gunboat have instantly got into action for salvation. When the ferry company sent representatives on board for consoling purposes, the officers have made very friendly expressions. If the ferry company would just have the hole fixed, they should not make any other claim. The officers even at the same time have consoled the passengers if who were so severely frightened. . . .

The friendly sentiments stretched over some weeks, while the Chinese workmen in the most primitive but ingenious manner put things to rights. Nor were the social amenities connected therewith overlooked.

Oil drums on deck on the undamaged side were filled with water to lift the jagged hole clear of the surface. This listed the ship to the point soup no longer stayed in the dish nor sleepers in their bunks except with tie ropes. Rivet holes were laboriously bored with an archaic type of drill called an "old man." Held to a metal arch bolted to adjacent holes, the drill was forced down a fraction of an inch at a time while a coolie turned it by a spike through a hole in its shank. Replacement plates were heated over a charcoal fire, then pounded into shape with a sledge hammer.

The completion of the repairs was celebrated by a Chinese banquet hosted by the ferry company, honoring *Tutuila*'s officers and repair company officials. There were *gambais* (bottoms up) in several types of liquid fire to wash down the pressed duck and ancient eggs, until in quintessence of hospitality, the hosts could no longer maintain a balance on the stools surrounding the large circular table.

Shortly thereafter, the repair company repeated this gesture of amity with the *Tutuila* and ferry company officers as guests, until these hosts in turn brought the party to a close by falling off their perches.

With this sort of one-upmanship, *Tutuila* scarcely could be expected not to finesse. Her officers followed up with a third celebratory bash in the little officers' club ashore. By that time, the ferry and repair company officials were practically in the category of *Tutuila* messmates. The heretofore limited common language had been enriched from just plain "*gambai.*" As a result of some earnest homework by the guests, "Downee hotch!" or, "Lookee aht, sotowmack! Heah it come!" rendered by the beaming Chinese, assumed part of the burden of communication.

The pity of it all was that obviously the guests were uninitiated in the subtle effects of western potables. The Americans were thus robbed of the opportunity to demonstrate their exquisite hospitality by falling out first. Although the chairs had backs, the Chinese soon found trouble staying on them, to the point the party ended in celestial confusion well before the dessert. If by this time, there was any better reason for avoiding collisions than the subsequent rounds of hospitality, no one aboard *Tutuila* could think of it.

In May of 1940, the *Tutuila* shifted downriver, from winter moorings to the little cove behind the rooster's comb pinnacle of rock abreast the Standard Oil building. Precious fuel oil was used to build up steam in the boilers, heretofore dead for many months. A brisk wind and strong chow-chow water combined to catch the ship's stern as she backed downstream. There was no great thump when she hit, but with a falling river, the horrible truth soon dawned; unless immediately lightened, perched on pinnacles fore and aft as she was, she would soon break in two.

During the following hectic days, as the river fell slowly, the crew of the 370-ton ship hauled off 168 tons of guns, armor, ammunition, stores, oil, water, machinery, chains, anchors, and lifelines, as everything movable went into lighters alongside or ashore. Through the amidships passageway a huge log reached across the main deck, secured by chains passing under the ship. The ends hung over junks on either side, which were flooded while supports were built under the log, then pumped out to support the sagging hull.

In a week, the capricious river had risen and the ship was free, and the crew was busy putting her back together again, veterans of one more round with the river dragons.

Luck was with *Tutuila* on 31 July 1941, when a Japanese bomb came close enough to do US$27,045.78 worth of damage. Admiral Shimada, CinC, China Seas Fleet, said so very sorry. The wardroom officers had lost their pride and joy, the motor skimmer and its outboard engine. But nobody had been more than scratched. It had been just "Another Day in the Life of a Chungking Gunboat."

The year 1941 had been a tough one for *Tutuila*. Almost daily Japanese overflights in the brilliant blue summer sky kept one question uppermost in mind: when would a misdirected hundred pounder produce a "regrettable incident?" With the approach of winter, clouds shrouded the hilltops and the air raids mercifully ceased. Life was passable on board as long as the little coal-burning tug alongside furnished steam. But on boiler cleaning day, everyone froze. The galley bean kettle went cold for lack of steam and candles lighted the clammy living quarters. Those not on duty sought out the enlisted men's club or the Chungking Club and hugged the open fires.

News of Pearl Harbor brought hopes they would be remembered and the ordeal soon over. On 19 January 1942, the great day came; forlorn, abandoned *Tutuila* was signed over to Colonel James McHugh, USMC, naval attache, and *Tutuila*'s ship's company, down to two officers and twenty-two men, commenced a remarkable hegira to the "outside."

Devil-discouraging firecrackers popped and sputtered in farewell as sorrowing sampan coolies and the foreshore fraternity of *mafus,* chair boys, *chinpings* (lantern bearers), and small urchins watched the U.S. Navy men depart for the last time from the south bank across to the mud island airport recently uncovered by a falling river.

The *Tutu*'s skipper, Lieutenant Commander W. A. Bowers, had written himself a set of orders "to take charge of the below listed officer and men, proceed and report to the nearest ship or station for duty." That "nearest" would turn out to be Trinidad, halfway around the earth, must have been well beyond his wildest expectations.

The little group flew via CNAC, the Sino-American commercial line, to Calcutta, thence rode across India by rail, signing chits for service and transportation as they went. At Bombay they found the SS *President Madison* in port and joined the load of Netherlands East Indies refugees on board. Via South Africa, *Madison* headed for Trinidad, where *Tutu*'s party arrived just three months and five days after their not too nostalgic last look at Chungking's chill winter fog and muddy paths. There, the group broke up to fight the war apart. But the four faithful Chinese steward's mates got no farther; they were dragooned into the service of Vice Admiral Jonas Ingram, Commander, U.S. Naval Forces South Atlantic, who knew solid gold when he chanced upon it.

The end of the war brought an interesting sequel. Former Skipper Bowers helped one of the steward's mates bring his family to the United States, where son Philip won his doctorate in physics. Things had come a long way for Chang Tung Sing since he had first timidly found his way aboard a Yangtze gunboat.

On 10 December, the *Mindanao* reached Manila from Hongkong. She had tacked to the very entrance of Swatow, 180 miles up the China coast, before being safely able to swing down toward Manila. She had picked up ten Japanese fishermen en route, the first—and for a long time the only—Japanese prisoners taken by the U.S. Navy. When the ship sailed into Manila, her traditional gunboat white and buff was gone—sailors had used swabs and varicolored paint to give her the appearance of a burned-out hulk.

The same day the *Mindanao* reached Manila, the Japanese began daily air raids which continued uninterrupted and unopposed.

On the first day of war, 8 December, MacArthur's proud air force of 256 light and heavy bombers, with pursuit aircraft in proportion, had been caught napping on the ground and largely destroyed. Thenceforth, Japanese planes came and went at will. On 10 December, two waves of enemy bombers wheeled unchallenged at 20,000 feet over Cavite Navy Yard, dropping bombs at will. Fire swept the yard, and by nightfall only the great stone-bastioned ammunition depot remained intact. All else was ashes. Life for the ships became one of scavenge and make-do.

On 26 December, the bomb-and-collision damaged destroyer *Peary* and the schooner yacht *Lanikai* left Manila Bay under cover of darkness, the last surface ships to escape the bomb-pounded focal point of American defense. From then on, for the few months the defense held out, only U.S. submarines came and went, bringing small quantities of antiaircraft ammunition and medicines and taking out valuable personnel no longer fulfilling useful functions locally.

The first casualties of the war aboard *Mindanao*[61] after her arrival at Manila were the two Spanish War era saluting guns, replaced by a .50 caliber machine gun. No one expected to fire any more salutes, or take pot shots at flocks of fly-fly ducks. Twelve Browning heavy machine guns had been received to replace the Lewis guns, but the old guns were kept and the Brownings added, to give the *Mindanao* a formidable light battery of twenty-four automatic weapons plus the 50-caliber.

Manila was declared an open city on Christmas Day and so was no longer available as a source of any kind of supplies. The ships were thenceforth strictly on their own, and the gunboat patrols off the Corregidor minefields were discontinued on 27 December to save fuel.

On 29 December the Japanese bombed Corregidor and ships in the anchorages nearby. The *Mindanao* was straddled and slightly damaged. Debris from ashore mixed with geysers of salt water from near misses flailed down on the gunboats. Their gunners banged away at the enemy nevertheless. Lieutenant David Nash, the *Mindanao*'s executive officer, visited an army battery some time later and was pleased to learn that on that rough day, things had been so bad the army guns had stopped firing and their crews had taken cover. But that little gunboat, blazing away despite the fact its shells were 4,000 feet short, so heartened them by its spirited gunfire that they resumed their own. In fact, the mutual appreciation syndrome even extended to such a holy elevation as food: when an army battery shot down two Japanese aircraft, the *Mindanao* sent over a half-dozen oven-hot raisin pies to augment their iron rations.

On 29 December, also, the gunboaters turned amphibious. By day, they worked ashore, setting up machine gun nests and beach defenses for a last stand. By night, they slept aboard, grateful for such simple comforts as a little rain water caught in oil drums, and a real bunk. And on that date, apparently considering the sea phase of the defense at an end, Commander Inshore Patrol moved into a Corregidor tunnel.

For a few days in January, night patrols once again were set up, while during daylight hours in good weather the men spent their time near foxholes ashore, with Japanese planes constantly overhead.

By early February, enemy shells began to crump into the anchorages off Corregidor. The *Luzon* was hit, but damage was slight. Sometimes there was a welcome chance to slap back: "This afternoon there was a classic aerial dogfight," wrote Nash on 9 February. "Four P-40s, last of the U.S. Army Air Force, operating from secret jungle airstrips, were jumped by four Zeros and the battle was on. In a flash a Jap was on a P-40's tail and heading parallel to us down the eastward side, low and at about three or four hundred yards. Then our machine guns—all 13 of them that could bear—cut loose on the Jap. The hail of tracers he was flying in was a joy to behold. Apparently it turned him back, for all of a sudden he faltered, his wings wagged and he turned away. . . ."

On 15 February, the last of the fuel oil, a pitiful 20,000 gallons, was split between the *Mindanao* and *Luzon*. By 21 March, boilers were run down to about thirty pounds pressure except on the day a ship had the ready duty. On 25 February there had been a call for that ready gunboat. Five small boats appeared Manila-ward, and *Mindanao* set out in pursuit. She opened fire at 5,000 yards and shots soon straddled the formation. Perhaps it was a trap, for she was soon straddled in turn by two shells from ashore. She zigzagged wildly as shots continued to fall close astern. To liven the action, three Japanese planes appeared astern in single file, but turned away when fired on.

From then on, the bombing intensified. As many as nine planes roamed constantly overhead in daylight and a new contingent took over at night. From the first of April, gunboat boilers were generally cold and dead. A kerosene-driven generator powered radios and chill boxes. Fifty-gallon drums filled with water stood on deck for fire fighting; with no steam the pumps were useless.

The word on 6 April was better: patrol east of Bataan! By 2:00 A.M. the *Oahu* and *Mindanao* were steaming in column when they sighted eleven small craft in the classic target position of being silhouetted against the moonlight. Apparently unaware of the gunboats, the Japanese plugged complacently along. After an hour the gunboats opened fire, but their joss was bad—clouds suddenly obscured the moon. Literally and figuratively the *Mindanao* was in the dark, but her starshells illuminated the targets for the *Oahu*. The gun flashes were good bait; almost immediately, shells from ashore started whistling around them. A fire in *Mindanao*'s pyrotechnic locker filled the bridge with alarming smoke but was soon put out. "We raised hell with them," Nash reported a Japanese officer had told them after their final capture. "Four boats sunk and every time we opened fire, they hit the deck."

By 9 April the end of Bataan clearly was very near. That afternoon, three army officers paddled out to the *Mindanao* in a banca. There were more ashore, they said. Sixty were picked up, the remnants of the 31st Infantry, of 1932 Shanghai fame. They were one step ahead of the enemy, without food for two days, and too dead tired to swim.

On 10 April the final move commenced. The gunboaters graduated from 3-inch popguns to heavy artillery. *Mindanao*'s crew, some fifty men, went to Fort Hughes on Caballo Island, which guarded Corregidor's eastern flank, and took over Battery Craighill. They fired 26 rounds from four model 1912 mortars under the tutelage of a

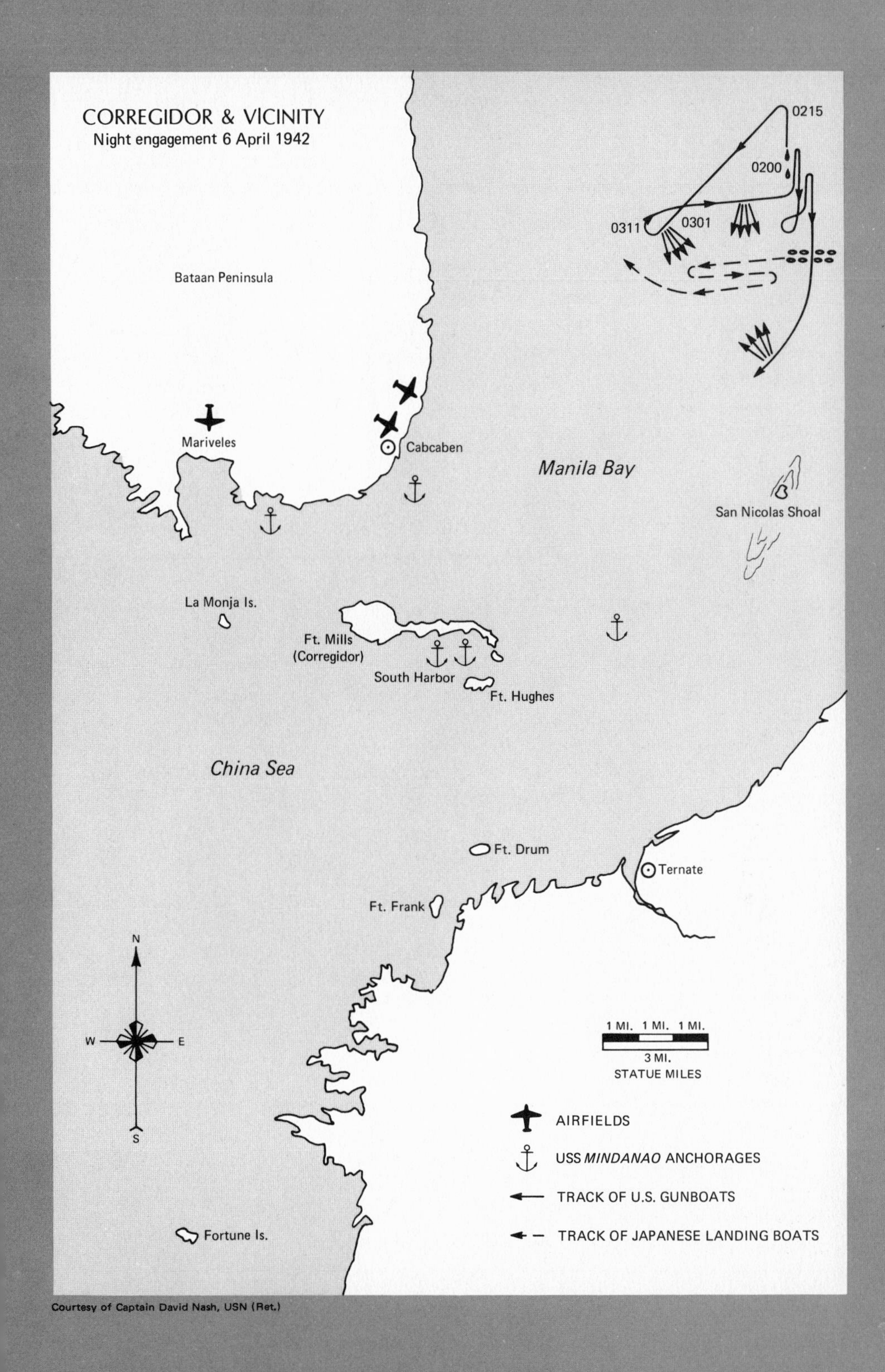

Courtesy of Captain David Nash, USN (Ret.)

couple of army coast artillerymen and then they were on their own. The first day they fired 88 rounds from the stubby old guns, which could elevate almost straight up.

Luzon's crew took over Battery Gillespie—two 14-inch disappearing rifles that rose above a parapet to fire, then sank back out of sight to reload. *Oahu*'s crew manned a battery of 155's. All hands moved into the casemates and set up housekeeping, with *Mindanao*'s galley supplying food for over a hundred men.

For the next two weeks, the duel between Japanese batteries and the Navy's only heavy artillerymen gradually heated up. From atop an observation post 177 steps straight up, three *Mindanao* officers took turns spotting the fall of shot. By 24 April they had forced the Japanese to bring up some heavy artillery of their own which, for two and a half hours, every two and a half minutes, sent a monster shell—some 240-mm—into the fort. One of the mortars was knocked out. One enemy shell dropped into a gun pit and circled it several times with a noise like an express train, but failed to explode. Ingenious gunboaters experimented in making antiaircraft weapons out of the 14-inch mortars, but the project failed; the clumsy pointing gear was not agile enough and the improvised fuzes were duds.

On 2 May, Lieutenant Nash climbed down from his high tower to carry out an unusual order—assume command of a sinking ship. Several ships already had been scuttled to avoid capture, and *Mindanao* was next. She had been hit, holed, and was in a sinking condition. Nash's orders were to take command at midnight and put her on the bottom of South Harbor. If he had time to hoist his first commission pennant, it did not fly long, because seven hours and twenty minutes later his orders had been carried out, the job was done, and the *Mindanao* was no more.

As the erstwhile skipper put it, "The ship was finished, but the crew wasn't." Along with the other gunboaters, they continued to fire on Japanese positions. Battery Craighill could fire in any direction and at the very last was in the anomalous position of firing into friendly territory—a final barrage laid on Corregidor to blast the Japanese landing. In their short stay, the Navy's amateur artillerymen had fired many more rounds than had gone down the bores of the old guns in their entire previous existence.

Before the situation in the Philippines grew desperate, it had been suggested to General MacArthur that the *Mindanao, Luzon,* and *Oahu* be allowed to attempt an escape south with volunteer crews. His staff replied that "these vessels are performing essential duties covering entrances to minefields, on patrols toward Manila and off the coasts of Bataan, the vital North Channel and the water adjacent to Corregidor . . . but that if the opportunity arises and their essential usefulness ceases, endeavor would be made to permit these naval units to seek safety to south."

That opportunity never arose. The gunboats fought to the very last, their death sentence starkly clear in MacArthur's dramatic "I intend to fight this army to destruction! The gunboats are necessary in the defense." And so they were. The last one, *Oahu,* was sunk on 6 May 1942, the last day the American flag flew over Corregidor.

On that date, Captain Kenneth M. Hoeffel, Commander Inshore Patrol, sent out a sort of radio epitaph:

ONE HUNDRED AND SEVENTY-THREE OFFICERS AND TWENTY-THREE HUNDRED SEVENTEEN MEN OF THE NAVY REAFFIRM THEIR LOYALTY AND DEVOTION TO COUNTRY, FAMILIES AND FRIENDS

In that number were the gunboat sailors from the *Oahu, Mindanao, Luzon,* and *Wake*. By noon it was all over. A white flag went up over Corregidor, where haggard men stumbled out of the dust-filled tunnels with their hands up in surrender. The Yangtze Patrol, many years and many miles from its birthplace, had fought its final action.

NOTES

1. William Crosby and H. P. Nichols, *The Journals of Major Samuel Shaw* (Boston: Metcalf and Company, 1847).
2. Earl Swisher, *China's Management of the American Barbarians* (New Haven: Far Eastern Publications, Yale University, 1951). Copyright 1955 by The Association for Asian Studies, University of Michigan, Ann Arbor, Michigan. Various passages reproduced by permission of the copyright holder. (A compilation of mid-nineteenth century communications between the emperor and his officials.)
3. Based on information kindly supplied by Mr. John S. Service, Center for Chinese Studies, University of California, Berkeley, California.
4. D. C. Boulger, *History of China* (New York: The Co-operative Publication Society, 1893), p. 307.
5. Thomas W. Blakiston, *Five Months on the Yangtze* (London: John Murray, 1862).
6. Tyler Dennett, *Americans in Eastern Asia* (1922; reprint ed., New York: Barnes and Noble, Inc., 1963), p. 159.
7. Swisher, *American Barbarians.*
8. Frank Hawks, *Narrative of the Expedition of an American Squadron to the China Sea and Japan* (New York: D. Appleton and Company, 1857), p. 304. (The expedition was under the command of Commodore M. C. Perry, USN.)
9. E. S. Maclay, "New Light on the 'Blood is Thicker than Water' Episode," U.S. Naval Institute *Proceedings* 40 (1914): 1085; J. W. Bellah, "The Grand Manner of Josiah Tattnall," *Shipmate* 28 (December 1965): 4.
10. John S. Littell, "Revolutionary Spirit on the Yangtze," *Asia* Magazine 27 (June 1927): 482.
11. *The Journals of Daniel Noble Johnson (1822–1863) USN,* (Washington, D.C.: Smithsonian Institution, 1959).
12. J. F. Bishop, *The Yangtze Valley and Beyond* (New York: G. P. Putnam's Sons, 1899).
13. Dennett, *Americans in Eastern Asia,* p. 673.
14. Hawks, *Narrative of American Squadron.*
15. Swisher, *American Barbarians.*
16. Swisher, *American Barbarians.*
17. Charles E. Clark, *My Fifty Years in the Navy* (Boston: Little, Brown & Co., 1917).
18. G. Carsalade duPont, *La Marine Français sur le Haut Yang tsé* (Paris: Academie de Marine, 1963).
19. Lieutenant Commander Glenn Howell, USN (Retired), "Ascent of the Min," U.S. Naval Institute *Proceedings* 65 (1939): 709. (Lieutenant Commander Howell was in the *Palos.* Also, from personal letters of Lieutenant Scott Umsted *(Monocacy).*

20. R. Logan Jack, *The Back Blocks of China* (London: Edward Arnold, 1904).
21. Robley Evans, *An Admiral's Log* (New York: D. Appleton and Company, 1910).
22. Much of this episode is based on Lieutenant A. B. Miller's detailed intelligence report. Record Group 45, the National Archives.
23. Thomas Woodrooffe, *River of Golden Sand* (London: Faber and Faber, 1936).
24. Parts of this chapter are derived from a personal letter from Captain George S. Gillespie, USN (Retired), to Rear Admiral Stanley M. Haight, USN (Retired), dated February 1969. Gillespie was skipper of the *Palos* in 1920.
25. Lieutenant Commander Glenn Howell, USN (Retired), "Chungking to Ichang," U.S. Naval Institute *Proceedings* 64 (1938): 1312.
26. "Report on Conditions of the Upper Yangtze," Record Group 80, the National Archives.
27. Alexander Hosie, *Three Years in Western China* (London: George Phelps and Son, 1890).
28. Elizabeth C. Enders, *Temple Bells and Silver Sails* (New York: D. Appleton and Company, 1925).
29. Rear Admiral B. O. Wells, USN (Retired), in a letter to the author.
30. Cornell Plant, *Glimpses of the Yangtze Gorges* (Shanghai: Kelly and Walsh, Ltd., 1926).
31. Part of this chapter is derived from the private letters of Rear Admiral Scott Umsted, USN (Retired) written from *Palos* to his family and friends, and kindly lent to the author.
32. H. B. Ellison, former editor of the *Washington Post,* quoted in *Donald of China* by Albert Selle (New York: Harper and Brothers, 1948), p. 225.
33. F. A. Sutton, *One-Arm Sutton* (New York: Viking Press, 1933).
34. Lieutenant Commander Glenn Howell, USN (Retired), "Opium Obligato," U.S. Naval Institute *Proceedings* 64 (1938): 1729.
35. Harold Isaacs, *The Tragedy of the Chinese Revolution* (New York: Atheneum Press, 1966), p. 112.
36. Vincent Sheean, *In Search of History* (London: Hamish Hamilton, 1935).
37. *China Year Book* (1928).
38. Lieutenant K. N. Gardner, USN, "The Beginning of the Yangtze River Campaign of 1926–27," U.S. Naval Institute *Proceedings* 58 (1932): 40.
39. Dennett, *Americans in Eastern Asia.*
40. This section is based on Smith's account of the Nanking outrage: Lieutenant Commander Roy C. Smith, Jr., USN, "Nanking, March 24, 1927," U.S. Naval Institute *Proceedings,* 54 (1928): 1, frequently quoted hereafter; and in the personal files of the same officer, now held by his son, Captain Roy C. Smith, USNR (Retired). Also extensively quoted are communications between U.S. Consul John K. Davis at Nanking and American Minister J. V. A. MacMurray at Peking, held in the Smith files.
41. Sheean, *Search of History.*
42. Summarized from various issues, between April and May 1927, of the Shanghai *North China Daily News, Peking and Tientsin Times, The New York Times,* and Baltimore *Sun.*
43. Sir Robert Hart, *These from the Land of Sinim* (London: Chapman and Hall, 1901).
44. Adapted in part from "Chinese Huntsman," Rear Admiral Kemp Tolley, USN (Retired), *Shipmate* 29 (June-July 1966); 12.
45. Adapted from "The Chameleon," Rear Admiral Kemp Tolley, USN (Retired), *Shipmate* 29 (October 1966): 2.
46. H. G. W. Woodhead, *The Yangtze and Its Problems* (Shanghai: Mercury Press, 1931).
47. Captain Walter E. Brown, USN, "Fiat Justitia, Ruat Caelum, Chinese Style," U.S. Naval Institute *Proceedings* 64 (1938): 1585.
48. Franz Schurmann and Orville Schell, *Imperial China* (New York: Random House, 1967).
49. Arthur H. Smith, *Chinese Characteristics* (New York: Fleming H. Revell Company, 1894).
50. The material following was originally published under the title "Three Piecie and Other Dollars Mex," *Shipmate* 28 (July 1965): 8.
51. Samuel Eliot Morison, *"Old Bruin," Commodore Matthew Calbraith Perry* (Boston: Little, Brown & Co., 1967), p. 348.

52. Adapted from an article by Vice Admiral T. G. W. Settle, USN (Retired), "Last Cruise of the *Palos*," *Shipmate* 24 (April 1961): 2.
53. Bishop, *Yangtze Valley*.
54. Courtesy of Rear Admiral Karl G. Hensel, USN (Retired), through whose family this ode has been passed down.
55. Claire Lee Chennault, *Way of a Fighter* (New York: G. P. Putnam's Sons, 1949).
56. Hamilton Darby Perry, *The Panay Incident, Prelude to Pearl Harbor* (New York: The Macmillan Co., 1969).
57. Based on articles by the author, "A Day in the Life of a Chungking Gunboat," *Shipmate* 30 (June-July 1967): 8; and "Yangpat—Shanghai to Chungking," U.S. Naval Institute *Proceedings* 89 (1963): 80.
58. William Edgar Geil, *A Yankee on the Yangtze* (London: Hodder and Stoughton, 1904).
59. Rear Admiral Kemp Tolley, USN (Retired), "The Strange Assignment of USS *Lanikai*," U.S. Naval Institute *Proceedings* 88 (1962): 71.
60. Quentin Reynolds, *Officially Dead, the Story of Commander C. D. Smith* (New York, Random House, Inc., 1945).
61. The executive officer of the *Mindanao*, Lieutenant David Nash, who was captured by the Japanese, kept a diary which he buried at Cabanituan prison camp, which was recovered after World War II ended. The following account of the last weeks of the Inshore Patrol, to which the river gunboats were attached, is based on that diary, held by the present Captain Nash, USN (Retired).

APPENDIX 1

CHARACTERISTICS OF SHIPS SERVING ON THE EAST INDIA AND ASIATIC STATIONS

Sources:

Marsh, Captain C. C., USN (Retired). *Official Records of the Union and Confederate Navies in the War of the Rebellion.* Navy Department Library, Series 2, Vol. 1. Washington, D.C., 1921.

Bauer, K. Jack. *Ships of the Navy, 1775–1969. Vol. 1, Combat Vessels.* Troy, New York: Rensselaer Polytechnic Institute, 1970.

Naval History Division, Navy Department. *Dictionary of American Naval Fighting Ships.* Washington, D.C.: Government Printing Office. Vol. 1, A-B, 1959; Vol. 2, C-F, 1963; Vol. 3, G-K, 1968; Vol. 4, L-M, 1969; Vol. 5, N-Q, 1971.

Jane's Fighting Ships, various issues.

Various characteristics of many of the ships changed markedly over the years since their completion. Data given below generally are of commissioning date, the latter being the year given. Length is overall unless followed by (pp), between perpendiculars, or (wl), waterline. Draft, as far as can be determined, is full load. Displacement, a good measure of the real "heft" of a ship, is given as full load displacement, as far as is known. Pdr—pounder; R—rifle; MLR—muzzle loading rifle; BLR—breech loading rifle; SB—smooth bore; CT—conning tower; TT—torpedo tube; MG—machine gun; cpl.—complement. The slant following gun bore diameter precedes caliber—the multiplying factor of bore diameter to arrive at barrel length. The term "sloop" generally indicates a warship with guns all on one deck, although the sail rig might be schooner, brig, ship, or hermaphrodite.

Alaska
Wooden screw-sloop built by Navy Yard, Boston, 1867. 2,400 tons. 11.5 kts. 250′ x 38′ x 16½′. Cpl. 273. One 11″ SB, six 8″ SB, one 60-pdr R. Two horizontal engines, single screw. Ship rig. Two stacks. Sold at Mare Island 1883.

Asheville
Gunboat built by Navy Yard, Charleston, 1918, for tropical duty. 1,760 tons. 12 kts. 241′ x 41′ x 11′. Cpl. 185. Three 4″/50, two 3-pdr, three 1-pdr. Sunk by Japanese south of Java, March 1942.

Ashuelot
Side wheel, double-ender iron hull gunboat built by Donald McKay, Boston, 1866. 1,030 tons. 11.2 kts. 255′ x 35′ x 9′. Cpl. 159. Four 8″ SB, two 60-pdr R. Full sail rig. Hit Lamock Rocks, China coast, and sank in 47 minutes, February 1883.

Augusta
Heavy cruiser, built at Newport News, Va., 1931. 9,050 tons. 32.7 kts. 600′ x 66′ x 23′. Cpl. 691/795. Nine 8″/55, eight 5″ AA. Four AC, 2 catapults. Three inch side armor, 2″ deck, 1½″ gunhouses. Sold for scrap, 1959.

HMS *Bee*
River gunboat. 1915. 625 tons. 14 kts. 237½′ x 36′ x 4′. Cpl. 65. One 6″ behind shield. (After 6″ omitted for flag configuration.) One 3″ AA, 10 MG. Twin screws in tunnels. Two stacks athwartships.

Benicia
Wooden screw-sloop, built at Portsmouth Navy Yard, N.H., 1864. Generally similar to *Alaska*. Sold at Mare Island, 1884.

Callao
Built at Cavite, P.I., 1888. Captured at Cavite 12 May 1898. 243 tons. 10 kts. 121′ x 18′ x 6½′. Cpl. 31. Four 3-pdr, two 1-pdr. Sold in 1923.

HMS *Cockchafer*
Similar to *Bee*, but with two 6″ guns.

Dakotah
Wooden screw-sloop built at Navy Yard, Portsmouth, Va., 1859. 1,300 tons. 11 kts. 198′ x 33′ x 13′. Two 11″ SB, 4 long 32-pdr. Two horizontal engines, single screw, one stack. Bark rig. Sold at Mare Island, 1873.

Detroit
Third rate steel cruiser, built at Baltimore, Md., 1893. 2,072 tons. 19.2 kts. 270′ x 37′ x 16½′. Cpl. 217. Nine 5″, six 6-pdr, 4 TT. Two-inch conning tower, ½″ deck. Sold, 22 December 1910.

Elcano (Originally *El Cano*)
Built of iron in Spain, 1885. Captured at Manila 1 May 1898. Comissioned in USN 1902, under Lieutenant Commander A. G. Winterhalter, who in 1915 became CinC Asiatic Fleet. 620 tons. 11 kts. 165′ x 26′ x 10′. Cpl. 103. Four 4″. Four 3-pdr. Sunk as a target 4 October 1928.

HMS *Emerald*
Built by Armstrong, 1926. 9,000 tons. 33kts. 570′ x 54½′ x 17′. Cpl. 576. Seven 6″/50, five 4″ AA, four 3-pdr, 16 TT. Three-inch side, 1″ deck.

General Alava
Built in Scotland, 1895. Captured at Manila, 1898. 1,390 tons. 10½ kts. 212′ x 28′ x 13′. Cpl. 76. One 6-pdr, two 3-pdr. Sunk as target, 17 July 1929.

Guam, renamed *Wake* 5 April 1941.
River gunboat, first of the "new six," built on similar plans in three sizes from material sent from United States and assembled by Kiangnan Dock and Engineering Works, Shanghai, 1927. 370 tons. 14.5 kts. 159½′ x 27′ x 6′. Cpl. 60. Two 3″/23 AA behind shields, eight 30 cal. MG. Two triple expansion engines, total 1,900 HP, twin shafts in tunnels, triple rudders. Captured at

Shanghai, 8 December 1941, and renamed HIJMS *Tatara*. Turned over to Nationalist Chinese post war and renamed RCS *Tai Yuan*. Captured by Red Chinese on collapse of Nationalists on mainland.

Hartford
Wooden screw-sloop built at Navy Yard, Boston, 1859. 2,900 tons. 9.5 kts. 225′ x 44′ x 18′. One 8″ MLR, twelve 9″ SB. Single screw which could be hoisted up for sailing. Ship rig. One stack. Foundered alongside pier at Portsmouth, Va., 1956.

Helena
Built at Newport News, Va., 1897. Steel gunboat. 1,390 tons, 13 kts. 215′(wl) x 40′ x 9′. Cpl. 175. Four 4″, four 6-pdr, four 1-pdr, 4 Colt MG, one 3″ field piece. Sold 7 July, 1934.

Houston
Sister ship of *Augusta*. Sunk by Japanese in Sunda Strait, N.E.I., 28 February 1942.

Isabel
Yacht built by Bath Iron Works, 1917, and taken over by Navy for use as a small destroyer during World War 1. 950 tons. 26 knots, 230′(wl) x 26′ x 8½′. Cpl. 99. Two 3″/50, two 3″/23 AA.

HMS *Ladybird*
Similar to HMS *Cockchafer*. Sunk by enemy aircraft off Libyan coast, 12 May 1941.

Luzon
Built under same circumstances as *Guam,* 1928. 560 tons. 16 kts. 211′ x 31′ x 6½′. Cpl. 82 (as flagship). Two 3″/50 AA behind shields, eight 30 cal. MG. Sunk 5 May 1942. Salvaged by Japanese and renamed HIJMS *Karatsu*.

Mindanao
Sister ship of *Luzon*. Allowed to sink after battle damage off Corregidor, 3 May 1942.

Monadnock
Double turreted monitor laid down 1874, launched 1883, completed 1896. 3,990 tons. 11.6 kts. Twin screws. 262′ x 55½′ x 14½′. Cpl. 156. Four 10″, two 4″, two 6-pdr. Nine-inch belt, 11″ barbettes, 7½″ turrets. Decommissioned 1921. First ironclad to round Cape Horn.

Monocacy #1
Sister ship of *Ashuelot*. Sold in 1904, at Shanghai, $11,325.

Monocacy #2
River gunboat constructed, then disassembled at Mare Island, then reassembled at Shanghai, 1914, by Kiangnan Dockyard. 204 tons. 13 kts. 165½′ x 24½′ x 2½′. Cpl. 47. Two 6-pdr, six 30 cal. MG. Sunk by demolition charges 10 February 1939, at sea off Shanghai.

Monterey
Double-turreted monitor, 1893. 4,084 tons. 13.5 kts. 261′ x 59′ x 14′. Cpl. 210. Two 12″, two 10″, six 6-pdr, four 1-pdr. Thirteen-inch belt, 14″ barbettes. Sold 1922.

New Orleans (Originally *Amazonas*)
Steel-protected cruiser, sheathed with teak below waterline, built by Elswick, England, for Brazil, 1897. Acquired by USN in 1898. 3,950 tons. 20.5 kts. 355′ x 44′ x 19′. Cpl. 300. Six 6″/50, four 5″/50. Sold at Mare Island, 1929. Had been re-equipped with U.S. guns in 1907.

New York, later *Saratoga,* later *Rochester*
Armored cruiser built by William Cramp & Sons, Philadelphia, 1893. 8,950 tons. 21 kts, later reduced to 16 by removal of one of three stacks and two boiler rooms. 380′(wl) x 64′ x 28′. Cpl. 525. Four 8″/45, ten 5″/50, two 3″/23 AA. Four-inch belt, 6″ deck, 7″ Ct, 6″ barbettes and turrets. Decommissioned 1932, hulk scuttled at Olongapo, P.I., 24 December 1941.

Noa
Flush deck "four piper" destroyer, built by Norfolk Navy Yard, 1921. 1,190 tons. 35 kts. 314′ x 31′ x 13′. Cpl. 122. Four 5″/51, one 3″/23 AA, twelve 21″ TT. Sunk in a collision off Palau, 12 September 1944.

Oahu
River gunboat, 1928. Built under same circumstances as *Guam.* 450 tons. 15 kts. 191′ x 28′ x 6½′. Cpl. 65. Two 3″/50 AA behind shields, eight 30 cal. MG. Sunk at Corregidor, 4 May 1942, enemy gunfire from Bataan.

Oregon
Battleship built by Union Iron Works, San Francisco, 1896. 10,288 tons. 16.5 kts. 351′ x 69′ x 24′. Cpl. 473. Four 13″/35, eight 8″/35, four 6″/30. Eighteen-inch belt, 15″ turrets, 3″ deck, 10″ CT. Lent to state of Oregon in 1922. Used as ammunition barge during World War II. Broken up for scrap in Japan, 1957.

Palos #1
Built by James Tetlow, Boston, 1866, for $128,000. Iron. One of 8 sisters; 4th rate screw tug. First U.S. warship to transit Suez Canal, 11/12 August 1870. 420 tons. 10 kts. 137′(pp) x 26′ x 10′. Sold 6 January 1893.

Palos #2
Sister ship of *Monocacy,* built under same circumstances, 1914. Sold to Ming Sun Industrial Co. at Chungking, China, for use as wood oil barge, 1937.

Pampanga
Iron gunboat built by Manila Ship Co. at Cavite, P.I., 1888. 243 tons. 11 kts. Twin screws. 121′ x 18′ x 6½′. Cpl. 31. Four 3-pdr, two 1-pdr. In 1900 commissioned and commanded by Lieutenant M. M. Taylor, who in 1931 became CinC Asiatic Fleet.

Panay
Sister ship of *Oahu,* built under same circumstances as *Guam,* 1928. Sunk 12 December 1937 by Japanese aircraft between Nanking and Wuhu.

Pennsylvania, later *Pittsburgh*
Armored cruiser built by William Cramp & Sons, Philadelphia, 1905. 15,138 tons. 22 kts. 504′ x 70′ x 27′. Cpl. 822. Four 8″/45, fifteen 6″/50. Six-inch belt, 6½″ turrets, 4″ deck, 9″ CT. Sold in 1931.

Pompey
Purchased from James and Charles Harrison, London, April 1898, for $111,929. 3,085 tons. 10.5 kts. 245′ x 33½′ x 16′. Cpl. 114, Four 6-pdr. Decommissioned in 1921.

Preston, William B.
Generally similar to *Noa.*

Quiros
Built by Hongkong and Whampoa Dock Co., Hongkong, for Spain in April 1895. Purchased by

the U.S. Army in the Philippines and transferred to Navy 21 February 1900. Composite, single-screw gunboat. 350 tons. 11 kts. 138′ x 22′ x 9′. Cpl. 44. Two 6-pdr, two 3-pdr, two 1-pdr, 2 Colt MG. Sunk 16 October 1923 by destroyer gunfire.

Sacramento
Steel gunboat built by William Cramp & Sons, Philadelphia, 1914, for tropical service. Very tall natural draft stack, coal or wood; later converted to oil. 1,140 tons. 12 kts. 210′(wl) x 40½′ x 12′. Cpl. 153. Three 4″/50, two 3-pdr, two 1-pdr. Transferred to Great Lakes as training ship. Carried large junk sail on mainmast en route United States from Far East. Known as "the galloping ghost of the China Coast."

Saginaw
Wooden side-wheel gunboat built at Mare Island, 1859, at a cost of $188,667. Twenty-foot paddle wheels, 6′ wide. 500 tons. 155′ long, 4½′ draft. Cpl. 50. One 150-pdr Dahlgren R, one 32-pdr, two 24-pdr MLR. Wrecked on Ocean Island, 1870.

Samar
Similar to *Pampanga*. Sold 11 January 1921, after collision in Yangtze in which she was damaged beyond economical repair.

Shenandoah
Built by Navy Yard, Philadelphia, 1863. Wooden screw-sloop, two horizontal engines. 228′(pp) x 38′9″ x 15′(max.). 10 kts. Cpl. 191. 1,375 tons. Two 11″ SB, one 8″ SB, three 20-pdr R, one 150-pdr R. Sold July 1887.

South Dakota, later *Huron*
Armored cruiser similar to *Pennsylvania*. Built by Union Iron Works, San Francisco, 1907. Sold 1930.

Stewart
Similar to *Noa*.

Swatara
Built by Navy Yard, New York, 11 May 1874. Wooden screw-sloop, ship rig. Single screw, compound engine. 11 kts. 216′(pp) x 37′ x 16′6″(mean). Cpl. 230. One 30-pdr R, one 8″ R, six 9″, one 60-pdr BLR, two 20-pdr BLR. Sold August 1891.

Tulsa
Similar to *Asheville*.

Tutuila
Sister ship of *Guam*. Turned over to Nationalist Chinese at Chungking, February 1942.

Villalobos
Named after Magellan's first lieutenant. Generally similar to *Quiros*. Launched by same yard, 1896. Captured at Cavite, P.I., by U.S. Army and transferred to Navy, commissioned March 1900. Sunk as target 9 October 1928. Played the semi-fictional role of the USS *San Pablo* in the motion picture *Sand Pebbles*.

Wachusett
Wooden screw-sloop built at the Navy Yard, Boston, 1826. 1,200 tons. 201′ x 34′ x 14′. Three 100-pdr MLR, four 32-pdr SB, two 30-pdr MLR, one 12-pdr MLR. Bark rig. Sold at Mare Island 1887.

HMS *Widgeon*
Built 1904. 180 tons. 13 kts. 165′ x 24½′ x 2½′. Two 6-pdr, four MG. Cpl. 35.

Wilmington
Similar to *Helena*. In 1941, renamed *Dover*. During World War II, trained 20,000 men and 2,000 officers for armed guard crews. By decommissioning, post-World War II, *Wilmington/Dover* had served under thirty commanding officers.

Wisconsin
Battleship built by Union Iron Works, San Francisco, 1901. 13,000 tons. 17 kts. 374′ x 72′ x 24′. Cpl. 700. Four 13″/35, fourteen 6″/40. 14″ belt. Two athwartships stacks. Sold in 1922.

Wyoming
Wooden screw-sloop built by Navy Yard, Philadelphia, 1859. 1,200 tons. 10 kts. 198′6″(pp) x 33′2″ x 14′10″(max.). Two 11″ SB, one 60-pdr MLR, three 32-pdr. Ship rig. Sold in 1892.

APPENDIX 2

COMMANDERS OF U.S. NAVAL FORCES IN THE FAR EAST
1835—1942

Name and Rank	*Dates of Command*
Commodore Edmund P. Kennedy	22 March 1835 —30 October 1837
Commodore George C. Read	14 December 1837 —13 June 1840
Commodore Lawrence Kearney	4 February 1841 —27 February 1843
Commodore Foxhall A. Parker	27 February 1843 —21 April 1845
Commodore James Biddle	21 April 1845 —6 March 1848
Commodore William B. Shubrick	6 March 1848 —13 May 1848
Commodore David Geisinger	13 May 1848 —1 February 1850
Commodore Philip F. Voorhees	1 February 1850 —30 January 1851
Commodore John H. Aulick	31 May 1851 —20 November 1852
Commodore Matthew C. Perry	20 November 1852 —6 September 1854
Commodore Joel Abbott	6 September 1854 —15 October 1855
Commodore James Armstrong	15 October 1855 —29 January 1858
Commodore Josiah Tattnall	29 January 1858 —20 November 1859
Commodore Cornelius K. Stribling	20 November 1859—23 July 1861
Commodore Frederick Engle	23 July 1861 —23 September 1862
Commodore Cicero Price	23 September 1862—11 August 1865
Rear Admiral Henry H. Bell	11 August 1865 —11 January 1868
Commodore John R. Goldsborough	11 January 1868 —18 April 1868
Rear Admiral Stephen C. Rowan	18 April 1868 —19 August 1870
Rear Admiral John Rodgers	19 August 1870 —15 May 1872
Rear Admiral Thornton A. Jenkins	19 August 1872 —12 December 1873
Rear Admiral Enoch C. Parrott	12 December 1873—12 January 1874
Commodore Edmund R. Calhoun	12 January 1874 —29 May 1874
Rear Admiral Alexander M. Pennock	29 May 1874 —24 June 1875
Commander R. F. R. Lewis	24 June 1875 —16 August 1875
Rear Admiral William Reynolds	16 August 1875 —12 August 1877
Commodore Jonathan Young	12 August 1877 —4 October 1877
Rear Admiral Thomas H. Patterson	4 October 1877 —11 September 1880
Rear Admiral John M. B. Clitz	11 September 1880—21 April 1883
Rear Admiral Peirce Crosby	21 April 1883 —30 October 1883

Name and Rank	*Dates of Command*
Commodore Joseph S. Skerrett	30 October 1883 —19 December 1883
Rear Admiral John Lee Davis	19 December 1883—22 November 1886
Rear Admiral Ralph Chandler	22 November 1886—11 February 1889
Rear Admiral George E. Belknap	4 April 1889 —20 February 1892
Rear Admiral David B. Harmony	20 February 1892 —7 June 1893
Rear Admiral John Irwin	11 June 1893 —11 December 1893
Commodore Joseph S. Skerrett	11 December 1893—1 September 1894
Rear Admiral Charles C. Carpenter	1 September 1894 —21 December 1895
Rear Admiral Frederick V. McNair	21 December 1895—3 January 1898
Commodore George Dewey	3 January 1898 —15 June 1899
Rear Admiral John C. Watson	15 June 1899 —19 April 1900
Rear Admiral George C. Remey	19 April 1900 —1 March 1902
Rear Admiral Frederick Rodgers	1 March 1902 —29 October 1902
Rear Admiral Robley D. Evans	29 October 1902 —21 March 1904
Rear Admiral Philip H. Cooper	21 March 1904 —11 July 1904
Rear Admiral Yates Stirling	11 July 1904 —23 March 1905
Rear Admiral William M. Folger	23 March 1905 —30 March 1905
Rear Admiral Charles J. Train	30 March 1905 —4 August 1906
Rear Admiral Williard H. Brownson	15 October 1906 —31 March 1907
Rear Admiral James H. Dayton	31 March 1907 —31 July 1908
Rear Admiral William T. Swinburne	31 July 1908 —17 May 1909
Rear Admiral Uriel Sebree	17 May 1909 —19 February 1910
Rear Admiral John Hubbard	19 February 1910 —16 May 1911
Rear Admiral Joseph B. Murdock	16 May 1911 —24 July 1912
Rear Admiral Reginald F. Nicholson	24 July 1912 —3 May 1914
Admiral Walter C. Cowles	3 May 1914 —9 July 1915
Admiral Albert G. Winterhalter	9 July 1915 —4 April 1917
Admiral Austin M. Knight	4 April 1917 —7 December 1918
Vice Admiral William L. Rodgers	7 December 1918 —1 September 1919
Admiral Albert Gleaves	1 September 1919 —4 February 1921
Admiral Joseph Strauss	4 February 1921 —28 August 1922
Admiral Edwin A. Anderson	28 August 1922 —11 October 1923
Admiral Thomas Washington	11 October 1923 —14 October 1925
Admiral Clarence S. Williams	14 October 1925 —9 September 1927
Admiral Mark L. Bristol	9 September 1927 —9 September 1929
Admiral Charles B. McVay	9 September 1929 —1 September 1931
Admiral Montgomery M. Taylor	1 September 1931 —18 August 1933
Admiral Frank B. Upham	18 August 1933 —4 October 1935
Admiral Orin G. Murfin	4 October 1935 —30 October 1936
Admiral H. E. Yarnell	30 October 1936 —25 July 1939
Admiral Thomas C. Hart	25 July 1939 —14 February 1942

SELECTED BIBLIOGRAPHY

PRINCIPAL DOCUMENTS

East India Squadron Letters to the Secretary of the Navy, 1841–1861, National Archives.

Asiatic Squadron Letters to the Secretary of the Navy, 1865–1885, National Archives.

Asiatic Fleet Reports, 1916–1923, National Archives, Record Group 45, Box 359.

Deck logs of ships, National Archives.

"U.S. Gunboats on the Yangtze: History and Political Aspects, 1842–1922." An annotated report of an address by E. Mowbray Tate. Published in the annual series "Studies on Asia," University of Nebraska Press, Lincoln, Nebraska, 1966.

American State Papers, Naval Affairs, Volumes 23–27, from 1794 to 1836. Pratt Library designation XJ33A2, Baltimore, Maryland.

"The Growth of the Yangtze Delta." Royal Geographic Society, North China Branch Journal, Shanghai, 1922, pp 21–36, 80 Volume 53.

Rosenberg, David Alan. "History of the Yangtze Patrol, A Study in American Imperialism, Part I. The Formative Years." Unpublished thesis prepared for partial satisfaction of requirement for Departmental Honors in History on Bachelor of Arts degree at American University, Washington, D.C., 1961. In U.S. Navy Department Library, Washington, D.C.

Record Groups 38 and 80, National Archives: General correspondence between Washington and Far East commanders.

Senate Miscellaneous Documents, 1837–1861, published semiannually by various printers, covering activities of all departments of government, including the Navy. National Archives.

Secretary of the Navy's Annual Report, 1862–1942, with only partial reports for 1933 and 1934. Government-bound document. Pratt Library designation XPL52A1, Baltimore, Maryland.

BOOKS

Augur, Helen. *Tall Ships to Cathay*. New York: Doubleday, 1951.

Basil, George C. *Test Tubes and Dragon Scales*. Chicago: John C. Winston Co., 1940.

Bishop, J. F. *The Yangtze Valley and Beyond*. New York: G. P. Putnam's Sons, 1899.

Blakiston, Thomas W. *Five Months on the Yangtze*. London: John Murray, 1862.

Borg, Dorothy. *The Unite‿ States and the Far Eastern Crisis*. Boston: Harvard University Press, 1964.

Boulger, D. C. *History of China*. New York: The Co-operative Publication Society, 1893.

Chennault, Claire Lee. *Way of a Fighter*. New York: G. P. Putnam's Sons, 1949.

Clark, Charles E. *My Fifty Years in the Navy*. Boston: Little, Brown and Co., 1917.

Clark, Grover. *Economic Rivalries in China*. New Haven: Yale University Press, 1932.

Clyde, Paul H., and Beers, Burton F. *The Far East*. Englewood, New Jersey: Prentice, Hall, Inc., 1948.

Collis, Maurice. *Foreign Mud*. London: Faber and Faber, 1946.

Crosby, William, and Nichols, H. P. *The Journals of Major Samuel Shaw*. Boston: Metcalf and Co., 1847.

Crow, Carl. *The Traveller's Handbook for China*. Shanghai: Kelly and Walsh, 1921.

Dennett, Tyler. *Americans in Eastern Asia*. New York: Barnes and Noble, Inc., 1963.

duPont, G. Carsalade. *La Marine Français sur le Haut Yang tsé*. Paris: Academie de Marine, 1963.

Elegant, Robert S. *The Center of the World*. London: Meuthen and Co., 1963.

Enders, Elizabeth C. *Temple Bells and Silver Sails*. New York: D. Appleton and Co., 1925.

Evans, Robley. *Admiral's Log*. New York: D. A. Appleton and Co., 1910.

Finney, Charles. *The Old China Hands*. New York: Paperback Library, Inc., 1963.

Fischer, Louis. *The Soviets in World Affairs*. Princeton: Princeton University Press, 1951.

Fitzgerald, C. P. *A Concise History of East Asia*. New York: Frederick A. Praeger, 1966.

Fitzgerald, C. P. *China, a Short Cultural History*. New York: Frederick A. Praeger, 1954.

Fontenoy, Jean. *The Secret Shanghai*. New York: Grey-Hill Press, 1939.

Geil, William Edgar. *A Yankee on the Yangtze*. London: Hodder and Stoughton, 1904.

Gill, William. *The River of Golden Sand*. London: John Murray, 1883.

Hart, Sir Robert. *These from the Land of Sinim*. London: Chapman and Hall, 1901.

Hawks, Francis L. *Narrative of the Expedition of an American Squadron to the China Sea and Japan*. New York: D. Appleton and Co., 1857.

Hersey, John. *A Single Pebble*. New York: Alfred A. Knopf, Inc., 1956.

Hobart, Alice Tisdale. *Within the Walls of Nanking*. London: Butler and Tanner, 1928.

Hosie, Alexander. *Three Years in Western China*. London: George Phelps and Son, 1890.

Isaacs, Harold. *The Tragedy of the Chinese Revolution*. New York: Atheneum Press, 1966.

Jack, R. Logan. *The Back Blocks of China*. London: Edward Arnold, 1904.

Kazakov, V. G. *Nemie Svidetelii*. Shanghai: 1936.

Komroff, M. *The Travels of Marco Polo*. New York: Liveright Publishing Corporation, 1950.

Leonard, Royal. *I Flew for China*. New York: Doubleday, Doran, 1942.

Lubbock, Basil. *The China Clippers*. Glasgow: James Brown and Sons, Ltd., 1925.

Maclellan, J. W. *The Story of Shanghai*. Shanghai: North China Herald, 1889.

MacGregor, David R. *The Tea Clippers*. London: Percival Marshall and Co., Ltd., 1952.

Miller, G. E. *Shanghai, The Paradise of Adventurers*. New York: Orsay Publishing House, 1937.

Morison, Samuel Eliot. *"Old Bruin," Commodore Matthew Calbraith Perry*. Boston: Little, Brown & Co., 1967.

Paullin, Charles Oscar. *American Voyages to the Orient, 1690–1865*. Annapolis: U.S. Naval Institute, 1970. A collection of articles from the U.S. Naval Institute *Proceedings*.

Perry, Hamilton Darby. *The Panay Incident, Prelude to Pearl Harbor*. New York: The Macmillan Co., 1969.

Plant, Cornell. *Glimpses of the Yangtze Gorges*. Shanghai: Kelly and Walsh, 1926.

Polley, C. E. and Hill, M. O. *USS Augusta under Fire*. Shanghai: USS *Augusta*, 1938.

Reynolds, Quentin. *Officially Dead*. New York: Random House, 1945.

Rodman, Hugh. *Yarns of a Kentucky Admiral*. Indianapolis: Bobbs-Merrill Co., 1928.

Selle, Albert. *Donald of China*. New York: Harper and Brothers, 1948.

Schurmann, Franz, and Schell, Orville. *Imperial China*. New York: Random House, 1967.

Scidmore, Eliza Rumahah. *China—The Long Lived Empire*. London: Macmillan and Co., Ltd., 1900.

Sheean, Vincent. *In Search of History*. London: Hamish Hamilton, 1935.

Smith, Arthur H. *Chinese Characteristics*. New York: Fleming H. Revell Company, 1894.

Stirling, Yates. *Sea Duty*. New York: G. P. Putnam's Sons, 1939.

Sutton, F. A. *One-Arm Sutton*. New York: Viking Press, 1933.

Swisher, Earl. *China's Management of the American Barbarians*. New Haven: Yale University; Far Eastern Publications, 1951. Copyright 1955 by The Association for Asian Studies, University of Michigan, Ann Arbor, Michigan.

Tompkins, Pauline. *American-Russian Relations in the Far East*. New York: Macmillan, 1949.

Tregear, T. R. *A Geography of China*. Chicago: Aldine Publishing Co., 1965.

Wetmore, W. S. *Recollections of Life in the Far East*. Shanghai: 1894.

Woodhead, H. G. W. *The Yangtze and Its Problems*. Shanghai: Mercury Press, 1931.

Woodrooffe, Thomas. *River of Golden Sand*. London: Faber and Faber, 1936.

PERIODICALS

Articles from U.S. Naval Institute *Proceedings:*

Brown, W. F., "Fiat Justitia, Ruat Caelum, Chinese Style." Vol. 64, November 1938.

Carter, A. F. "The Upper Yangtze River." Vol. 43, February 1917.

Eller, E. M. "United States Disaster in China." Vol. 75, July 1949.

Eyre, James K., Jr. "The Civil War and Naval Action in the Far East." Vol. 68, November 1942.

Gale, Esson M. "The Yangtze Patrol." Vol. 81, March 1955.

Gardner, K. N. "The Beginning of the Yangtze River Campaign of 1926–1927." Vol. 58, January 1932.

Gulliver, Louis J. "The Yangtze U.S. Gunboats." Vol. 68, September 1942.

Howell, Glenn. "The Battle of Wanhsien." Vol. 53, May 1927.

———. "Operations of the United States Navy on the Yangtze River—September 1926 to June 1927." Vol. 54, April 1928.

———. "Captain Plant." Vol. 55, March 1929. (Excellent illustrations.)

———. "Hwang Tsao." Vol. 64, August 1938.

———. "Chungking to Ichang." Vol. 64, September 1938.

———. "Army-Navy Game; Or, No Rules of the Road." Vol. 64, October 1938.

———. "Opium Obligato." Vol. 64, December 1938.

———. "Ascent of the Min." Vol. 65, May 1939.

Hutchins, Charles T. "Why This Chaos in China?" Vol. 53, April 1927.

Jacobs, V. F. G. "Port of Call." Vol. 65, February 1939.

Johnson, Felix L. "Naval Activities on the Yangtze." Vol. 53, April 1927. (Excellent collection of photographs.)

———. "The Asiatic Station." Vol. 58, May 1932.

Johnson, Nelson T. "Ten Books on China." Book Review Section. Vol. 81, February 1955.

Kendall, David W. "The Navy in the Orient in 1842." Vol. 68, May 1942.

Lee, J. A. "Between Wars in the Far East." Vol. 65, January 1939.

Manning, G. C. "Yangtze." Vol. 60, February 1934.

Okumiya, Masatake, assisted by Roger Pineau, "How the *Panay* Was Sunk." Vol. 79, June 1953.

Pfaff, Roy. "Sea Duty on the Yangtze." Vol. 59, November 1933.

Pineau, Roger. "*U.S.S. Noa* and the Fall of Nanking." Vol. 81, November 1955.

Sheehan, J. M. "From the Side Lines." Vol. 65, January 1939.

———. "Nanking." Vol. 69, September 1943.

———. "The Gorges of the Yangtze Kiang." Vol. 69, November 1943.

Smith-Hutton, H. H. "Lessons Learned at Shanghai in 1932." Vol. 64, August 1938.

Smith, Roy C., Jr. "Nanking, 24 March 1927." Vol. 54, January 1928.

Sutliff, R. C. "Duty in a Yangtze Gunboat." Vol. 61, July 1935. (Excellent illustrations.)

Swanson, H. J. "The Panay Incident: Prelude to Pearl Harbor." Vol. 63, December 1967.

Tolley, Kemp. "YangPat—Shanghai To Chungking." Vol. 89, June 1963.

———. "Divided We Fell." Vol. 92, October 1966.

Wharton, W. S. "Our Chinese Navy." Vol. 51, January 1925.

Winslow, Cameron McR. "Action on the Yangtze." Vol. 63, April 1937.

Articles from *Shipmate* Magazine:

Bellah, J. W. "The Grand Manner of Josiah Tattnall." Vol. 28, December 1965.

Settle, T. G. W. "Last Cruise of the *Palos*." Vol. 24, April 1961.

Tolley, Kemp. "Three Piecie and Other Dollars Mex." Vol. 28, July 1965.

———. "Chinese Huntsman." Vol. 29, June-July 1966.

———. "The Chameleon." Vol. 29, October 1966.

———. "A Day in the Life of a Chungking Gunboat." Vol. 30, June-July 1967.

INDEX

Edited by Louise Gerretson.

Designed by Edward Martin Wilson.

Composed in nine-point Optima with two points of leading by Monotype Composition Company, Baltimore, Maryland.

Printed offset on sixty-pound Natural Hopper Bulkopaque, by Universal Lithographers, Cockeysville, Maryland.

Photograph section printed Vintone-Gravure on seventy-pound Mead Moistrite Matte by the Vinmar Lithographing Company, Baltimore, Maryland.

Bound in Columbia Mills' Bolton Buckram by L. H. Jenkins, Inc., Richmond, Virginia.